Common Flowering Plants
of the Northeast

Common Flowering Plants

of the Northeast)

their natural history and uses

DONALD D. COX

STATE UNIVERSITY OF NEW YORK
COLLEGE OF ARTS AND SCIENCES AT OSWEGO

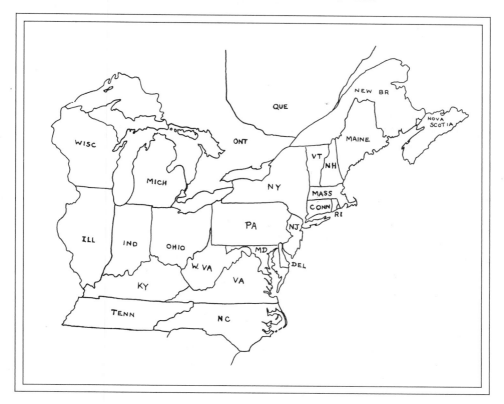

KEY TO THE FAMILIES OF PLANTS BY MILDRED E. FAUST

ILLUSTRATIONS BY SHIRLEY PERON AND RUTH SACHIDANANDAN

State University of New York Press, Albany

Published by
State University of New York Press, Albany

For information, address State University of New York Press, State University Plaza,
Albany, N.Y., 12246

Library of Congress Cataloging in Publication Data
Cox, Donald D.
   Common herbaceous flowering plants of the Northeast.
   Bibliography: p. 395
Includes index.
   1. Botany—Northeastern States—Ecology.   2. Wild
plants, Edible—Northeastern States—Identification.
3. Medicinal plants—Northeastern States—Identification.
4. Plants—Identification.   I. Title.   II. Title: Flowering
plants of the Northeast.
QK117.C828   1984   582.13'0974   83-24242
ISBN 0-87395-889-6
ISBN 0-87395-890-X (pbk.)

Dedicated to the Memory of
Deborah

Acknowledgements

In a project that has taken several years to complete it is impossible to acknowledge everyone who has contributed. However, I would like to express special appreciation to the following people: Dr. Mildred Faust for generosity with her time and botanical expertise; Shirley Peron and Ruth Sachidanandan for their cooperation, enthusiastic field work, and talent; Dr. Alan Davies for the loan of *The Dispensatory of the United States of America*; Shelby Marshall, Gerry Smith, and Diann Jackson for "holding the fort"; Judith Block, Production Editor at State University of New York Press; and finally to Barbara, Cindy, and Lisa for tolerance, patience, and love.

CONTENTS

Woodlands

Open Fields

Wetlands

Beaches

Salty Soils

Key to Symbols

 Food uses
℞ Medicinal and other Information
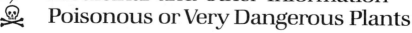 Poisonous or Very Dangerous Plants

Introduction

Area and Plants Included

   This publication presents an ecological ordering of common herbaceous flowering plants that occur in northeastern North America, with aids for their field identification. No attempt has been made to include every herbaceous plant in the northeast. The plants included are the ones most often observed in the following five common environmental situations: (1) woodlands, (2) roadsides and open spaces, (3) wetlands, (4) beaches and sandy areas, and (5) areas of high concentrations of salt. Each of these habitat types will be described in detail. In addition, information has been included on food and medicinal uses for many species. In recent years there have been numerous excellent publications dealing with the medicinal or food value of natural wild species of plants. Most of these have concentrated on one or the other of these specializations, but few have included both.

   Since species of herbaceous plants have wide ranges of tolerance for environmental variation, it is not unusual for some to be observed in more than one setting. For this reason some of the included species are listed in two or even three of the major habitats identified above. These species are listed for each habitat in which they may occur, but are illustrated only for the habitat in which they are most commonly observed. For example, *Spiranthes cernua* (Autumn Ladies' Tresses) may occur in woods, open spaces, or bogs. It is thus listed in sections 1, 2, and 3, but is illustrated only in section 1, woodlands. Similarly, *Potentilla anserina* (Silver-weed) normally occurs on sandy soils and beaches. However, the author has observed it growing vigorously in areas occupied by halophytes (salt tolerant plants). Therefore, silver-weed is listed in sections 4 and 5 but is illustrated only in section 4, beaches and sandy areas.

   Certain groups of herbaceous flowering plants have been omitted entirely in the preparation of this publication, e.g., Poaceae (the grasses), Cyperaceae (the sedges), and Juncaceae (the rushes). In addition, it has not been the intent of the author to include every genus and species of some large families of plants such as Brassicaceae (the Mustard family), Lamiaceae (the Mint family), and Asteraceae (the Aster family). Other families of plants that normally occur in the northeast may have been omitted either because the plants are small and in-

1

conspicuous, and rarely encountered, or they have a very limited distribution.

The map on the title page gives the approximate minimum geographic area in which all the included species may be found. It should be noted, however, that every species included here does not occur throughout the area depicted. A certain amount of overlapping of the ranges of species from contiguous areas is inevitable. Some southern species occur at the northern edge of their range in the southern portion of our map, while some northern species occur at the southern edge of their distribution in the northern portion of the map. For example, *Delphinium tricorne* (Dwarf Larkspur), which is common in the southern states of the map, does not ordinarily extend north of Pennsylvania. Similarly, *Epilobium augustifolium* (Fireweed) and *Cornus canadensis* (Dwarf Cornell) are abundant in the northern portion of our map, but are found only at higher elevations in the southern portion.

This publication deals mainly with herbaceous plants. These are plants that do not have woody tissues. Many plants that do have woody tissues and are technically shrubs, have the physical appearance of low-growing herbaceous plants. For example, *Epigaea repens* (Trailing Arbutus), *Arctostaphylos uva-ursi* (Bearberry), and *Vaccinium macrocarpon* (Large Cranberry) could all be mistaken for herbaceous plants. These and several other low-growing shrubs have been included.

Food and Medicinal Uses of Plants

The information included here on food and medicinal uses of wild plants is of historical interest only, and is not intended as recommendations for use by the reader. When the population of the United States was restricted mainly to the eastern seaboard, and could be counted in thousands rather than in hundreds of millions, plants could be collected for food and medicine without damage to plant populations. This is no longer true. Many species of wild plants are threatened today because of habitat destruction. If a species that is already threatened must also bear collection pressure from humans, its chances of survival are greatly diminished.

The collection of wild plants for home remedies is a practice that should never be engaged in without first consulting a professional botanist or an experienced herbalist. The shapes and sizes of plant leaves and stems vary according to the environmental conditions under which they grow. Futhermore, sometimes unrelated plants have similar fruits or leaves. For example, the fruits of Poison Ivy (*Rhus radicans*), which may cause serious dermatitis if handled, could be

2

mistaken by the novice for those of the relatively harmless Mistletoe (*Phoradendron flavescens*). (The berries of Mistletoe are toxic if eaten.) Both of these plants have small, white, berry-like fruits, and both are found growing in trees. Although plants are the source of many useful drugs, some people are highly allergic to various wild plant juices. Therefore, the misidentification of a specimen to be used in a home remedy could be dangerous and maybe even fatal.

The same reasoning can be used with regard to the collection of wild plants for food. An individual who is not familiar with plants should never attempt to collect wild food plants without the counsel of one who knows the local vegetation. In several plants the leaves are edible when very young, but are toxic in the mature plant. In addition, the young leaves of many plants are similar, making identification difficult at this stage, even for the expert. The literature is replete with descriptions of the unpleasant consequences of misidentification and subsequent consumption of wild plants.

Plant Names

For us to talk about a given plant, it must have a name, and in order for a name to be meaningful, it must consistently refer to the same plant. In 1753 a Swedish botanist named Carolus Linnaeus wrote a book in which he named every known plant in the world. The system he used for naming plants is accepted by botanists today. It is a binomial system, in which the name of each plant species has two parts (for example the name for May Apple is *Podophyllum peltatum*): the first part designates the genus to which the plant belongs (*Podophyllum*), and the second part (*peltatum*) identifies a particular species of that genus. These two words comprise the scientific or Latin name for that species. The advantage of the scientific name is that no other plant in the world has that name. If you mentioned *Podophyllum peltatum* to a Russian, Japanese, Indian, or Ugandan botanist, he would recognize it as a discrete entity and never confuse it with another plant.

Many plants have common or popularized names that are widely accepted. However, there are several problems associated with the use of common names. In the first place, some plants have more than one common name. For example, May Apple is not the only popular name for *Podophyllum peltatum*. In different areas it may be known as Mandrake, Wild Lemon, Raccoon Berry, Duck's Foot, Hog Apple, or Wild Jalap. Secondly, the same common name is often applied to different plants. Swamp Loosestrife is a common name for the entire genus of *Decodon* and also for *Lysimachia terrestris*.

3

In this publication the scientific or Latin name is given for each species along with one or more common names. The first common name given is the one that seems to be most widely recognized. Most of the common names are those given in Fernald's *Gray's Manual of Botany*, Eighth Edition.Most of the scientific names follow the classification given in Gleason and Cronquist's *Manual of Vascular Plants of Northeastern United States and Adjacent Canada*.

Early botanists were known as herbalists and their main concern was finding plants that could be used to treat human ailments. Thus, the earliest books on plants, called herbals, consisted of plant descriptions with their medicinal uses. Consequently, many early botanists were physicians. Modern physicians receive little training in medicinal botany, but lay practictioners of herbal medicine carry on the traditions of the herbalists, and a significant body of literature on herbal medicine is available today. In this book the word "herbology" is used to refer to this literature.

Many common names for plants date back to a medieval concept known as the "Doctrine of Signatures." This idea held that when a plant, or some part of it, resembled an organ or structure of man, the plant was good medicine for that organ or structure. For example, a common spring flowering plant in our area has a popular name and the genus name of *Hepatica*. This name is derived from a Greek word meaning "liver," which the three-lobed leaves of this plant are supposed to resemble. Other common names for Hepatica are Liver Leaf and Liverweed, and in herbal medicine it is recommended for hepatitis, gallstones and other liver ailments. Although there is no basis in modern American medicine for many of the remedies proposed by herbalists (including the use of Hepatica for liver problems), herbal medicine is widely practiced in many parts of the world and may deserve more serious consideration in the United States.

How To Use This Book

This book has been compiled primarily as a field guide for the identification and ecology of common herbaceous flowering plants in the northeastern United States and adjacent Canada. It has been written to be useful to the interested layman as well as the student of field botany. A suggested procedure for using the book is outlined below.

Suppose you have discovered a flowering plant that is unknown to you and you wish to identify it.

(1) Making use of the included dichotomous key, identify the family of the unknown plant specimen.

(2) Refer to the type of habitat in which the unknown plant was found.

(3) Locate the family of the unknown specimen. The families are alphabetically arranged.

(4) In the section under this family find an illustration that matches the appearance of the unknown plant. The common name and the Latin name of the species are given with each illustration.

(5) Confirm the identification by reading the complete description of the specimen in the text. The genera of a given plant family that occur in a specific habitat are listed and illustrated alphabetically under that family name.

The Nature and Use of Plant Keys

The function of a plant key is to make easier the identification of unknown plants. The most commonly used type of key is the dichotomous key. A dichotomous key is based on the assumption that any collection of plants can be divided into two groups by some observable characteristic, i.e., one group has the characteristic the other group does not. The construction of a dichotomous key is made more difficult by the fact that plants are highly variable in their expression of vegetative traits. For example, two individuals of the same species grown under different environmental conditions may have quite different vegetative characteristics. Consequently, the parts of plants that are most reliable in the construction of keys are the flower parts. In most instances these parts are discernable with the unaided eye, but sometimes a simple magnifying lens is useful.

A dichotomous key can be best illustrated with a small group of plants. Consider the following six specimens:

Plant A—10 white petals, 4 stamens, leaves opposite
Plant B—5 white petals, 10 stamens, leaves alternate
Plant C—3 white petals, 6 stamens, leaves alternate
Plant D—6 blue petals, 6 stamens, leaves opposite
Plant E—5 blue petals, 5 stamens, leaves opposite
Plant F—3 blue petals, 3 stamens, leaves alternate

A key for these plants is given below.

A. Plants with petals and stamen in numbers divisible by 4 or 5
  B.   Petals Blue .... Plant E
  B.   Petals White
      C.   Leaves Opposite .... Plant A
      C.   Leaves Alternate .... Plant B
A. Plants with petals and stamen in numbers divisible by 3 or 6
  D.   Petals White .... Plant C
  D.   Petals Blue
      E.   Leaves Opposite .... Plant D

      E.   Leaves Alternate .... Plant F

Two general observations can be made on the nature of dichotomous keys from the above example.

(1) The two statements for each dichotomous division begin the same number of spaces from the left margin of the page.

(2) The two statements for each division usually begin with the same word followed by a statement expressing contrasting conditions, e.g.

        B. Petals Blue

        B. Petals White

The sample key given above is for a very limited group of plants and is thus an oversimplification. The larger the group of plants the greater the spaces in the key between paired contrasting statements. The plant key to families given in the following pages is valid for most of the herbaceous flowering plants in the northeastern United States and adjacent Canada.

Key to the Families of Plants

A. Plants green

    B. Land plants

    (This includes plants of swamps and bogs as well as uplands, woods and fields.)

        C. Flowers showy

      CC. Flowers *not* showy

      (This includes those with small flowers which may not have all of the parts or are not clear).

    BB. Aquatic plants

PLANTS GREEN · LAND PLANTS · FLOWERS SHOWY
MONOCOTYLEDONS

Flower parts in 3's and multiples of 3's; leaves parallel veined; petals and sepals often joined in a tube, both the same color.

(A number of exceptions. Example: LILIACEAE, flower parts in 4's in MAIANTHEMUM; sepals differ from petals and leaves are net-veined in Trillium)

perianth ⎰ corolla (petals) — stamen (anther, filament)
        ⎱ calyx (sepals) — pistil (stigma, style, ovary)

a. HYPOGYNOUS or PERIGYNOUS
    (perianth below or around the ovary)

   b. ACTINOMORPHIC
    (perianth regular)

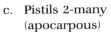

     c. Pistils 2-many
        (apocarpous)   actinomorphic   hypogynous   perigynous

apocarpous

       d. Flowers pink, in umbels;
            leaves narrow, basal .... BUTOMACEAE

     dd. Flowers white, in panicles or racemes;
            leaves oval, linear or sagittate .... ALISMATACEAE

                umbel   panicle   raceme

   cc. Pistil 1
         (syncarpous)

syncarpous

     d. Flowers in heads or fleshy spikes   head   spike

spathe
spadix

       e. Inflorescence (spadix) surrounded or not
            by a leafy involucre (spathe); leaves var-
            ious .... ARACEAE

       ee. Inflorescence in spherical heads, stami-
            nate heads above the pistillate ones;
            leaves narrow, clasping .... SPARGANIACEAE

     dd. Flowers axillary or in various inflorescences;
            sepals 3, petals 3, often joined; stamens 6; leaves
            various .... LILIACEAE

bb.   Zʏɢᴏᴍᴏʀᴘʜɪᴄ, flowers blue
      (perianth irregular)

zygomorphic c.   Flowers in a spike ....   Pᴏɴᴛᴇᴅᴇʀɪᴀᴄᴇᴀᴇ
             cc.   Flowers one or a few in a folded spathe .....   Cᴏᴍᴍᴇʟɪɴᴀᴄᴇᴀᴇ

  aa.   Eᴘɪɢʏɴᴏᴜs
        (perianth above the ovary)

epigynous

      b.   Aᴄᴛɪɴᴏᴍᴏʀᴘʜɪᴄ
         (perianth regular)

actinomorphic c.   Stamens 6 ....   Aᴍᴀʀʏʟʟɪᴅᴀᴄᴇᴀᴇ
             cc.   Stamens 3 ....   Iʀɪᴅᴀᴄᴇᴀᴇ
      bb.   Zʏɢᴏᴍᴏʀᴘʜɪᴄ; stamens 2 or 1
         (perianth irregular) ....   Oʀᴄʜɪᴅᴀᴄᴇᴀᴇ
zygomorphic

PLANTS GREEN • LAND PLANTS • FLOWERS SHOWY
DICOTYLEDONS

Flower parts in 4's or 5's; leaves net-veined.
(A number of exceptions. Examples: Pʀɪᴍᴜʟᴀᴄᴇᴀᴇ, flower parts in 7's in Trientalis;
Bᴇʀʙᴇʀɪᴅᴀᴄᴇᴀᴇ, flower parts in 3's in Caulophyllum.)

perianth ⎡corolla (petals) ——————— stamen (anther, filament)
        ⎣calyx (sepals) ——————— pistil (stigma, style, ovary)

  a.   Pᴏʟʏᴘᴇᴛᴀʟᴏᴜs or Aᴘᴇᴛᴀʟᴏᴜs ....aa. page 17
      (petals separate or none)

      b.   Hʏᴘᴏɢʏɴᴏᴜs or Pᴇʀɪɢʏɴᴏᴜs ....bb. page 15
         (perianth below or around the ovary)

         c.   Aᴄᴛɪɴᴏᴍᴏʀᴘʜɪᴄ ....cc. page 15
           (perianth regular)   hypogynous   perigynous
polypetalous
actinomorphic   d. Leaves opposite, whorled or basal ......dd. page 12

9

opposite whorled  basal leaves

e. Leaves deeply divided or compound

    f. Stamens many

        g. Cell sap yellow-orange;
           pistil one .... PAPAVERACEAE

        gg. Cell sap clear; mostly more
            than one pistil

           h. Perigynous; leaves with
               stipules .... ROSACEAE

stipule⟍ hh. Hypogynous; no stipules .... RANUNCULACEAE

    ff. Stamens 4–10 (12); pistil one

        g. Petals and sepals 4; stamens 6 (2 + 4) .... BRASSICACEAE
                                       (Cruciferae)

        gg. Petals 5 or more

           h. Petals 5

              i. Leaves lobed .... GERANIACEAE

              ii. Leaves compound, of 3 leaflets .... OXALIDACEAE
           hh. Petals 6 or more; flowers white, bluish, green
              .... BERBERIDACEAE

ee. Leaves simple, entire or toothed

    f. Pistils 2-many

        g. Stamens 3–10

           h. Pistils 2 .... SAXIFRAGACEAE
           hh. Pistils 4–5 .... CRASSULACEAE

        gg. Stamens many

      h. Flowers white; stipules ....   Rosaceae

    hh. Flowers yellow; no stipules ....   Ranunculaceae

  ff.   Pistil one

     g. Pistil stalked; cell sap milky ....   Euphorbiaceae
    gg. Pistil sessile; cell sap clear

      h. Styles more than one (or deeply divided)

       i. Leaves with punctate (oil) dots;
           stamens grouped (9-many) ....   Hypericaceae

        ii. Leaves without dots; stamens as many or twice
            the petals

          j. Sepals 2; styles 3, joined or not at base...........   Portulacaceae

         jj. Sepals 4–5; styles 2–5

            k.   Leaves basal
             l. Leaves with sticky hairs ....   Droseraceae
            ll. Leaves mostly smooth ....   Plumbaginaceae

          kk. Leaves not basal,
             without sticky hairs

            l. Plant prostrate with whorled leaves ....   Aizoaceae
            ll. Leaves opposite ....   Caryophyllaceae

   hh. Style one

      i. Flower parts in 5's–7's; leaves whorled ....   Primulaceae

     ii. Flower parts 4–6 (7) or many

       j. Tall or sprawling plant; flowers magenta ....   Lythraceae

      jj. Low plant; flowers various colors

        k. Flowers yellow; leaves small........................   Cistaceae

kk. Flowers not yellow;
leaves basal or whorled

l. Leaves pitcher-shaped; flowers maroon....... SARRACENIACEAE

ll. Leaves oval;
flowers white or pink.......................... ERICACEAE

dd. Leaves alternate

alternate

e. Plant woody or shrubby

f. Pistils 2 or more; stamens mostly more than 10

g. Vine with oval-round leaves .... MENISPERMACEAE

gg. Upright plant; leaves simple or compound .... ROSACEAE

ff. Pistil 1; stamens 3–10; leaves simple

raceme g. Leaves toothed, oval

h. Inflorescence a raceme .... CLETHRACEAE

hh. Inflorescence umbellate .... RHAMNACEAE

umbel

gg. Leaves entire, small; flowers terminal

head h. Flowers in purple heads .... EMPETRACEAE

hh. Flowers separate, yellow .... CISTACEAE

ee. Plants herbaceous

f. Leaves lobed or compound

g. Stamens mostly more than 10

h. Stamens joined around the pistil .... MALVACEAE

hh. Stamens separate

i. Cell sap yellow-orange;
sepals 2 .... PAPAVERACEAE

12

      ii.   Cell sap clear; sepals 4–5

           j.   Stipules present;
               perigynous ....   ROSACEAE

          jj.   No stipules;
               hypogynous

               k. Sepals usually falling as flower opens
                  ....   BERBERIDACEAE

              kk. Sepals persistent ....   RANUNCULACEAE

  gg. Stamens 3–10 (12)

      h. Petals 4; stamens 6 (2 + 4) ....   BRASSICACEAE
                                              (Cruciferae)

    hh. Petals 5–9

        i.   Petals 6–9 ....   BERBERIDACEAE
        ii.   Petals 5; stamens 10

            j.   Flowers yellow or pinkish; leaflets 3
                ....   OXALIDACEAE
           jj.   Flowers blue-reddish-purple; leaves
               lobed ....   GERANIACEAE

ff. Leaves simple

  g. Pistils 2-many (some joined at base)

    h. Stamens the same number or twice the petals

        i.   Pistils 4–5 ....   CRASSULACEAE

        ii.   Pistils 2 ....   SAXIFRAGACEAE

    hh. Stamens usually many

        i.   Stamens joined into a tube around the
            pistil ....   MALVACEAE

        ii.   Stamens separate

j.   Stipules present;
     perigynous .... ROSACEAE

jj.   No stipules;
     hypogynous .... RANUNCULACEAE

gg. Pistil one

    h. Stamens mostly many
       (see also Phytolaccaceae)

        i. Leaves pitcher-shaped;
          stigma broad .... SARRACENIACEAE

       ii. Leaves roundish; stamens joined in a tube
         .... MALVACEAE

  hh. Stamens 3–12

      i. Corolla lacking or not obvious; sepals or
        bracts petal-like

        j. Stipules surrounding stem nodes .... POLYGONACEAE

        jj. Stipules, if present, not surrounding
          nodes

            k. Cell sap milky;
              pistil stalked .... EUPHORBIACEAE

            kk. Cell sap clear; sepals white or green
              .... PHYTOLACCACEAE

      ii. Corolla present

        j. Flowers lavender;
          calyx tubular .... PLUMBAGINACEAE

        jj. Flowers yellow or white; sepals and pet-
          als each 4; stamens 6 .... BRASSICACEAE
                                        (Cruciferae)

cc. ZYGOMORPHIC (perianth irregular)

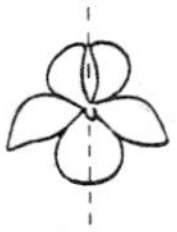

zygomorphic d. Leaves deeply cut or compound

e. Stamens many; pistils 1–3;
flowers blue .... RANUNCULACEAE

ee. Stamens 5–10; pistil 1; various colors

f. Leaves with stipules; sepals mostly 5

petiole

g. A gland on the petiole; stamens opening
by pores; flowers slightly zygomorphic   CAESALPINIACEAE

gg. No gland; flowers quite zygomorphic .   FABACEAE
(Leguminosae)

ff. Leaves without stipules; sepals 2, petals 4, stamens 6 .... FUMARIACEAE

dd. Leaves simple

e. Leaves mostly heart-shaped with stipules ...   VIOLACEAE

ee. Leaves ovate or narrow

f. Leaves entire; flowers rose-purple ....   POLYGALACEAE

ff. Leaves toothed; flowers axillary

g. Leaves with a few teeth; flowers
greenish ....   VIOLACEAE

gg. Leaves with many teeth;
flowers yellow ....   BALSAMINACEAE

bb. EPIGYNOUS
(perianth above the ovary)

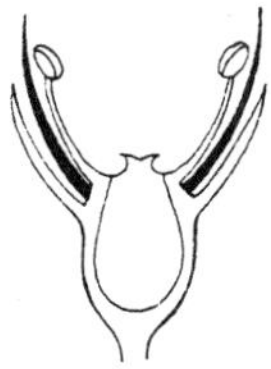

epigynous

actinomorphic c. ACTINOMORPHIC (perianth regular)
– – – cc. page 17

d. No leaves; stems spiny, ovate, fleshy; petals
many ....   CACRACEAE

15

dd. Leaves present

　　e. Leaves opposite or whorled

opposite 　　　　whorled

　　　　f.　Flower parts 3 or 5;
　　　　　　woody or herbaceous

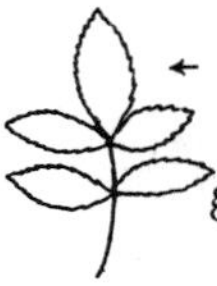

　　　　　　g.　Leaves compound .... ARALIACEAE

　　　　　gg.　Leaves simple

　　　　　　　h.　Epiphyte growing on
　　　　　　　　　trees .... LORANTHACEAE

　　　　　　hh.　Not epiphytic

　　　　　　　　i.　Leaves heart-shaped, 2; sepals 3 ... ARISTOLOCHIACEAE

　　　　　　　　ii.　Leaves ovate, fleshy .... PORTULACACEAE

　　　　ff.　Flower parts in 2's or 4's or multiples

　　　　　　g.　Pistil cylindric or tiny globoid with hooked
　　　　　　　　bristles .... ONAGRACEAE

　　　　　gg.　Pistils globoid, smooth .... MELASTOMATACEAE

　　ee. Leaves alternate or basal

alternate 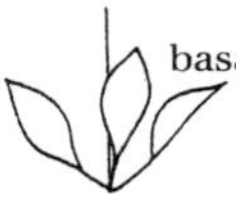　　　　basal

　　　　f.　Leaves compound

umbel

　　　　　　g.　Plant woody, with thorns

　　　　　　　h.　Flowers in umbels .... ARALIACEAE

　　　　　　hh.　Flowers single or few .... ROSACEAE

　　　　　gg.　Plants herbaceous

spike　raceme　　h.　Flowers in spikes or racemes .... ROSACEAE

16

hh. Flowers in umbels; sepals minute or
lacking

     i.   Leaves whorled or single ....   ARALIACEAE

    ii.   Leaves several; outer flowers of um-
bels may be zygomorphic ....   APIACEAE
(Umbelliferae)

  ff. Leaves simple; entire or toothed,
petals 2–5

    g. Sepals 2; styles 3;
leaves blunt at tip ....   PORTULACACEAE

  gg. Sepals 3–5; style 1; leaves acute

    h. Petals 2 or 4 ....   ONAGRACEAE

    hh. Petals lacking; sepals 3–5, whitish   SANTALACEAE

cc. ZYGOMORPHIC (perianth irregular)

    d. Woody vine; leaves roundish ....   ARISTOLOCHIACEAE

  dd. Herbaceous; upright; leaves ovate ....   ONAGRACEAE

aa. SYMPETALOUS (petals joined
or appearing so); corolla
often deeply lobed

hypogynous   perigynous

  b. HYPOGYNOUS or PERIGYNOUS...bb. page 21

  c. ACTINOMORPHIC .... cc. page 20

      d. Leaves opposite or whorled ..... dd. page 19

sympetalous
actinomorphic e. Cell sap milky; 2 pistils joined by the stigma

    – f.   Flowers in umbels, sepals reflexed.........   ASCLEPIADACEAE

    –ff. Axillary and terminal racemes; sepals up-
right ....   APOCYNACEAE

ee. Sap clear; leaves simple

    f. Leaves entire

        g. Fertile stamens fewer than petals (2 or 4)

           h. Pistil 4-lobed ....   LAMIACEAE (Labiatae)

          hh. Pistil 2-lobed or not lobed ....   SCROPHULARIACEAE

       gg. Stamens the same number as the petals (4–7)

          h. Evergreen trailing plant with blue axillary flowers ....   APOCYNACEAE

         hh. Non-evergreen, upright

            —i. Stigmas 2-parted

               — j. Inflorescence circinate; pistil 4-lobed ....   BORAGINACEAE

               jj. Not circinate; pistil not lobed .....   GENTIANACEAE

            ii. Stigma single or 3-parted

               —j. Stigma 3-parted; leaves sessile ....   POLEMONIACEAE

              —jj. Stigma 1; leaves with petioles or sessile if whorled ....   PRIMULACEAE

   ff. Leaves toothed

      g. Pistil 4-lobed

        — h. Style from center of lobes ....   LAMIACEAE (Labiatae)

        hh. Style from tip of lobes ....   VERBENACEAE

     gg. Pistil 2-lobed or not lobed

h. Flowers umbellate; leaves evergreen; stamens 10 .... ERICACEAE

hh. Not umbellate; stamens 2 or 4 .... SCROPHULARIACEAE

dd. Leaves alternate or basal

e. Inflorescence circinate

f. Leaves lobed or divided .... HYDROPHYLLACEAE

ff. Leaves entire or shallowly toothed; pistil 4-lobed .... BORAGINACEAE

ee. Inflorescence not circinate

f. Plant a twining vine .... CONVOLVULACEAE

ff. Plant upright or sprawling

g. Leaves deeply cut or compound

h. Leaflets 3, palmate; flowers white .... GENTIANACEAE

hh. Leaflets more than 3, pinnate .... POLEMONIACEAE

gg. Leaves simple

h. At least some leaves toothed or lobed

i. Stamens many in a ring .... MALVACEAE

ii. Stamens 5, separate .... SOLANACEAE

hh. Leaves entire or wavy

i. Petals reflexed; leaves basal .... PRIMULACEAE

ii. Petals upright

j. Leaves alternate .... ERICACEAE

jj. Leaves basal

19

                    k. Style 1 ....   PLANTAGINACEAE

                    kk. Styles 5 ....   PLUMBAGINACEAE

cc. ZYGOMORPHIC
(perianth irregular, some only slightly so)

    d. Leaves opposite

zygomorphic e. Woody vine: leaves compound ....   BIGNONIACEAE

       ee. Herbaceous; leaves simple

          f. Flowers reflexed when in fruit; stamens
            4 ....   PHRYMACEAE

          ff. Flowers upright; 2 or 4 functional stamens

             g. Style from center of 4-lobed ovary ....   LAMIACEAE
                                                  (Labiatae)

           gg. Style from top of ovary

              h. Ovary 4-lobed ....   VERBENACEAE

            hh. Ovary 2-lobed or not lobed ....   SCROPHULARIACEAE

    dd. Leaves alternate or basal

       e. Leaves basal or none, flowers one or few ....   LENTIBULARIACEAE

       ee. Leaves alternate, some also with basal leaves

          f. Inflorescence circinate; hairs stiff   BORAGINACEAE

          ff. Flowers in spikes, racemes or heads; hairs soft
          ....   SCROPHULARIACEAE

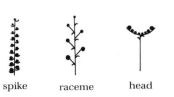

spike   raceme   head

bb. EPIGYNOUS
(perianth above the ovary)

c. ACTINOMORPHIC
(perianth regular)

actinomorphic

epigynous

    d. Leaves opposite or whorled

        e. Flowers in heads or compact spikes

            f. Lobes of the corolla 3 or 5; stamens joined by anthers; disc (center) usually 5 parted corolla; rays (outer) if present 3–5 lobed .... **ASTERACEAE (Compositae)**

           ff. Lobes of corolla mostly 4; stamens separate

               g. Outer flowers in a head; somewhat zygomorphic .... **DIPSACACEAE**

bract

              gg. 4 broad white bracts outside the flowers .... **CORNACEAE**

        ee. Flowers axillary or terminal, not in heads or spikes

            f. Lobes of the corolla 5; style 1 .... **CAPRIFOLIACEAE**

           ff. Lobes of corolla 4 (3); styles or stigmas 2 or 4; pistil 2-lobed or 2-jointed .... **RUBIACEAE**

alternate

    dd. Leaves alternate or basal

basal

        e. Plant evergreen, reclining; 4-parted corolla recurved .... **ERICACEAE**

        ee. Plant not evergreen

            f. A vine with tendrils; leaves palmately lobed .... **CUCURBITACEAE**

tendril

           ff. Plant upright

g. Flowers in heads (They may be all disc flowers, or ray flowers or both in the head) Ray flowers may have 3–5 divisions .....   ASTERACEAE (Compositae)

ray

disc

gg. Flowers axilliary, in spikes or racemes, often bell-shaped ....   CAMPANULACEAE

cc. ZYGOMORPHIC (perianth irregular)

zygomorphic

perfoliate

d. Leaves opposite, some perfoliate

e. Lobes of the corolla 5; flowers often axillary   CAPRIFOLIACEAE

—ee. Lobes of the corolla 4 (5); flowers terminal   DIPSACACEAE

dd. Leaves alternate

e. Woody vine; flowers yellow, green, brown or purple ....   ARISTOLOCHIACEAE

ee. Herbaceous; flowers white, blue or red   LOBELIACEAE

PLANTS GREEN · LAND PLANTS · FLOWERS *NOT* SHOWY
MONOCOTYLEDONS

– a. Flowers in spikes; leaves strap-shaped from ground
....   TYPHACEAE

aa. Flowers in heads or thick spikes

–b. Flowers in spherical heads with staminate ones above the pistillate ones; leaves narrow, clasping ....   SPARGANIACEAE

– bb. Flowers mostly in thick spikes (spadix) with or without a surrounding leaf-like spathe; leaves various ....   ARACEAE

DICOTYLEDONS

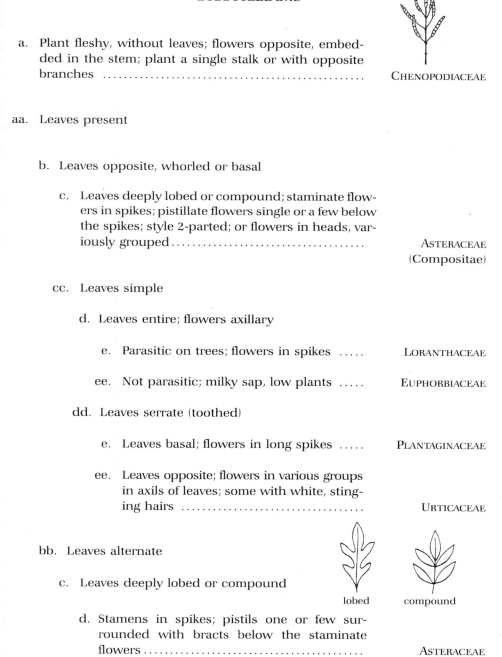

a. Plant fleshy, without leaves; flowers opposite, embedded in the stem; plant a single stalk or with opposite branches .... CHENOPODIACEAE

aa. Leaves present

   b. Leaves opposite, whorled or basal

      c. Leaves deeply lobed or compound; staminate flowers in spikes; pistillate flowers single or a few below the spikes; style 2-parted; or flowers in heads, variously grouped................................... ASTERACEAE (Compositae)

     cc. Leaves simple

        d. Leaves entire; flowers axillary

           e. Parasitic on trees; flowers in spikes ..... LORANTHACEAE

          ee. Not parasitic; milky sap, low plants ..... EUPHORBIACEAE

       dd. Leaves serrate (toothed)

           e. Leaves basal; flowers in long spikes ..... PLANTAGINACEAE

          ee. Leaves opposite; flowers in various groups in axils of leaves; some with white, stinging hairs .... URTICACEAE

 bb. Leaves alternate

    c. Leaves deeply lobed or compound

lobed   compound

       d. Stamens in spikes; pistils one or few surrounded with bracts below the staminate flowers.. ASTERACEAE (Compositae)

dd. Flowers in leafy panicles or spikes; Leaves palmately compound, leaflets narrow .... MORACEAE

cc. Leaves simple

  d. Leaves lobed or toothed

    e. Pistil more than one

      f. Pistils 5 (joined at base); one-sided racemes CRASSULACEAE

      ff. Pistils 2–3; somewhat woody vine MENISPERMACEAE

    ee. Pistil 1 (there may be 2 or more styles or stigmas)

spike

      f. Thick spikes; leaves thick .... AMARANTHACEAE

      ff. Thin spikes, panicles, or axillary flowers

        g. One or more axillary burs with hooked
prickles; a large coarse plant .... ASTERACEAE
(Compositae)

        gg. Plant otherwise

          h. Leaves broadly ovate with white
stinging hairs; flowers small, crowded
.... URTICACEAE

          hh. Leaves oval or mostly triangular;
flowers axillary or in spikes .... CHENOPODIACEAE

  dd. Leaves entire, simple

    e. Leaves broadly ovate or roundish

      f. Plant creeping; spike of white flowers ..... SAURURACEAE

      ff. A vine or shrubby plant

        g. Stipules surrounding stem POLYGONACEAE

       gg. No stipules; a woody vine ....   MENISPERMACEAE

  ee. Leaves narrow

     f. Woody; leaves tiny ....   EMPETRACEAE

   ff. Herbaceous

      g. Pistil stalked; cell sap milky ....   EUPHORBIACEAE

    gg. Pistil sessile; sap not milky

       h. Stipules surrounding stem nodes .....   POLYGONACEAE

     hh. No stipules around nodes

        i. Flowers in axils, either a few or in racemes; some with spiny bracts ..   CHENOPODIACEAE

       ii. Flowers in heads near tip; plant with few branches ....   ASTERACEAE
                                            (Compositae)

PLANTS GREEN • AQUATIC PLANTS
MONOCOTYLEDONS = M
DICOTYLEDONS = D

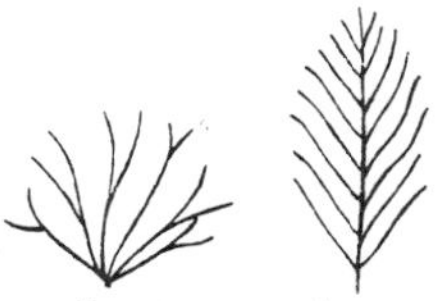

a. Leaves finely divided or compound

  b. Leaves opposite, palmately divided; flowers axillary   palmate   pinnate

    c. Flowers peduncled ....   NYMPHAEACEAE–D

   cc. Flowers sessile ....   CERATOPHYLLACEAE–D
    peduncle
  bb. Leaves whorled

    c. Leaves pinnately divided; flowers in a spike   HALORAGACEAE–D cc. Leaves variously divided; flowers few, yellow   LENTIBULARIACEAE–D

aa. Leaves simple, entire (filamentous, strap-shaped or ovate)

    b. Plants tiny, ovate-thin or fleshy-floating; with or
       without roots; seldom flowering ....   LEMNACEAE–M

    bb. Leaves small or large; flowers on peduncles

        c. Flowers in short spikes or umbels; some with
          leaves both under water and floating ....   NAJADACEAE–M

        cc. Flowers single on peduncles

          d. EPIGYNOUS; leaves short, whorled or long,
            strap-shaped, basal; flowers white   HYDROCHARITACEAE–M

          dd. HYPOGYNOUS

            e. Leaves broad, floating; petals 3-many;
              stamens often many ....   NYMPHAECEAE–D

            ee. Leaves narrow, submerged; flowers
              yellow ....   PONTEDERIACEAE–M

PLANTS *NOT* GREEN · DICOTYLEDONS

a. Stems twining upon other plants, yellow; flowers in small
    groups ....   CONVOLVULACEAE

aa. Stems upright; plants small; leaves none or reduced to
    scales; flowers and stems similar in color, white, tan,
    reddish and purplish

    b. POLYPETALOUS (petals separate);
       ACTINOMORPHIC (perianth regular) ....   ERICACEAE

    bb. SYMPETALOUS (petals united);
       ZYGOMORPHIC (perianth irregular)
       flowers single or grouped ....   OROBANCHACEAE

Plants of Woodlands

When this country was first settled by Europeans, the most abundant type of natural vegetation in northeastern North America was forest vegetation. Although there were swamps and undoubtedly many open areas caused by natural phenomena such as windfall and fires, forests were the main features of the landscape. Fossil pollen studies have suggested that these forests persisted with little change in floristic composition for thousands of years before settlement. A number of significant changes have taken place since. At first trees were cut to clear the land for crops and grazing. Then commercial timber harvesters worked their way westward, from Maine to Minnesota, frequently accompanied by devastating fires which destroyed millions of acres of forestland and took many human lives. Less than half of the original commercial forest remains today, with only a few small isolated patches of virgin timber.

Climatically, all of the Northeast continues to be potential forestland, and any open space, left undisturbed, will be invaded by and eventually dominated by trees. For example, an uncultivated cornfield will produce a rich crop of annual weeds the first year. Over a period of several years, shrubs and tree seedlings will become established. Trees will eventually develop an overtopping canopy and shade the open field species, which will be replaced by plants that can survive in lower light concentrations. The trees resulting from these successional changes will not be replaced by other species, and they therefore become the climax vegetation.

The climax forest is made up of four vertical levels of vegetation: (1) The dominant trees that make up the canopy, (2) A layer of smaller trees that grow to the underside of the canopy, but usually do not become a part of it, (3) A shrub zone, and (4) a rich strata of herbaceous plants that carpet the forest floor. The latter are the plants listed in the following section.

The environmental conditions on the forest floor depend on the nature of the canopy. Under a dense canopy that permits little direct sunlight, it will be moist and cool with an herbaceous flora that requires these conditions. Under an open canopy where an abundance of light reaches the forest floor it will be dryer and warmer, allowing a different population of herbaceous plants to develop. There has been no attempt to separate the types of forest conditions in the listing of families because there are so many intermediate situations. The ecological description of each species will identify it as one of dry or moist woods.

The following species are sometimes observed in woods but they also occur in other types of habitats. These plants are described and illustrated in the section that describes the habitat where they are believed to be most common.

Amaryllidaceae, the Amaryllis Family
   Leucojum aestivum, Snowflake
Apiaceae, the Parsley Family
   Zizia aurea, Golden Alexanders
Apocynaceae, the Dogbane Family
   Apocynum androsaemifolium, Spreading Dogbane
   Vinca minor, Periwinkle
Araceae, the Arum Family
   Symplocarpus foetidus, Skunk Cabbage
Asteraceae, the Aster Family
   Antennaria plantaginifolia, Plantain-leaved Pussytoes
   Aster linariifolius, Stiff Aster
   Bidens bipinnata, Spanish Needles
   Erigeron philadelphicus, Philadelphia fleabane
   Lactuca biennis, Tall Blue Lettuce
   Lactuca canadensis, Wild Lettuce
   Rudbeckia laciniata, Green-headed Coneflower
   Rudbeckia triloba, Thin-leaved Coneflower
Balsaminaceae, the Touch-me-Not Family
   Impatiens biflora, Spotted Touch-me-not
   Impatiens pallida, Pale Touch-me-not
Bignoniaceae, the Trumpet Creeper Family
   Campsis radicans, Trumpet Creeper
Brassicaceae, the Mustard Family
   Alliaria officinalis, Garlic Mustard
   Arabis lyrata, Lyre-leaved Rock Cress
   Cardamine bulbosa, Spring Cress
   Cardamine pratensis, Cuckoo Flower
Campanulaceae, the Harebell Family
   Campanula rotundifolia, Harebell
   Specularia perfoliata, Venus's Looking-glass
Commelinaceae, the Spiderwort Family
   Commelina communis, Asiatic Dayflower
Ericaceae, the Heath Family
   Arctostaphylos uva-ursa, Bearberry
Euphorbiaceae, the Spurge Family
   Euphorbia corollata, Flowering Spurge
Fabaceae, the Bean Family
   Lupinus perennis, Wild Lupine
Gentianaceae, the Gentian Family
   Gentiana crinita, Fringed Gentian
   Gentiana andrewsii, Closed Gentian
   Sabatia angularis, Rose Pink

Lamiaceae, the Mint Family
   Glecoma hederacea, Ground Ivy
   Lycopus virginicus, Virginia Bugleweed
   Satureja vulgaris, Wild Basil
   Scutellaria lateriflora, Mad-dog Skullcap
   Stachys hispida, Rough Hedge-nettle
   Trichostema dichotomum, Blue Curls
Liliaceae, the Lily Family
   Smilacina trifolia, Three-leaved Solomon's Seal
   Veratrum viride, False Hellebore
Lobeliaceae, the Lobelia Family
   Lobelia cardinalis, Cardinal Flower
   Lobelia inflata, Indian Tobacco
   Lobelia siphilitica, Great Lobelia
Orchidaceae, the Orchid Family
   Cypripedium reginae, Showy Lady's Slipper
   Habenaria dilatata, Leafy White Orchis
   Listera australis, Southern Twayblade
   Spiranthes cernua, Nodding Lady's Tresses
Ranunculaceae, the Crowfoot Family
   Caltha palustris, Marsh Marigold
   Ranunculus septentrionalis, Swamp Buttercup
   Trollius laxus, Globe-flower
Rosaceae, the Rose Family
   Fragaria vesca, Wood Strawberry
   Potentilla simplex, Common Cinquefoil
Rubiaceae, the Madder Family
   Houstonia caerulea, Bluets
Saxifragaceae, the Saxifrage Family
   Chrysoplenium americanum, Golden Saxifrage
Scrophulariaceae, the Figwort Family
   Chelone glabra, Turtlehead
   Penstemon digitalis, Foxglove Beardtongue
   Veronica arvensis, Corn Speedwell
   Veronica officinalis, Common Speedwell
   Veronica serpyllifolia,Thyme-leaved Speedwell
Solanaceae, the Nightshade Family
   Physalis heterophylla, Clammy Ground Cherry
   Solanum nigrum, Black Nightshade
Urticaceae, the Nettle Family
   Boehmeria cylindrica, False Nettle
   Urtica dioica, Stinging Nettle
Verbenaceae, the Vervain Family
   Verbena urticifolia, White Vervain
Violaceae, the Violet Family
   Viola odorata, Sweet Violet
   Viola sororia, Woolly Blue Violet

Hypoxis hirsuta
Yellow Stargrass

1

Amaryllidaceae
Amaryllis Family

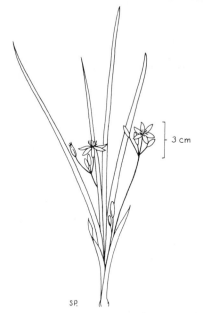

3 cm

S.P.

A perennial with a short, solid, roundish, bulb-like rhizome. The flowers are self-incompatible and cross-pollination is effected mainly by small bees of the genus *Halictus*. The flowers are frequently visited also by

Flowers yellow, to 2 cm. wide (¾ in.), in a terminal cluster, often of 3, on a leafless stem; corolla of 6 divisions, petals 3, petal-like sepals 3, stamens 6; leaves basal, long and narrow, grass-like, hairy, to 35 cm. long (14 in.); flowering stem hairy, to 15 cm. high (6 in.).

April–August

the Meadow Fritillary Butterfly. The seed capsule splits irregularly, releasing several glossy black seeds covered with concentric rows of sharp, wart-like projections.

The 4 species of this genus native to the Northeast have similar growth habits but can be distinguished from one another by seed characteristics. All have long narrow leaves and can be mistaken for grasses in the absence of flowers. *H. micrantha*, not illustrated, is very similar to *H. hirsuta* but has brown seeds with blunt projections. It occurs in sandy pine-lands along the Atlantic coastal plain from Virginia to Florida and Texas.

Yellow Stargrass (*H. hirsuta*) is found in open woods and dry meadows from Maine to Manitoba, south to Florida and Texas.

Osmorhiza claytoni
Hairy Sweet Cicely*

2

Apiaceae or Umbelliferae
Parsley Family

Also Sweet Jarvil

A perennial with spindle-shaped and fibrous roots. The flower clusters consist of a mixture of staminate flowers and those with both stamens and pistils. The anthers mature before the stigmas, reducing the possibility of self-pollination. Fertilization is usually the result of cross-pollination by bees and flies. The bristly seeds disperse by clinging to the coats of passing animals.

This is a genus of about 15 species native to North America, South America, and eastern Asia. The genus name is derived from Greek words meaning "odor" and "root" and

refers to the anise or licorice odor. Of the species native to North America, *O. claytoni* and *O. longistylis* are the most abundant in the Northeast. In an earlier system of classification, they were listed as *Washingtonia claytoni* and *W. longistylis*.

A related species, *O. longistylis*, Smooth Sweet Cicely or Aniseroot (see drawing of fruit), is very similar but its styles are longer than the petals, and it often has a smooth stem. Smooth Sweet Cicely is most abundant in the states south of New England and New York.

Flowers white, tiny, in clusters of umbels, at the ends of stems and branches, styles shorter than petals; leaves mostly alternate, twice ternately divided, lower ones with petioles, upper ones sessile, leaflets toothed; stems soft-hairy, slender, to 90 cm. high (3 ft.).

May–June

Hairy Sweet Cicely is found in moist woods from Nova Scotia and Quebec to Saskatchewan, south to North Carolina, Alabama, and Oklahoma.

✗ The roots of all the native species of this genus are edible and contain anise oil. Those that are too strong to be eaten can be used as flavoring.

℞ The root has been used to relieve mucus congestion, upset stomach, intestinal gas, and to improve the appetite.

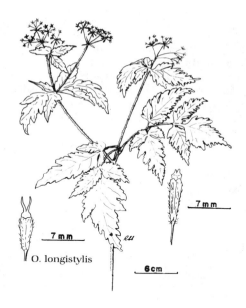

O. longistylis

Arisaema triphyllum
Jack-in-the-pulpit*

3

Araceae
Arum Family

*Also Indian Turnip

Flowers enclosed in a brown or purple-brown spathe, the "pulpit," with over-arching pointed flap; staminate or pistillate flowers or both borne at base of club-shaped spadix or "Jack"; fruit a cluster of shiny, bright red berries; leaves 1 or 2, long stalked, with 3 leaflets, overtopping the spathe, to 90 cm. high (3 ft.).

April–July

A perennial with an enlarged underground solid bulb. Staminate and pistillate flowers are usually on different plants, and cross-pollination is effected by beetles, small flies, and gnats. The spathe withers with age exposing the cluster of 1–3 seeded scarlet berries. The seeds are dispersed by the Ring-necked Pheasant, Wild Turkey, Wood Thrush, and other birds and mammals that eat the berries.

This is a somewhat variable species with three varieties based on spathe and leaflet characteristics. These are recognized by some botanists as distinct species. A closely

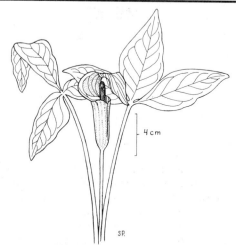

related species, *A. dracontium*, Green Dragon, not illustrated, has 7–13 leaflets and a spadix that protrudes considerably beyond the spathe. It occurs throughout the Northeast but is more common in the western part.

Jack-in-the-pulpit is found in moist to wet

woods from Nova Scotia and Quebec to Minnesota, south to Florida and Louisiana. ✖ The enlarged underground base can be dried and ground into a flour. American Indians pounded the turnip shaped root to a pulp and allowed it to dry for several weeks before grinding it into flour. *Caution*: The fresh root contains calcium oxalate crystals which, if eaten, pierce the tongue and mucous membranes of the mouth causing an intense burning sensation. The crystals are broken up by thorough drying.

℞ Some tribes of American Indians used the powdered root in water to induce sterility. Other tribes applied the powder to the head and temples as a cure for headache. In herbology the powdered root is recommended for croup, whooping cough, asthma, rheumatism and pains in the chest. The juices of the leaves and roots may cause dermatitis in sensitive individuals.

Aralia nudicaulis
Wild Sarsaparilla

4

Araliaceae
Ginseng Family

6 cm

Flowers greenish-white, on a leafless stalk, usually in 3 globular, terminal clusters, each to 5 cm. wide (2 in.); 5 long, white stamens per flower give a fuzzy appearance to each cluster; fruit a blue-black berry; the flower is overtopped by a single, 3-parted basal leaf with each part pinnately divided into 3–5 finely toothed leaflets, to 40 cm. high (16 in.).

May–July

A. racemosa, Spikenard, neither illustrated, are herbaceous species with leafy stems to 2 m. high (6½ ft.). *A. hispida* is characterized by sharp slender spines on the lower part of its stem. *A. spinosa*, Hercules' Club, not illustrated, is a large shrub or small tree with thorny stems and leaves.

Wild Sarsaparilla is found in moist or dry woods from Newfoundland to British Columbia, south to Georgia, Missouri, and Colorado.

✖ The aromatic rhizomes of the herbaceous species described above may be used to make a tea or as a source of flavoring for root beer. They are not related to true Sarsaparilla (*Smilax officinalis*), which is a native of South America. Indians of the Northeast are said to have used the rhizome of Wild Sarsaparilla for food on long marches and hunting trips.

℞ In herbology, Spikenard and Wild Sarsaparilla have similar uses. The boiled and

A perennial with a long, horizontal, aromatic rhizome. Each berry usually contains 5 elliptical, blue-gray, finely wrinkled seeds. This species is not a major wildlife food plant but the berries are occasionally a late summer or autumn source of food for the White-throated Sparrow, Wood Thrush, foxes, skunks, and chipmunks. These consumers probably serve as agents of seed dispersal.

This is a genus of Asia and North America represented by 4 woodland species in the Northeast. *A. hispida*, Bristly Sarsaparilla, and

powdered rhizome of Wild Sarsaparilla was used by several groups of American Indians to make a cough medicine. The rhizome of Spikenard was used, sometimes with the bark of White Pine, to make a popular nine-teenth-century commercial cough syrup. This species was also used by Cherokee Indians and white residents of the southern Appalachian Mountains for backache and rheumatoid arthritis.

Panax quinquefolium
Ginseng*

5

Araliaceae
Ginseng Family

Also Sang

Flowers greenish-white, very small, in a terminal umbel, petals 5, stamens 5; in fruit a cluster of bright red berries; leaves in a whorl of 3, palmately compound, long stalked; leaflets usually 5, toothed, pointed; stems unbranched, smooth, to 40 cm. high (16 in.).

June–July

A perennial with a large, thick, often forked root. Three types of flowers are produced: pistillate, staminate, and those with both pistils and stamens. Self-pollination is rare, thus the Lily-of-the-valley odor of the staminate flowers may serve as an attractant for insect pollinators. Each of the bright red berries is about 1 cm in diameter (½ in.) and contains 2–3 gray, rough-coated seeds.

This species was once fairly abundant in eastern North America, but is now rare mainly as a result of collections for folk medicine, especially for the Chinese market. It is used as a substitute for Chinese Ginseng (*P. pseudoginseng*) which is a panacea in Chinese folk medicine. Although it is grown commercially for export, the wild roots bring the best prices from Ginseng dealers. It is protected by law in some parts of the Northeast.

This is a genus of seven species native to eastern North America and eastern Asia. The genus name is derived from Greek words that mean "all healing" and refers to supposed medicinal properties. The two-forked root was imagined to resemble a human body and is the source of the name Ginseng, which is a corruption of the Chinese Jin-chen, meaning man-like.

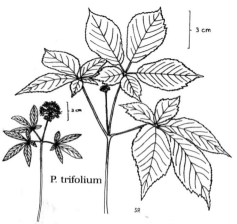

P. trifolium

A closely related species, *P. trifolium*, Dwarf Ginseng or Ground-nut (see illustration), is much smaller, to 20 cm. high (8 in.); flowers April–May, has white flowers followed by yellow berries; leaves commonly with 3 leaflets; a globular tuberous root. Dwarf Ginseng is found in moist woods from Nova Scotia and Quebec to Minnesota, south to Pennsylvania and Iowa, and along the mountains to Georgia.

Ginseng is found in moist cool woods from Quebec to Manitoba, south to Georgia and Oklahoma.

✖ The roots of both Ginseng and Dwarf Ginseng are edible, raw or cooked, but they should be gathered for food only in emergencies. The leaves are reputed to make a palatable tea.

℞ Chinese Ginseng has been used for thousands of years in Chinese folk remedies for a variety of ills and as an aphrodisiac. Although American Ginseng is an acceptable

substitute with similar medicinal properties, it has never achieved the level of use
in western medicine as it has in the Orient.

Recent investigations by Chinese, Japanese
and Russian scientists have demonstrated
that it is a tonic and a stimulant.

Aristolochia durior **Dutchman's Pipe***	**6**	*Aristolochiaceae* **Birthwort Family**

*Also Pipe-vine

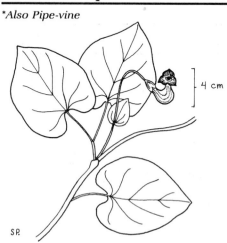

S.P.

A perennial, woody, high-climbing vine.
Plants of this genus have evolved a very interesting method of assuring cross-pollination. The flower has a peculiar odor that
attracts certain small gnats and flies. These
enter the throat of the flower tube easily but
downward pointing hairs prevent them from
escaping. In the base of the tube the stigmas
are receptive and can be pollinated, but the
anthers have not yet opened. After 2–3 days
the anthers mature and the insects are
dusted with fresh pollen. The hairs in the
throat of the tube wilt, and the insects escape to repeat the process with another
flower.

The cylindrical seed capsule is up to 10
cm. long (4 in.) and 5 cm. wide (2 in.). It

4 cm

*Flowers brown-purple, on long stalks from
leaf axils; petals none, calyx tube curved or
pipe-shaped with 3 petal-like lobes; leaves
alternate, broadly heart-shaped, finely
hairy on underside, length and width about
the same; stems twining on other vegetation, to 20 m. long (67 ft.).*

May–June

splits into 6 sections, releasing numerous
flattened heart-shaped seeds, each with a
soft appendage that may function to attract
ants as agents of dispersal.

This species is often cultivated as a climbing trellis plant but is more popular in Europe than in North America. A related
species, *A. tomentosa*, Pipe Vine, not illustrated, is similar but has yellowish calyx
lobes and densely hairy petioles and calyx
tubes. It occurs in moist wooded floodplains from North Carolina to Illinois, south
to Florida and Texas.

Dutchman's Pipe is found in moist upland woods from Pennsylvania and West
Virginia south to Georgia and Alabama.

The genus name is derived from Greek
words that mean "best childbirth," referring
to a supposed usefulness in easing delivery,
thus the name Birthwort. The plants of the
genus have been used for a variety of ailments including indigestion, nausea, snakebite, and impotency. However, some species
of the genus, and perhaps all, are poisonous
to humans and livestock, and they should
not be used in home remedies.

Asarum canadense — Wild Ginger
7
Aristolochiaceae — Birthwort Family

Flowers reddish-brown, to 4 cm. wide (1.5 in.), cup-shaped with 3 pointed lobes; a single flower arising from the juncture of 2 leaf stalks, often resting on the ground; leaves 2, basal, heart-shaped, to 30 cm. high (12 in.).

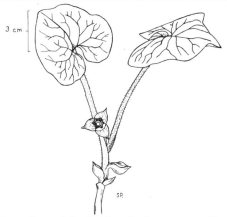

April–May

A perennial with a slender, aromatic rhizome that bears 2 leaves and a flower annually. The flowers are often buried, or partially so, in the leaf litter of the forest floor. Although they are frequently visited by fungus gnats and early flesh flies, fertilization is chiefly the result of self-pollination. The seed capsule splits irregularly to release many transversely ridged, glossy, green or brown seeds. In some European species the seeds are dispersed by ants but further investigation is needed to establish them as agents of dispersal for *A. canadense*.

Wild Ginger is found in rich, moist woods from New Brunswick and Quebec to Minnesota, south to North Carolina and northern Alabama.

✗ The rhizome is edible and has the taste of ginger. In colonial times it was ground and used as a substitute for commercial ginger. A ginger flavored candy was made by boiling the rhizome in a sugar syrup. These plants are usually not abundant, and if local extinction is to be avoided they should not be collected for food or for home remedies.

℞ In folk medicine the rhizome has been used for heart palpitations, chest pains, stomach cramps and intestinal gas. The American Indians used it as a contraceptive taken by women, and as a solution that was poured directly into the ear for infections. Early American settlers mixed the powdered rhizome of Wild Ginger with the powdered barks of Alder, Bayberry, and Black Oak to form a mixture that was applied to the teeth to prevent decay. Wild Ginger contains aristolochic acid, which has antibacterial properties.

All of the above uses can be better served by the local pharmacy.

Aster cordifolius — Heart-leaved Aster
8
Asteraceae or Compositae — Aster Family

A perennial with a short horizontal rhizome and dense clusters of fibrous roots. Its pollen grains are indistinguishable from those of the goldenrod (*Solidago spp.*), as with other asters, and it is insect-pollinated.

Each flower head produces numerous oblong, light brown, 2-ribbed, seed-like fruits with parachutes of hairs that aid in dispersal by air currents.

This is a highly variable species with sev-

36

4 c m

Flower heads with blue or violet rays, to 18 mm. wide (¾ in.), on stalks, in dense, branched, terminal clusters; leaves alternate, lower ones heart-shaped with long petioles, sharply toothed, to 15 cm. long (6 in.); stems erect, smooth below the flower cluster, to 1.2 m. high (4 ft.).

August–October

eral varieties based on leaf and flower variations recognized by some botanists. It is believed to hybridize with at least two other species and some of the varieties may be hybrids. A single plant may bear more than 100 flower heads, each with 10–20 ray flowers.

Heart-leaved Aster is found in dry or moist woods, thickets, and clearings from Nova Scotia and Quebec to Minnesota, south to Georgia and Missouri. For additional information on this genus, see *A. novae-angliae.*

Aster macrophyllus **Large-leaved Aster**	**9**	*Asteraceae or Compositae* **Aster Family**

4 cm

Flower heads with violet to lavender rays, to 2.5 cm. wide (1 in.), in a branched terminal cluster; stalks of flower heads with sticky glands; leaves alternate, heart-shaped, basal leaves very numerous, to 20 cm. wide (8 in.), stem leaves smaller; stems coarse, purplish, hairy, to 1.2 m. high (4 ft.).

July–October

A perennial with a thick creeping rhizome that produces flowering stems and dense clumps of basal leaves. Each flower head bears several smooth, longitudinally ribbed, seed-like fruits with parachutes of hairs that aid in dispersal by air. The very large and abundant basal leaves appear early in the spring and remain a conspicuous aspect of the forest floor throughout the summer and autumn.

This is a highly variable species that has been segregated by some botanists into several geographic varieties. Older systems of classification listed some of these as separate species. The flower head has 10–20 ray flowers and numerous disk flowers that turn purplish-brown as the season advances. Forms with white ray flowers are rarely observed.

Large-leaved Aster is found in dry or moist woods from Nova Scotia and Quebec to Manitoba, south to West Virginia and Indi-ana and along the mountains to Georgia.
✗ The young basal leaves can be cooked as a spinach-like green. The older leaves are less palatable, since they soon become tough and leathery, but they can still be eaten as a survival food. Some tribes of American Indians enjoyed the leaves boiled with fish.

Eupatorium rugosum
White Snakeroot

10

Asteraceae or Compositae
Aster Family

Flower heads showy white, in dense clusters at the tips of stems and upper axillary branches; leaves opposite, oval, coarsely toothed, long petioled; stems smooth, stiff, to 1.2 m. high (4 ft.).

July–October

A perennial with a knobby rhizome and fibrous roots. The flowers are self-incompatible and cross-pollination, often effected by bees, is necessary for the production of seeds. Each flower head produces 20–30 oblong, black, longitudinally ridged seeds, with tufts of barbed bristles that aid in dispersal by wind.

This is a highly variable species with several recognized varieties based on the shape of the bracts surrounding the flower heads, and the amount and location of hairs on the stems and leaves. An older name for the species, *E. urticaefolium*, refers to the similarity of the leaves to those of *Urtica*, the genus of the nettles. This is perhaps the most delicate and attractive species of the genus, and its conspicuous white flowers brighten many late summer and autumnal woodlands of northeastern North America.

A closely related species, *E. aromaticum*, not illustrated, is very similar but is not as tall, averaging about 80 cm. (32 in.). It occurs in sandy areas along the Atlantic Coast from Massachusetts to Florida and Mississippi. Another similar species, *E. luciae-brauniae*, not illustrated, is more delicate, slender, to 60 cm. high (2 ft.), and occurs around sandstone outcrops in eastern Kentucky.

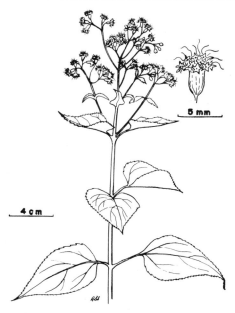

5 mm

4 cm

White Snakeroot is found in moist woods, clearings and shady damp pastures from New Brunswick to Saskatchewan, south to Florida and Texas.

☠ The stems and leaves of White Snakeroot contain a poisonous alcohol known as tremetol. Poisoning of cattle and sheep has resulted from eating this plant, most often when there is a shortage of other forage. Eating the plant fresh, or dried in hay, causes trembling, difficulty in breathing, collapse and often death. The toxic alcohol is passed on to humans in milk and milk products, and causes a disorder known as milk sickness. This was a dreaded disease in colonial times, resulting in death for 10–25% of those who were poisoned. Modern processing methods have greatly reduced the danger of poisoning.

Helianthus decapetalus **Thin-leaved Sunflower** **11** ***Asteraceae or Compositae*** **Aster Family**

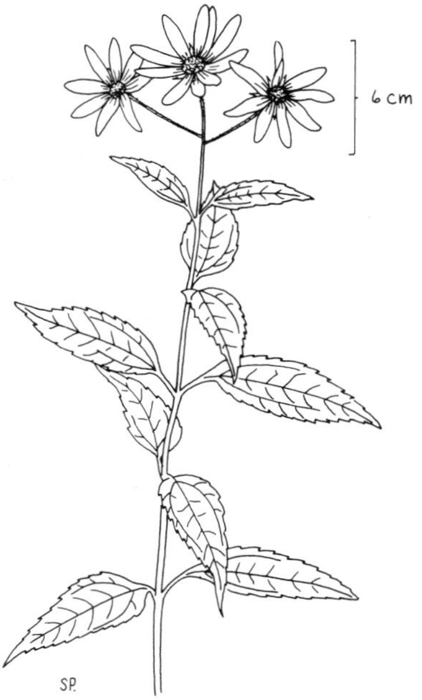

Flower heads with 8–15, often 10, yellow rays, usually several on terminal branches, to 7.5 cm. wide (3 in.); leaves opposite, upper ones sometimes alternate, toothed, with a sharp tip, broadest near base, tapering abruptly to a winged petiole; stems slender, hairy on upper part, to 1.5 m. high (5 ft.).

August–October

A perennial with a thick rhizome and fibrous roots. The flower head produces numerous smooth, dark seeds, each with 2 scale-like awns that are soon shed. The value of this genus as food for wildlife is moderate in the Northeast but is outstanding in the prairies and other parts of the West. The part of the plant that is eaten is mainly the seeds.

This is a genus of about 60 species, most of which are native to North America. Some of the species are difficult to distinguish and hybridization may take place between *H. decapetalus* and several others. There are about 18 species in the Northeast and numerous others in the West. Sunflower is the state flower of Kansas.

Thin-leaved Sunflower is found in moist open woods and woodland borders and along stream banks from Quebec to Minnesota, south to Georgia and Missouri.

✕ The American Indians were using the sunflowers for food long before the arrival of Europeans. The seeds of the larger species were roasted and ground into a meal that was used to make bread-like foods and to enrich soups. The meal was mixed with bone marrow to make a concentrated food for traveling. The unopened flower buds were cooked as a vegetable.

Common Sunflower, *H. annuus*, not illustrated, is grown commercially today in many parts of the world. The seeds are used for chicken feed, wild bird feeding stations, and human snacks. They contain much oil, which is extracted for use in cooking and soap making. The roasted seed shells can be used to make a drink that is reputed to taste very much like coffee.

Helianthus divaricatus
Woodland Sunflower

12

Flower heads with yellow rays, 1 to few at stem tip, to 5 cm. wide (2 in.); leaves opposite, sessile or short petioled, tapering to a long pointed tip, hairy, to 15 cm. long (6 in.); stems hairy in upper part, usually branched at tip, to 1.5 m. high (5 ft.).

July–September

A perennial with a creeping rhizome and fibrous roots. Each of the showy flower heads produces numerous smooth, longitudinally ridged black or slightly mottled seeds. These are a source of food for numerous upland gamebirds, songbirds, and small mammals.

A related species, *H. hirsutus*, Stiff-haired Sunflower, not illustrated, is similar but has slightly larger flower heads and leaves that are more densely hairy. Stiff-haired Sunflower usually does not occur north of Pennsylvania in the Northeast.

Woodland Sunflower is found in dry open woods, woodland borders, and shady roadsides from Quebec to Saskatchewan, south to Florida and Louisiana.

For information on ecology and food uses see *H. decapetalus*, Thin-leaved Sunflower.

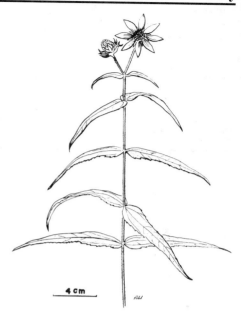

4 cm

Helianthus strumosus
Pale-leaved Wood Sunflower

13

Flower heads with yellow rays, few to several at the tips of stems and branches, rays 8–15; leaves opposite, upper ones sometimes alternate, usually with short petioles, light green on underside, hairy; stems usually branched in upper part, often smooth, with a white bloom, to 2 m. high (6½ ft.).

July–September

A perennial with an elongate branched rhizome, sometimes forming slender tubers, and woody roots. Each flower head produces numerous smooth dark seeds. As with other members of the genus, the seeds are eaten by several species of birds and small mammals.

H. strumosus is one of several very closely related and similar species. Sometimes they grade into one another and are difficult to identify. Hybridization may occur when their

6 cm

40

ranges overlap. In some characteristics *H. strumosus* is intermediate between *H. hirsutus*, Stiff-haired Sunflower, and *H. decapetalus*, Thin-leaved Sunflower.

Pale-leaved Wood Sunflower is found in dry, open woods and thickets from Quebec to Minnesota, south to Florida and Texas.

For information on ecology and food uses see *H. decapetalus*.

| *Prenanthes altissima* Tall White Lettuce* | **14** | *Asteraceae or Compositae* Aster Family |

*Also Rattlesnake Root

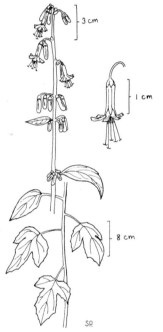

SP.

A perennial with milky juice and rounded tuberous roots. It produces spiny pollen grains similar to those of Dandelion (*Taraxicum officinale*). Each of the numerous heads bears 5–6 narrow, brown, prominently ribbed seeds with tufts of creamy white hairs that aid in dispersal by wind.

This genus includes about 25 species of North America, Europe, and Asia. The genus name is derived from Greek words that mean

Flower heads greenish-white, nodding, with protruding styles, usually surrounded by 5 bracts; leaves alternate, lower ones with long petioles, variable in shape from triangular or arrow-shaped to 3–5 lobed; stems hairy at base, slender, often tinged with purple, to 2 m. high (6½ ft.).

July–October

"drooping flowers" and refers to the clusters of nodding flower heads. Eleven native, mostly woodland, species grow in eastern North America. The flower heads of American species have fewer flowers than their European relatives, and this characteristic has been used by some botanists to place them into a separate genus, *Nabalus*.

Tall White Lettuce is found in moist woodlands from Newfoundland and Quebec to Manitoba, south to Georgia and Louisiana. A variety with cinnamon colored parachutes of hairs on the seeds occurs from Indiana to Missouri and Louisiana.

℞ The tuberous roots of all members of this genus are intensely bitter, and this has probably contributed to their use in home remedies (perhaps based on the idea that if it tastes bad it must be good for you). *The Dispensatory of the United States of America* (1851) reported that a North Carolina physician had successfully treated a dozen cases of rattlesnake bite by internal administration of a tonic made from the roots. In more recent herbology, the tonic is recommended for diarrhea and dysentery.

41

Woodlands

Senecio aureus
Golden Ragwort*

15

Asteraceae or Compositae
Aster Family

*Also Squaw-weed
Ray flowers yellow or orange, flower heads
few-many, on long stalks, in flat topped
terminal cluster; basal leaves long petioled,
heart-shaped at base, toothed; stem leaves
alternate, smaller, pinnately lobed, upper
ones sessile; stems not branched below
flower-head clusters, to 1 m. high (40 in.).

April–August

A perennial with a branched creeping rhizome. Pollination is probably by bees and fertile seeds are produced by both ray and disk flowers. The seed-like fruits are windborne with the aid of a parachute of white hairs. The plants of this genus contain toxic pyrrolizidine alkaloids which give them a bitter taste and are believed to be an adaptation to protect them from grazing animals. Two European species (S. jacobaea and S. vulgaris, neither illustrated), both of which have been introduced into North America, are fed upon by a moth larva that has developed the ability to isolate and store, in its body, the poisonous substances.

S. aureus is a variable species of moist woods, woody swamps, bogs, and wet meadows ranging from Labrador to Minnesota, south to Georgia and Arkansas.

☠ The plant has been used to promote the menstrual flow. It is said to have been used as a tea by the women of Catawba American Indians to ease the pains of childbirth and to hasten labor.

As indicated above, however, most of the species of the genus contain toxic alkaloids which have caused illness and death to livestock in many parts of the world. Cattle have most often been poisoned, with horses and sheep to a lesser extent. From Africa and the West Indies there have been reports of human poisoning.

Silphium perfoliatum
Cup-plant

16

Asteraceae or Compositae
Aster Family

A tall coarse perennial with resinous juice. The disk flowers are bisexual but the pistils are sterile. Each of the 20–30 ray flowers produces a very thin seed-like fruit with a well developed marginal wing which aids in dispersal by wind. Plants with hairy stems occurring in Virginia, West Virginia, and North Carolina are sometimes listed as S. connatum, but they may be a variety of S. perfoliatum.

This is a genus with about 8 species native to the northeast. Two of these, S. perfoliatum and S. laciniatum, Compass Plant, not illustrated, are occasionally cultivated as border plants. The name Compass Plant refers to the tendency of the lower leaves to orient themselves, but not reliably so, in a north-south direction.

Cup-plant is found in moist woods and thickets, along stream banks, and in prairies

42

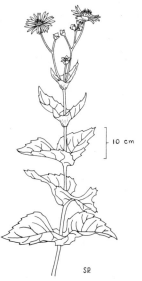

10 cm

S.P.

from Ontario to South Dakota, south to Georgia and Louisiana.

Flower heads with numerous yellow rays, few to several in branched, terminal clusters; leaves opposite, coarsely toothed, upper ones joined at base forming a cup around the stem; stems square, smooth, often branched above middle, to 2.4 m. high (8 ft.).

July–September

✘ When the flower heads are broken off, a drop of juice is exuded. After it hardens this can be used as a natural chewing gum. Several species of the genus exhibit this property. North American Indians and early settlers are said to have used the gum from these plants to clean their teeth and keep them white.

℞ In herbology, the root is reported to be a tonic and a diaphoretic that has been used for fevers, ulcers, liver problems and feebleness. Chewing the gum is said to sweeten the breath and soothe the nerves.

**Solidago bicolor
Silverrod***

17

**Asteraceae or Compositae
Aster Family**

*Also White Goldenrod

6 mm

4 cm

R.U.

Ray flowers silvery white, flower heads in a long, cylindrical, terminal cluster, with leaf-like bracts in lower part; leaves alternate, hairy, tapering at base and tip, lower ones toothed, upper ones smaller, entire; stems often unbranched, hairy, to 80 cm. high (32 in.).

July–October

A perennial with a persistent stem base and a fibrous root system. This is the only member of the genus with white ray flowers. Pollination is by insects but small amounts of pollen may become airborne. However, it is not a hayfever plant. It is not often listed as a wildflower, but it is a very attractive plant and never a troublesome weed.

The seed-like fruits have tufts of white hair that aid in wind dispersal, and they are eaten by several species of songbirds. The vegetation is eaten sparingly by rabbits and other small mammals. The abundance of goldenrods in North America is indicated by their selection as state flowers in Alabama, Kentucky, and Nebraska, but the intended species is not always clear.

Silverrod is found in dry, thin woods and open sandy or rocky areas from Nova Scotia and Quebec to Minnesota, south to Georgia and Arkansas.

For additional information see *S. caesia* and *S. sempervirens*.

Solidago caesia
Blue-stem Goldenrod

18

Flower heads with 3–5 yellow petal-like ray flowers, in axillary clusters with subtending leaves extending beyond the cluster; leaves alternate, sessile, lance-shaped with serrated margins; stems smooth, bluish, glaucous, to 1 m. tall (40 in.).

August–October

A perennial with a short thick rhizome and many fibrous roots. Pollination is effected mainly by bees and short-tongued flies. The seed-like fruits are hairy, longitudinally ribbed, and have a parachute of hairs at one end to aid in wind dispersal.

The goldenrods include about 100 species confined mainly to North America and concentrated in the Northeast. As wildlife food their uses are low in comparison to their abundance. Among the documented consumers are the Ruffed Grouse, Common Goldfinch, Junco, Cottontail Rabbit, Wood Rat, and White-tailed Deer.

Blue-stem Goldenrod is a species of open dry or shady moist woods ranging from Nova Scotia and Quebec to Wisconsin, south to Florida and Texas.

℞ In general the goldenrods are not important hayfever plants. Although most people who are sensitive to ragweed pollen show almost the same sensitivity to goldenrod pollen, the latter seldom gets into the air in more than minute quantities. Goldenrods are often blamed by hayfever sufferers

because their showy flowers are abundant during the flowering time of ragweeds. There are occasions, however, when goldenrod pollen may be of local importance. In dry windy weather at the end of the ragweed season, some species of goldenrods may be at their peak and their pollen grains in the atmosphere may outnumber those of ragweed.

For additional information see *S. sempervirens*.

44

*Also Papoose-root, Squaw-root

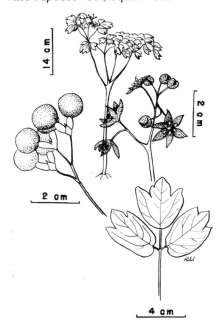

2 cm

4 cm

A perennial with a thick, knobby rootstock. The flower clusters appear in spring before the leaves are fully developed. For a given plant, the stigma matures before the anthers, thus cross-pollination, chiefly by bumblebees and other bees, is assured. Unlike most flowering plants, the seeds at maturity are not enclosed in the ovary but are borne berry-like on individual stalks. This

Flowers greenish yellow to brownish maroon, in branching terminal clusters, 6 petal-like, pointed sepals; the ovary splits early exposing the seeds which, when mature, resemble dark blue berries on thick stalks; leaf 1, sessile, 3-ternately compound, with leaflets resembling those of Meadow Rue; stems smooth, glaucous when young, to 90 cm. high (3 ft.).

April–June

has been interpreted as an adaptation for dispersal by birds.

This is the only species of the genus in North America. One other species is located in eastern Asia.

Blue Cohosh is a climax forest species ranging from New Brunswick to Manitoba, south to Alabama and Missouri.

℞ The dried rootstock has a record of use among several tribes of American Indians, and later by early settlers, to induce the menstrual flow and to hasten childbirth. Thus the common names Papoose-root and Squaw-root. In herbology, it is reported to be useful also in the treatment of epilepsy, to increase the flow of urine and perspiration, and to rid the body of intestinal worms.

Livestock do not usually eat the plant, but children have been poisoned by eating the berry-like seeds. The rootstock, the leaves, and the seeds all contain the alkaloid methylcytisine. Sensitive individuals may develop dermatitis from handling the plant.

*Also Rheumatism-root

A perennial with a dense, fibrous root system. The pear-shaped seed capsule is about 3 cm. long (1¼ in.) and contains numerous finely wrinkled, glossy, dark brown seeds. At maturity the capsule opens by splitting halfway around near the top, forming a hinged lid. Each seed has a fleshy appen-

dage on one side that is thought to be an adaptation for dispersal by ants.

The flowers resemble somewhat, but are more robust than, the fragile blossoms of Bloodroot. In addition to the native *J. diphylla*, the genus includes one other species that is a native of Manchuria in eastern China. The genus was named for Thomas

Flowers white, solitary, on long leafless stalks, to 2.5 cm. wide (1 in.); petals usually 8, stamens 8, sepals 4, shedding early; leaves basal, on long stalks, with 2 irregularly lobed or toothed leaflets, flowering stem to 20 cm. high (8 in.), elongating in fruit to 40 cm. (16 in.).

April–May

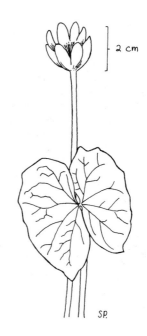

2 cm

SP.

Jefferson, third president of the United States, in recognition of his interest in horticulture and botany.

Twinleaf is found in moist woods, most often in limestone areas, from Ontario to Wisconsin, south to Maryland and Alabama. ℞ The roots have been used as a stimulant and a tonic and are said to be beneficial for chronic rheumatism. Large doses cause vomiting. It has been used as a gargle for throat irritations.

NOTE: This plant is usually not abundant and should not be collected. Its greatest value is the natural beauty it imparts to the woodland habitat.

Podophyllum peltatum
May Apple*

21

Berberidaceae
Barberry Family

Also Mandrake, Wild Lemon, Raccoon Berry, Hog Apple, Wild Jalap
Flowers white, 3–6 cm. wide (1-2.5 in.), 6–9 petals, unpleasant odor; a single, terminal, nodding flower, between two leaf stalks; mature fruit fleshy, many-seeded, yellow (July–Sept.); non-flowering stem with single, radially lobed, peltate leaf, to 50 cm. high (20 in.).

April–May

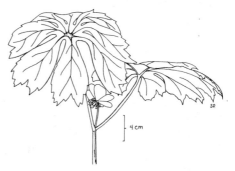

SP.

4 cm

A perennial with a creeping rhizome and fibrous roots. Pollination is by early bees that probably are attracted by the plentiful supply of pollen, and it may occasionally be self-pollinated. The fruit is commonly 5 cm. long (2 in.) at maturity and contains numerous large, light brown, finely wrinkled seeds. These fall to the ground when the fruits decay or they may be ingested by animals that eat the fruits.

May Apple is a colonial species often oc-curring in dense stands to the complete exclusion of other herbaceous plants. It is found in moist woods, shady meadows, roadsides, and forest margins from Quebec to Minnesota, south to Florida and Texas. ✘ The ripe fruit has a sweet strawberry-like flavor and the pulp can be eaten raw or made into marmalade or jelly. As a flavor additive, the juice of the ripe fruit can be added to lemonade or Madeira wine. (The green fruit and the seeds should not be eaten).

℞ May Apple is described in herbology as a remedy for gall bladder disfunctions, kidney stones, constipation, and intestinal worms. The rhizome contains a resin called podophyllin which has a powerful influence on living tissue. It is important in modern medicine in the treatment of venereal warts. One group of American Indians is reported to have boiled the whole plant in water then sprinkled the liquid on their potato plants to repel insects.

Caution: Podophyllin contains a compound known as podophyllotoxin which has been identified as a teratogen (a substance that may cause death or deformity in a developing embryo). Thus it is potentially dangerous to pregnant women. Podophyllin is a strong purgative that has been used to treat chronic constipation. However, a dose of as little as 5 grains can cause death.

Cynoglossum virginianum
Wild Comfrey

22

Boraginaceae
Borage Family

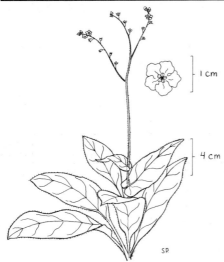

I cm

4 cm

SP

A perennial with most of its leaves on the lower part of the stem. Each flower produces 4 gray or light brown, bristly seeds that disperse by adhering to the coats of passing animals.

A related and somewhat similar species, *C. officinalis*, Hound's Tongue, not illustrated, is an introduced weed of open fields

Flowers pale blue, scattered, on 1–4 terminal branches; corolla saucer-shaped, 5-lobed, with small appendage at base of each lobe closing throat of flower; cauline leaves alternate, sessile, clasping, decreasing in size toward tip of stem; basal leaves petioled, tapering at base; stems hairy, unbranched, to 80 cm. high (32 in.).

May–June

with a leafy stem tip and reddish-purple flowers. Hound's Tongue is a summer blooming biennial throughout southern Canada and the United States. It is thought to be responsible for livestock poisoning in England.

Wild Comfrey is found in open upland woods often on south-facing slopes from southern New England and New York to Illinois, south to Florida, Louisiana and Oklahoma.

℞ Plants of this genus may cause contact dermatitis if handled by sensitive individuals. They contain alkaloids that have caused the development of tumors of the liver and pancreas in rats.

Mertensia virginica
Virginia Bluebells*

23

Boraginaceae
Borage Family

*Also Virginia Cowslip, Roanoke-bells
Flowers blue, trumpet-shaped, nodding, in
clusters at the ends of stems and
branches; flowers to 2.5 cm. long (1 in.);
stems smooth, branched or unbranched,
erect, to 60 cm. high (2 ft.).*

April–May

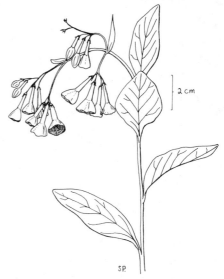

2 cm

S.P.

A perennial with soft stems and prominently veined leaves. Pollination is by bumblebees, honeybees, and other small wild
bees. The flowers are often pirated by bumblebees that bite holes in the bases of the
corolla tubes and steal the nectar, rather
than gather it in the conventional way. Carried to excess, this practice could prevent
seed formation and eventually result in the
extinction of the species.

The beautiful blue bells of this species are
sometimes grown as ornamentals in flower
gardens and borders. It is a favorite for wildflower gardens and adjusts well to cultivation. However, it does not transplant well
and should be initiated from seeds. When
observed in dense colonies, especially in its
natural habitat, it provides a spectacular
sight. Mutant white forms are sold by flo-
rists and are sometimes observed in the wild.

This genus is restricted to the cooler regions of the Northern Hemisphere, and consists of about 40 species. Only 3 species are
native to the Northeast, but 21 other North
American species occur in the western
states. Of the 3 northeastern species, *M. virginica* has the most southerly station and
the greatest north-south range of
distribution.

Virginia Bluebells is found in moist woods,
along stream banks, and in river bottoms
from Ontario to Minnesota, south to Alabama and Missouri.

Dentaria diphylla
Two-leaved Toothwort*

24

Brassicaceae or Cruciferae
Mustard Family

Also Crinkleroot

A perennial with a rough, continuous,
horizontal rhizome. Nectar produced by the
flowers attracts, as pollinators, small bees
(including honeybees) and flies. The erect,
slender seedpod splits to release numerous,
slightly wrinkled, green to brown seeds.

The genus name is from a Latin word
meaning tooth and refers to the tooth-like
projections on the rhizomes of some species. There are 7 similar woodland species
recognized in the Northeast. Some botanists
believe that there may be only 3 true species. The rhizomes of all species are eaten
to a limited extent by White-footed Mice.

A closely related species, *D. laciniata*, Cut-
leaved Toothwort or Pepperroot, is very similar but has leaves in a whorl of 3, each leaf
divided into 3 slender, toothed, segments;
stems hairy above leaves; rhizome consisting of a series of tuberous sections (see illustration). Its ecology and food uses are
similar to *D. diphylla*.

Two-leaved Toothwort is found in rich,

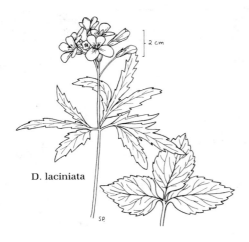

D. laciniata

Flowers white, becoming pink with age, in a terminal cluster, petals 4; leaves 2, nearly opposite, 3 parted, each leaflet rounded with course teeth; stems smooth, unbranched to 35 cm. high (14 in.).

April–June

moist woods from New Brunswick to Minnesota, south to Georgia and Alabama.

✕ The rhizomes of the several species of this genus are edible. They can be added fresh to salads or ground in vinegar as a substitute for horseradish. These plants should be picked in moderation and only where they are abundant. In all instances the terminal growing portion should be left undisturbed.

| *Campanula americana* **Tall Bellflower** | **25** | *Campanulaceae* **Harebell Family** |

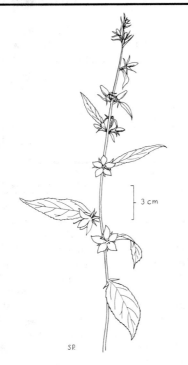

Flowers blue, to 2.5 cm. wide (1 in.), solitary or in clusters in axils of upper leaves and terminal leaf-like bracts; corolla with 5 spreading lobes, long style curving downward then upward at the tip; leaves alternate, toothed, lanceolate, tapering at base, to 15 cm. long (6 in.); stems erect, usually unbranched, to 1.8 m. high (6 ft.).

July–September

A slender annual with a flower cluster to 40 cm. long (16 in.). The most frequent insect visitors to the flowers are honeybees and bumblebees. The seed capsule opens by 3–5 round pores near the top from which the numerous smooth, glossy, dark brown, slightly winged seeds may be thrown as the stem sways in the wind.

The genus name is derived from a Latin word that means "a little bell" and refers to the shape of the corolla of most species. However, the name is not appropriate for Tall Bellflower since its corolla is flat and star-shaped. There are about 8 species of the genus in the Northeast, 4 of which are native to Europe and Asia. *C. americana* is probably the most widespread member of the genus in eastern North America.

Tall Bellflower is found in moist woods and thickets and along streambanks from Ontario to Minnesota, south to Florida, Alabama and Oklahoma.

Linnaea borealis
Twinflower

26

Caprifoliaceae
Honeysuckle Family

Flowers pink to white, funnel-shaped, 5-lobed, nodding, fragrant, in pairs on slender stalks; leaves opposite, roundish, unevenly toothed; stems prostrate and trailing, with short, erect branches, to 15 cm. high (6 in.).

June–August

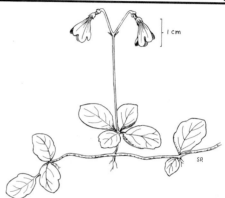

A perennial, semi-woody evergreen of North America, Europe and Asia. Each flower produces a small, dry, 1-seeded fruit enclosed by persistent bracts. It is eaten in the East by Ruffed Grouse and White-tailed Deer, and in the western mountains by Mule Deer and Bighorn Sheep. Although reproduction is mainly by the creeping stem, herbivores may contribute to the spread of the smooth green to brown seeds.

It is common in the northeastern Canadian evergreen forests but is less abundant and often restricted to higher elevations farther south. This delicate little vine is very attractive for rock gardens and planters. It is available commercially from American nurseries and landscape suppliers. In some systems of classification this species is listed as *L. americana.*

This plant was first dedicated to Linnaeus, the founder of modern systematic botany, by J. F. Gronovius, his teacher. Linnaeus described the plant as similar to himself in being "lowly, insignificant [and] flowering for a brief space." In his best known portrait, he is shown holding a flowering branch of Twinflower.

Triosteum perfoliatum
Wild Coffee*

27

Caprifoliaceae
Honeysuckle Family

**Also Feverwort, Horse Gentian, Tinker's Weed*

A coarse perennial with a thick stem. The flowers are rather inconspicuous in comparison with the yellow-orange berries that appear from August to September. Each berry contains 3 strongly ribbed, triangular seeds. The genus name is derived from Greek words that mean "three bones" in reference to the seeds.

This is a small genus with 2 species native to North America. A few others occur in Japan and the Himalaya Mountains. *T. perfoliatum* is probably the most widespread

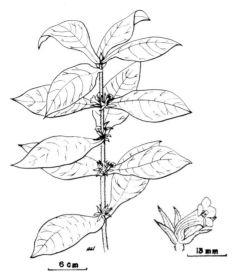

6 cm

13 mm

Flowers greenish to brownish, 3 or 4 per leaf axil of upper leaves; corolla tubular, with 5 unequal lobes; fruit an orange berry; leaves opposite, middle and upper pairs joined around stem, hairy on underside, to 22.5 cm. long (9 in.); stems sticky-hairy, unbranched, to 1.2 m. high (4 ft.).

May–June

ilar but usually has only one yellowish flower per leaf axil and narrow tapering leaf bases.

Wild Coffee is found in moist and dry woods, thickets, and late successional fields from New Brunswick and Quebec to Minnesota, south to Georgia, Alabama and Missouri.

✖ The dried and ground ripe berries can be used as a substitute for coffee.

℞ The 1851 edition of *The Dispensatory of the United States of America* lists *T. perfoliatum* as a medicinal plant. The root is the medicinal part with cathartic, emetic, and diuretic properties.

in North America. A closely related species, *T. angustifolium*, not illustrated, is very sim-

---

Arenaria lateriflora
Grove Sandwort

28

Caryophyllaceae
Pink Family

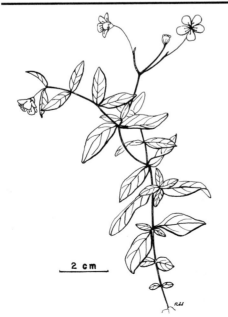

2 cm

Flowers white, to 12 mm. wide (½ in.), 2–5 per axillary or terminal cluster; sepals 5, petals 5, stamens 10, pistil 1; leaves opposite, oval, sessile, to 2.5 cm. long (1 in.); stems weak, branched, minutely hairy, to 30 cm. long (12 in.).

May–August

A perennial with a branched, thread-like, creeping rhizome, often forming extensive mats. The oval to oblong seed capsule splits from the top into 3 sections, releasing numerous shiny, black, kidney-shaped seeds. A small appendage on each seed may be an adaptation for dispersal by ants.

This is a genus with 200 or more species of worldwide distribution, including several that are cultivated as mat or border plants. There are 11 species in eastern North America, all but one native to the region. *A. lateriflora* is a highly variable species with a wide distribution also in Europe and Asia.

In North America, Grove Sandwort is found in moist woods, in clearings, and on sandy or gravelly beaches and shores from Newfoundland to Alaska, south to Maryland, Ohio, Missouri and southern California.

Silene virginica
Fire Pink*

29

Caryophyllaceae
Pink Family

Also Catchfly

Flowers bright red, on long stalks, in a branching terminal cluster; petals 5, 2-lobed, sepals united into a sticky tube with 5 lobes; stem leaves opposite, sessile, to 15 cm. long (6 in.), basal leaves spatula-shaped with narrowly winged stalks, to 10 cm. long (4 in.); stems unbranched, sticky-glandular, to 60 cm. high (2 ft.).

April–June

A colorful perennial often with weak or prostrate stems. This is the most widespread of the 3 species of this genus in the Northeast that have bright red flowers. Most members of this genus are pollinated by bees, butterflies, or moths. The seed capsule opens by 6 teeth to release numerous globular, finely pebbled seeds.

This is a very large genus represented in North America by about 54 species. In the Northeast, 18 species have been recognized, including several that have been introduced from Europe. The genus name is derived from the Greek word for "saliva" and refers to the sticky nature of the stems in several species. It is not an important wildlife food genus, but the plants are eaten by a few species of birds and small mammals.

Fire Pink is found in moist or open dry woods from New England and Ontario to Minnesota, south to Georgia and Oklahoma. ℞ The root has been used in the treatment of intestinal worms.

Chenopodium hybridum
Maple-leaved Goosefoot

30

Chenopodiaceae
Goosefoot Family

An annual, native of North America, Europe and Asia. Pollination is by wind, but this is not known to be an important hay-fever plant. Each plant produces a great number of seeds which are a source of food for numerous species of birds and small mammals. It is sometimes a weed in cultivated fields but can be controlled by clean cultivation.

There are about 20 species of this genus

52

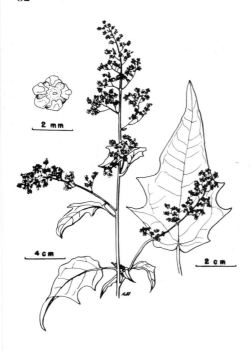

Flowers greenish, tiny, in branching, terminal and axillary clusters; seeds shiny black, lens-shaped; leaves bright green, oval to triangular in general outline, sometimes heart-shaped at base, with 2–5 large teeth on each side, tapering to a pointed tip, to 15 cm. long (6 in.); stems smooth, branched, to 1.5 m. high (5 ft.).

July–October

within our range. Most of these are herbaceous weeds with inconspicuous flowers and were introduced from Europe. C. hybridum is one of about 6 species native to the Northeast. A variety that occurs mostly west of the Mississippi River has seeds that are considerably larger than those of the plants in the eastern states.

Maple-leaved Goosefoot is found in open woods, thickets, clearings, and open shady areas from Nova Scotia and Quebec to the Yukon, south to Virginia, Kentucky and Texas.

Tradescantia virginiana
Spiderwort

31

Commelinaceae
Spiderwort Family

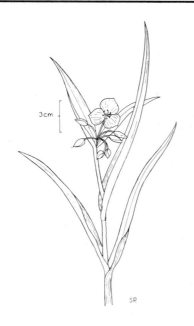

Flowers blue or purple, to 5 cm. wide (2 in.), in a terminal cluster subtended by 2 long leaf-like bracts; petals 3, sepals 3, hairy; stamens 6; leaves 1–4, alternate, long and narrow, forming basal sheaths around the stem, to 35 cm. long (14 in.); stems smooth, unbranched, succulent, to 45 cm. high (18 in.).

April–July

A perennial with very showy but short-lived flowers. The flowers open only for a morning and are pollinated by Queen Bumblebees that visit for the plentiful pollen. The petals then wilt and the flower fades. The seed capsule splits into 3 sections, releasing 3–6 oval to oblong, brown seeds. The genus was named in honor of J. Tradescant, a botanical explorer and gardener for King Charles I of England.

Most of the species of this genus in eastern North America occur west of the Appalachian Mountains. T. virginiana is one of about 4 similar species that occur east of the mountains. It is easily propagated by cuttings and is sometimes cultivated as a house or garden ornamental. Hairs on the stamens are often used in biology classes to study plant cells and the movement of protoplasm. A related species, *T. ohiensis*, not illustrated, is very similar but has sepals and flower stalks without hairs.

Spiderwort is found in moist woods, meadows, prairies, and roadsides from southern New England to Minnesota, south to Georgia and Missouri.

✖ The young leaves and stems can be added to salads or cooked as greens. All of the several species of the genus are edible but *T. virginiana* is reputed to be the best.

Cornus canadensis
Bunchberry*

32

Cornaceae
Dogwood Family

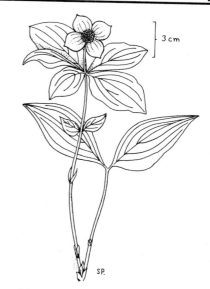

3 cm

SP.

*Also Dwarf Cornel, Crackerberry, Pudding-berry

Flowers white, 4 large petal-like bracts surrounding a central cluster of small greenish flowers; fruit a crowded cluster of red berries; leaves in single whorl of 4 or 6, a few centimeters below the showy bracts; stems unbranched, to 25 cm. high (10 in.).

May–July

A perennial with a slender branching rhizome, of North America and eastern Asia. The large white bracts function like petals in attracting insects to the small inconspicuous flowers. Common visitors include bees, bee-flies, bluebottle flies, and beetles. The large, smooth, light brown seeds, one in each berry, are dispersed by animals that eat the berries. These include the Sharp-tailed Grouse in the Great Lakes region, and songbirds such as the Veery-thrush and Warbling Vireo in the Northeast.

This is the smallest species of a relatively large genus of shrubs and small trees. Possibly the best known species is *C. florida*, Flowering Dogwood, not illustrated, which is commonly planted as a lawn ornamental. *C. canadensis* is sometimes planted in wild plant gardens where it is attractive and grows well.

A similar species, *C. suecica*, not illustrated, has smaller bracts, a central cluster of dark purple flowers, and opposite leaves. Its range includes northern Europe and Asia and northern North America, south to Nova Scotia.

Bunchberry is found in shady woods and bogs from Greenland to Alaska, south to New Jersey and Indiana, reaching the southern limit of its range in the mountains of West Virginia.

✖ The mature red berries can be eaten raw or cooked but are tasteless. Those of *C. suecica* have a slightly sour taste, and Linnaeus reported their use by Laplanders in making a pudding-like dish.

The more common Bunchberry can be used as a survival food, and the berries can be picked without damage to the plant. However, according to several reports, they are not likely to win any prizes as gourmet delights.

Sedum ternatum
Wild Stonecrop

33

Crassulaceae
Orpine Family

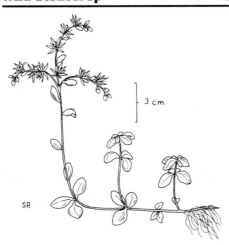

3 cm

S.P.

Flowers white, usually in a 3-branched terminal cluster, to 12 mm. wide (½ in.); petals usually 5, sharp pointed; leaves usually in whorls of 3, those of the flowering branches opposite or alternate, succulent, to 2.5 cm. long (1 in.); stems creeping, usually with one erect flowering branch and several non-flowering ones, to 20 cm. high (8 in.).

April–June

This is a very large and complex genus of over 300 species. The genus name is a Latin word, and there are two interpretations for its application to this genus. By one translation it means "to sit," referring to the way many species cling to rocks and walls or to their spreading growth habit. By another translation it means "to assuage," alluding to supposed healing properties of the Houseleek (*Sempervivum*), a plant very similar to the stonecrops, to which the name was applied by early Roman writers.

A perennial native to eastern North America. Each flower produces several small brown to black seeds with about 16 longitudinal ridges. These are small enough to become windborne in a moderate breeze. It is an attractive addition and grows very well in rock gardens. The name stonecrop is descriptive of the way it clings to rocks and crevices.

Wild Stonecrop is found in rocky woods, especially in limestone areas, from southern New England and New York to Michigan, south to Georgia and Arkansas.

Chimaphila maculata
Spotted Wintergreen*

34

Ericaceae
Heath Family

2 cm

*Also Spotted Pipsissewa
Flowers pink to white, waxy, fragrant, in terminal clusters of 1–5; petals 5, rounded, stamens 10; leaves whorled, lanceolate, with scattered teeth, white along veins on upper surface; stems unbranched, to 25 cm. high (10 in.).

June–August

An evergreen perennial with a long creeping rhizome. The sweet-scented flowers are visited by several species of flies and bees which effect pollination. The seedpod splits from the top into 5 sections, releasing numerous tiny, irregular dust-like seeds.

This is a small genus with only 2 species in the Northeast. They are of limited use as wildlife food plants because they usually do not occur in sufficient numbers, especially in the northern parts of their ranges. The genus name is Greek and means "to love winter," referring to the evergreen growth habit of these plants.

Spotted Wintergreen is found in upland dry woods, often in sandy soil, from New Hampshire to Ontario and Michigan, south to Georgia and Alabama.

✕ Neither Spotted Wintergreen nor Pipsissewa (C. umbellata) are important food plants but both are described as pleasant and refreshing woodland nibbles. C. umbellata was at one time used as an ingredient in root beer.

℞ Most herbals credit Spotted Wintergreen and Pipsissewa with similar properties. A solution made from the fresh plant is recommended as a diuretic and for its beneficial influence on the urinary tract. The name "pipsissewa" is said to be a Cree Indian term referring to the plant's ability to dispel kidney or bladder stones. Some American Indian tribes reportedly crushed the whole plant and applied it to swellings of the legs and feet.

| *Chimaphila umbellata*
Pipsissewa* | **35** | *Ericaceae*
Heath Family |

*Also Prince's Pine

Similar to C. maculata but leaves without white markings on upper surface, widest above the middle and with numerous fine teeth.

June–August

A perennial with a creeping rhizome, native of North America, Europe and Asia. One variety of this species extends into the western part of North America.

Pipsissewa is found in dry woods, preferring sandy soil, from Nova Scotia to Alaska, south to Georgia and California.

For further ecological, food, and medicinal information see *C. maculata*.

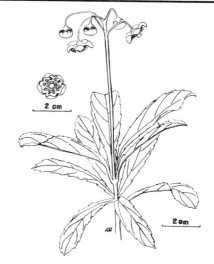

2 cm

2 cm

Epigaea repens
Trailing Arbutus*

36

Ericaceae
Heath Family

*Also Mayflower, Ground Laurel

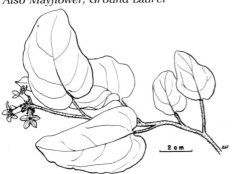

2 cm

A perennial, evergreen, creeping shrub with a tough rootstock. Many plants have unisexual flowers, and cross-pollination is commonly effected by early queen bumble-bees. At maturity the seed capsule splits into 5 sections, exposing a white pulpy interior in which numerous brown, pitted seeds are embedded. Ants are attracted to this tissue and carry it, along with the seeds, to their nests as food for their young. Only the soft tissue is eaten and the seeds are effectively dispersed.

When the very fragrant flowers appear in early spring, the leaves from the previous summer are still present. They are often leathery, insect-eaten and weather-beaten. The new leaves develop in June and are a

Flowers pink or white, in small axillary or terminal clusters; corolla tubular with 5 spreading lobes; leaves alternate, oval, rounded or heart-shaped at base; stems prostrate, branched, hairy to 60 cm. long (2 ft.).

March–May

lighter shade of green.

This attractive little plant has been so extensively collected that it has almost disappeared from some areas. Its collection and sale should be discouraged, and in some states it is protected by law. It is the state flower of Massachusetts and the floral emblem of Nova Scotia.

There are only 2 known species of this genus. The other one occurs in Japan and eastern Asia. The genus name is a Greek word meaning "upon the ground" and refers to the trailing habit of the stem.

Trailing Arbutus is found in rocky or sandy woods from Newfoundland to Saskatchewan, south to Florida and Alabama.

✘ ℞ The flowers are said to make a thirst quenching nibble or an additive for salads, and the leaves reputedly have been used in the treatment of kidney stones and diseases of the urinary system. Neither use justifies the disturbance of this plant in its natural habitat.

Gaultheria hispidula
Creeping Snowberry*

37

Ericaceae
Heath Family

*Also Moxieplum, Maidenhair-berry, Capillaire

2 cm

Flowers white, tiny, bell-shaped, nodding, corolla with 4 lobes; fruit a white berry; leaves alternate, oval, dark green, to 1 cm. long (½ in.); stems creeping, hairy when young, to 40 cm. long (16 in.).

May–June

A very leafy, prostrate, mat-forming, evergreen shrub. It produces flowers and fruits abundantly only in the northern part of its range. The pulpy white berry may be up to 10 mm. long (⅜ in.) and it encloses numerous shiny, triangular, cream-colored seeds. These are dispersed by birds and small mammals that eat the berries.

The genus is named in honor of Jean-Francois Gaultier, an early eighteenth-century amateur botanist and physician to the royal governor of French Canada. It includes only 2 species in the Northeast, both native to North America. In an earlier system, *G. hispidula* was classified as *Chiogenes hispidula*.

Creeping Snowberry is found in damp acid woods, frequently on rotting logs, mossy evergreen forests, and bogs from Newfoundland to British Columbia, south to Pennsylvania and Michigan and at higher elevations to North Carolina.

✕ The wintergreen-flavored berries are edible and are especially tasty with cream and sugar. With the addition of pectin they can be made into delicious jams and preserves. This is the favorite use of them in Newfoundland, where the plant grows in great abundance and is called Capillaire and Maidenhair-berry.

A tea made from the dried leaves has the flavor of wintergreen. Creeping Snowberry received high praise as a northwoods tea plant from Henry David Thoreau in *Walden*. ℞ Since *G. hispidula* contains oil of wintergreen it can be assumed to have the same medicinal properties as *G. procumbens*.

Gaultheria procumbens
Wintergreen*

38

Eriaceae
Heath Family

Also Teaberry, Checkerberry, Ivry-leaves Flowers white, sometimes tinged with pink, bell-shaped, nodding, corolla 5-lobed; fruit a bright red berry; leaves alternate, oval, to 5 cm. long (2 in.), clustered at tip of stem, firm and glossy dark green; stems erect, smooth, to 15 cm. high (6 in.).

July–August

A low evergreen shrub with a long creeping rhizome. The fruit ripens in August-September and remains on the plant, sometimes increasing in size, throughout the winter. It contains numerous irregularly angled, glossy brown seeds. The aromatic, wintergreen-flavored berries are eaten by the Ruffed Grouse, Wild Turkey, Black Bear, and White-footed Mouse, all of which may serve as agents of dispersal.

The genus includes about 150 species of worldwide distribution with greatest abundance in the mountainous regions of South America. *G. shallon*, Salal, not illustrated, is a related species common on the Pacific Coast. *G. procumbens* is the most abundant representative of the genus in the Northeast.

Wintergreen is found in dry or moist acid woods and sometimes in bogs from New-

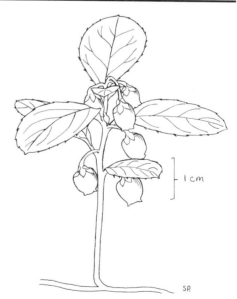

foundland to Manitoba, south to Georgia and Alabama.

✕ The bright red berries are very tasty. The young tender shoots, while they are still reddish in spring and early summer, are very spicy. Later they become tough but still have

58

the wintergreen flavor and can be eaten. The dried leaves can be used to prepare a wintergreen flavored tea. Moderation should be observed in collecting this plant, and under no circumstances should the rhizome be disturbed.

℞ The use of Wintergreen as a medicinal plant is the result of the oil of wintergreen in its leaves. The chief component of this oil in *G. procumbens* is methyl salicylate, which is related to the medicinal compound in aspirin. Thus, the former uses of this plant in the treatment of rheumatism and headaches and to bring down fevers were not without some effectiveness.

The protracted internal use of oil of wintergreen, however, causes severe stomach disorders as a side effect. This stimulated the Bayer company to conduct a search for a drug that would be as effective as methyl salicylate but without the side effects. In 1899 they rediscovered acetylsalicylic acid and named it aspirin. It is still not known how these drugs act to relieve pain.

Monotropa hypopithys
Pinesap*

39

Ericaceae
Heath Family

*Also False Beechdrops

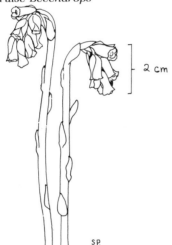

2 cm

S.P.

Flowers light red, tinged with yellow, 3–10 in a terminal cluster; flowers nodding, seed capsules becoming erect; leaves alternate, scale-like; stems often numerous, tan, yellowish or reddish, turning black with age, to 30 cm. high (1 ft.).

June–September

A fleshy parasite or saprophyte with a dense ball of matted fibrous roots. It grows in very close association with a soil fungus on which it is probably parasitic. The flowers are fragrant and mainly self-pollinated. The seedpod splits into 4–5 sections to re-lease a great number of tiny, irregular, thread-like seeds that are dispersed by wind.

This is a small genus in which most botanists recognize two highly variable species. Both are native to North America but are also widely distributed in Europe and Asia. In some systems of classification the genus is included in Pyrolaceae, the Wintergreen Family. In an older system, *M. hypopithys* is listed as *Hypopithys lanuginosa*. The name Pinesap apparently refers to the belief that the plant is parasitic on the roots of pine trees.

Pinesap is found in woodlands with thick humus layers from Newfoundland to British Columbia, south to Florida and Mexico.

Monotropa uniflora
Indian Pipe*

40

Ericaceae
Heath Family

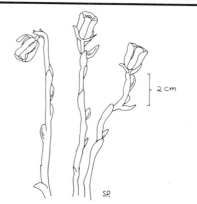

2 cm

S.R.

Also Corpse Plant, Convulsion-root, Fits-root

Flowers waxy-white or pale pink, solitary, nodding at first, then becoming erect as seed capsule matures; leaves alternate, scale-like, white; stems often numerous, fleshy, white, the whole plant turning black with age, to 25 cm. high (10 in.).

June–September

A fleshy, white, sometimes pink tinged parasite or saprophyte arising from a dense mat of fibrous roots. The roots are closely associated with a soil fungus on which they are parasitic. It is not clear whether the fungus gets something in return from the roots.

The stigma is covered with very sticky mucus, some of which is transferred, on contact, to the body of an insect visitor. When the insect subsequently comes into contact with an anther the smooth dry pollen adheres to the mucus and presumably is transferred to the stigma of the next flower visited. The oval seed capsule splits into 4–5 sections to release numerous minute, thread-like seeds that are dispersed by air currents.

Of the two species of this genus, *M. uniflora* is the most common and most widespread in North America. Its waxy white plants stand out conspicuously among the browns of the forest floor. Their unusual appearance has resulted in such descriptive common names as Fairy Smoke, Corpse Plant and Ghost Flower.

Indian Pipe is found in moist woodlands from Newfoundland to Alaska, south to Florida, Mexico and California.

� The plant is reported to be edible but tasteless.

℞ As the names Convulsion-root and Fits-root suggest, this plant has been used for spasms and other nerve disorders. In the nineteenth century, the juice of the stem was used as a remedy for inflamed eyes. *Note:* This is a small plant and not overly abundant. Its food and medicinal values are insufficient to justify collecting it for these uses.

Pyrola elliptica
Shinleaf

41

Ericaceae
Heath Family

An evergreen perennial with a creeping rhizome. The fragrant flowers produce pollen grains in clusters of 4's, or tetrads, and are visited by several species of bees and bee-like flies that use the upturned tip of the style as a landing place. Each chamber of the 5-celled seed capsule splits from the base releasing numerous dust-like seeds which are dispersed by air currents.

This is the most common of the approximately 7 species of this genus in the North-east. The genus name is the diminutive of the genus name for pear, referring to a not obvious similarity between the leaves of the two plants. *P. elliptica* is sometimes called Wild Lily-of-the-valley, as is *Maianthemum canadense*, but these plants are quite dissimilar.

A closely related species, *P. rotundifolia*, Round-leaved Pyrola, not illustrated, is very similar but has rounder shiny leaves, leaf blades about the same length as the pe-

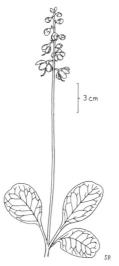

Flowers white with green veins, nodding, in a loose cluster of 7–15 at the top of a leaf-less stalk, petals 5, style protruding, bending downward then curving upward at tip; leaves basal, elliptical, rounded at tip, not shiny, leaf blade longer than its stalk; flowering stem to 30 cm. high (1 ft.).

June–August

from Newfoundland and Quebec to British Columbia, south to West Virginia, Indiana and Iowa.

℞ The leaves have been used to make a plaster for bruises and wounds, and to relieve pain. This kind of dressing has been called a shin plaster, thus the name Shinleaf. The plants of this genus were valued as wound healers by the American Indians. They have mild astringent properties and have been used to make a gargle for mouth and throat irritations, a wash for the eyes, and a vaginal douche.

tioles, and petals without green veins. It is widely distributed in North America as well as Europe and Asia.

Shinleaf is found in dry or moist woods

Amphicarpa bracteata
Hog Peanut

42

Fabaceae or Leguminosae
Bean Family

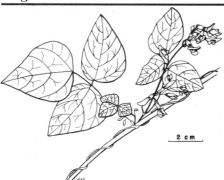

A perennial genus with only one species in North America. Two kinds of flowers are produced: aerial ones as described above, and basal ones near the ground, or just beneath the surface, that have no petals and do not open. The basal flowers are self-pollinated and viable seeds are borne by both kinds of flowers. At maturity the aerial pods split into 2 twisted sections, releasing the mottled light and dark brown seeds.

Flowers pale purple, to 12 mm. long (½ in.), in stalked clusters in upper leaf axils; corolla pea-like, 2–15 flowers per cluster; fruit a bean-like pod, to 3.8 cm. long (1½ in.), with 3–4 seeds; leaves alternate, pinnately compound, with 3 leaflets, leaflets to 10 cm. long (4 in.); stems thread-like, twining, to 1.5 m. long (5 ft.).

August–September

With the subterranean seedpods this species plants its own seeds. However, these are transported to other areas by small mammals that store them as a winter food supply. Both the aerial and the underground seeds are eaten by the Ruffed Grouse, Bobwhite Quail and Ring-necked Pheasant. The common name suggests that the underground pods are sought after also by hogs.

The genus name is derived from Greek words that mean "two kinds of fruits" and refers to the aerial and basal seedpods. A.

bracteata is a highly variable species with at least 2 well recognized varieties. These are classified as separate species, *A. monoica* and *A. pitcheri*, by some botanists. *A. pitcheri* has larger leaves, coarsely hairy stems and leaf stalks, and is more common in the western part of the range.

Hog Peanut is found in moist woods and woodland borders, especially near streams, from Nova Scotia and Quebec to Manitoba, south to Florida and Texas.

✘ According to E. Gibbons and G. Tucker the seeds of both the aerial pods and the underground ones are edible but the larger subterranean ones are best. The latter are high in protein and were well known to American Indians, who often collected them by robbing the stores of voles and field mice. They removed the tough shell by soaking the pods in warm water or in a mixture of water and hardwood ashes. The beans were then eaten raw or boiled. The roots were also eaten by the American Indians.

This plant was once cultivated for food in southeastern areas of the United States.

Apios americana
Groundnut*　　　**43**　　　**Fabaceae or Leguminosae**
Bean Family

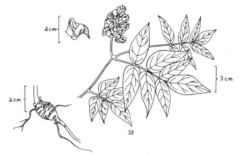

*Also Wild Bean, Potato Bean
Flowers brown-purple or maroon, in dense clusters on axillary stalks, fragrant; fruit a bean-like pod, several seeded, coiled after the seeds are released; leaves alternate, compound pinnate, 5–7 leaflets; stems climbing and twining extensively, to 3 m. long (10 ft.).

July–September

A perennial with a slender rhizome bearing numerous tuber-like enlargements. A notable work on seeds has reported that plants of this species in southeastern Canada are triploid and thus sterile. Other reports indicate that the flowers may be self-incompatible and cross-pollinated by bees.

There are about 7 species of this genus, two in eastern North America, the others in eastern Asia. The genus name is from a Greek word meaning "a pear" and refers to the shape of the tuber. Rootlets of the tubers have nitrogen-fixing nodules.

Groundnut is found in damp woods and thickets, low areas, and along the margins of swamps and ponds from Quebec to Minnesota, south to Florida and Texas.

✘ Both the seeds and the tubers are edible but, since the seeds are scarce, the tubers are much more important. The tubers were held in high regard by American Indians. During their first winter in New England, according to early American historians, the Pilgrims relied heavily upon these tubers as a source of food. The value of this plant to early settlers is suggested by a law, reputed to have been passed in 1654, that forbade Indians from digging groundnuts on English land. Captain John Smith wrote about the use of groundnuts as a food by the settlers in Virginia.

The tubers may be prepared in the same manner as potatoes. However, they should be eaten while still hot, since they are much less tasty when cold. Since the tubers take at least 2 or 3 years to grow to usable size, this plant has little potential as a cultivated food plant.

62

Desmodium canadense
Showy Tick-trefoil

44

Fabaceae or Leguminosae
Bean Family

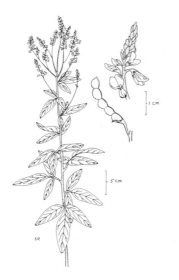

Flowers pink to violet, pea-like, in dense terminal clusters; fruit a flat, sticky, 3–5 segmented pod; leaves alternate, compound with 3 lanceolate leaflets; stems branched toward the tops, hairy, to 1.8 m. high (6 ft.).

July–August

Members of this genus are important successional pioneers that provide good cover for wildlife and protection of the soil from erosion in burned over or denuded areas. In addition, they enrich the soil through the spherical nitrogen-fixing nodules on their roots. The seeds may provide up to 10% of the diet of the Bobwhite Quail and are eaten in smaller quantities by the Wild Turkey and small mammals.

The 22 species that occur in the Northeast are most common southward and are predominantly woodland plants. They are all native perennial herbs with flat, jointed, sticky seedpods and alternate, 3-parted leaves.

Showy Tick-trefoil is found in open woods, woodland borders, clearings, and stream margins from Nova Scotia to Alberta, south to South Carolina and Arkansas.

℞ Various species have been used in folk medicine to reduce fever and to treat dysentery and liver diseases, and as poultices for acne, ulcers, and eye diseases.

A perennial with a long, slender, brown rootstock. The 2 lower petals form a boatlike structure that encloses the pistil and the stamens under slight pressure. When a relatively heavy insect, such as a bee, lands on the flower, the increased pressure causes the pistil and stamens to burst free in a cloud of pollen. The stigma comes into contact with its body as a bee is dusted with pollen. The seedpods adhere tenaciously to clothing or animal coats for efficient dispersal.

Lespedeza repens
Creeping Bush-clover

45

Fabaceae or Leguminosae
Bean Family

A perennial with a creeping stem and nitrogen-fixing root nodules. Two types of flowers are produced, those without petals that never open and are self-pollinated, and petal-bearing ones that are cross-pollinated by insects. Hard, single-seeded fruits are borne by both types of flowers. Seeds from plants of this genus comprise up to 50% of the diet of Bobwhite Quail. In some areas this genus provides up to 25% of the diet of White-tailed Deer.

The Lespedezas are high in protein and

Flowers pink to purple, in a cluster of 2–8, on a long terminal or axillary stalk that is longer than the subtending leaf; leaves alternate, with 3 leaflets; leaflets oval, to 13 mm. long (½ in.); stems trailing, soft-hairy, to 70 cm. long (28 in.).

June–September

are valued as forage and hay crops. They are good soil stabilizers and, since they add nitrogen to the soil, are useful as green manure crops. The genus was named in honor of V. M. de Cespedes, an eighteenth-century governor of Florida. There are about 12 species native to the Northeast, and several others, introduced as forage or cover crops, have escaped cultivation.

Creeping Bush-clover is found in open woods, thickets, and open fields from southern New England to Wisconsin, south to Florida and Texas.

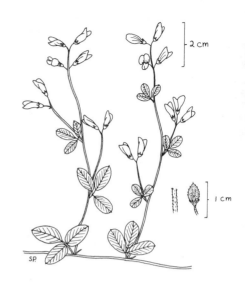

Corydalis flavula
Yellow Corydalis*

46

Fumariaceae
Fumitory Family

*Also Yellow Harlequin, Yellow Fumewort

Flowers yellow, to 8 mm. long (⅓ in.), with a short rounded spur, in short clusters; leaves alternate, finely twice pinnately dissected, pale green; stems slender, smooth, freely branched, sprawling, to 35 cm. long (14 in.).

April–May

A biennial with slender, beaked, often drooping seed capsules. The 3 upper petals of the corolla curve upward at the tip forming a crest that attracts insect pollinators. The seed capsule splits into 2 sections, releasing several shiny black seeds. The seeds have small soft appendages that may serve to attract ants as agents of dispersal.

There are 6 native species of this genus in eastern North America. Five of these are yellow flowered, and C. flavula is the most common in our area. A pink-flowered species C. sempervirens, Pale Corydalis, not illustrated, occurs in the same type of habitat but ranges northward to Alaska and flowers from May to September.

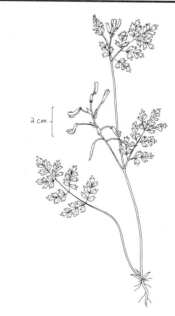

Yellow Corydalis is found in moist or dry rocky woods from New England and On-

64

tario to Minnesota, south to Alabama, Louisiana and Oklahoma.

☠ Some species of this genus are rich in toxic alkaloids and their consumption has caused losses in livestock. *C. flavula* is suspected to be poisonous.

Dicentra canadensis | 47 | *Fumariaceae*
Squirrel Corn* | | Fumitory Family

Also Turkey Corn, Stagger Weed

D. cucullaria

Flowers white or pinkish, heart-shaped, nodding, on a leafless stalk; leaves basal, pinnately dissected into fine segments; flowering stalk and leaves to 25 cm. high (10 in.).

April–May

Since these contain toxic alkaloids, it is unlikely that they are important in the diet of either squirrels or turkeys.

A similar and related species, *D. eximia*, Wild Bleeding Heart, not illustrated, has pink to purple flowers, is taller, to 40 cm. high (16 in.), and flowers from May to September. It ranges from New Jersey and Pennsylvania south to North Carolina and Tennessee. Like the introduced species *D. spectabilis*, Bleeding Heart, not illustrated, *D. eximia* is commonly cultivated as an ornamental.

Squirrel Corn is found in rich moist woods from Nova Scotia and Quebec to Minnesota, south to North Carolina and Tennessee.

☠ In herbal medicine the dried tubers are characterized as tonic, diuretic, and alterative. They have been used to treat chronic skin disorders, syphilis, and some menstrual problems.

A perennial with a rootstock bearing clusters of small tubers that resemble grains of yellow corn. These unusual small wildflowers have a slight fragrance of hyacinths. Each seedpod contains 10–20 smooth, shiny, black seeds, each of which has a ragged appendage that has been interpreted as an adaptation for dispersal by ants.

It may occur in the same area with Dutchman's Breeches, the 2 species often forming dense and extensive stands. Both species may be picked in moderation without endangering their survival, as long as their roots are not disturbed. However, when they are picked they soon wilt; thus their best use is the beauty they impart to an undisturbed woodland environment.

The common names Squirrel Corn and Turkey Corn refer to the shape and color of the tubers rather than their food utilization.

All the species of this genus are listed as poisonous plants. Several toxic alkaloids have been isolated from both the leaves and the tubers. There have been losses of cattle reported in Virginia and Indiana resulting from grazing the tops of these plants in woodland pastures, in spring, when other forage is scarce.

Although there has been no reported loss of life in humans from the use of these plants, they should not be experimented with in home remedies.

Dicentra cucullaria
Dutchman's Breeches*

48

Fumariaceae
Fumitory Family

*Also Little Blue Staggers
Flowers white, with 2 long, yellow-tipped spurs, 4–8 on a leafless stalk; flowes resembling upside-down pantaloons; leaves almost indistinguishable from those of Squirrel Corn; flowering stems and leaves to 25 cm. high (10 in.).

April–May

A perennial with a rootstock bearing clusters of white tubers. The plants are normally self-incompatible and cross-pollination is effected mostly by honeybees, bumblebees, and bee-like flies. Honeybees visit the flowers for pollen only, since their tongues are too short to reach the deeply recessed nectar. As in Squirrel Corn, each of the 10–20 seeds in the seedpod is shiny black and has a ragged appendage that may be a source of food for ants.

The common name, Little Blue Staggers, probably refers to the toxic alkaloids found in the leaves and tubers. These are reported to be in greater concentration in *D. cucullaria* than in *D. canadensis*. In view of this, it is remarkable that the tubers of *D. cucularia* have been reported as a part of the diet of the Pine Mouse.

Dutchman's Breeches is found in rich moist woods from Quebec to Minnesota, south to Georgia, Alabama, and Kansas.

For further information see *D. canadensis*.

Geranium maculatum
Wild Geranium*

49

Geraniaceae
Geranium Family

*Also Cranesbill, Alum Root, Chocolate Root
Flowers rose-pink to purple, in terminal clusters of 1-to-several, to 4 cm. wide (1½ in.), petals 5, stamens 10; fruit with a long pointed beak; stem leaves 2, opposite, deeply 5-parted, basal leaves similar, with long petioles; stems often branched above leaves, hairy, to 60 cm. high (2 ft.).

April–June

A perennial with a short thick rhizome and fibrous roots. The anthers mature one at a time and are shed in sequence. The stigma does not become receptive to pollen until the last anther has fallen. Thus, self-pollination is almost impossible and cross-fertilization is effected by honeybees and other smaller bees. At maturity the long seed capsule splits suddenly from the base into 5 sections. These curl upward with a snap, throwing the brown, pitted seeds outward, sometimes for several feet. The coiled sections of the capsule may remain attached to the beak.

Its large colorful flowers make this one of the most attractive of the spring flowers.

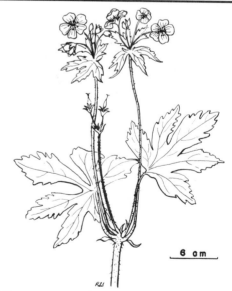

6 cm

The urge to make it part of a wildflower bouquet should be resisted however, because it loses its attractiveness soon after being picked. Although it is not a major wildlife food plant, its seeds are eaten by several species of birds and small mammals.

This is a very large genus of worldwide distribution. There are about 12 species in northeastern North America but only 4 are native to the region. The others are introduced from Europe and Asia. The familiar potted geraniums are members of this family but belong to the genus *Pelargonium* and are native to South Africa.

Wild Geranium is found in dry or moist woods, shady roadsides and rocky meadows from Maine to Manitoba, south to Georgia and Alabama.

℞ The rhizome has a high tannin content, which has led to its extensive use as an astringent, thus the common name Alum Root. When the rhizome is dried and ground, it yields a purplish-brown powder from which the common name Chocolate Flower is derived. This powder has been used for dysentery and diarrhea and to make a gargle for sore throat and ulcers of the mouth. American Indians and early settlers used it to treat thrush, a fungal disease of the mouth in children. Some tribes of American Indians sprinkled the powdered rhizome on wounds to stop bleeding.

Geranium robertianum
Herb Robert*

50

Geraniaceae
Geranium Family

---

*Also Red Robin

Flowers pink, usually in clusters of 2's, from upper axils; fruit with a long pointed beak; leaves opposite, divided into 3–5 leaflets, each leaflet finely pinnately dissected, the terminal one stalked; stems freely branched, hairy, often reddish and sprawling, to 60 cm. long (2 ft.); plant with a strong odor.

May–September

This is one of the latest blooming geraniums in the Northeast, and it often remains green into the winter months. With exposure to the cold nights of autumn, the stems and leaves turn bright red. A characteristic feature is the disagreeable odor of the bruised plant. It is smaller and the leaves are more finely dissected than those of *G. maculatum*. An older system of classification placed it in the genus *Robertiella*.

Herb Robert is found in moist woods, shady roadsides, and woodland borders from Newfoundland to Manitoba, south to West Virginia, Indiana, and Missouri.

℞ As with Wild Geranium, Herb Robert is an astringent. The dried herbage has been used for diarrhea, upset stomach, and internal bleeding. A poultice made from the leaves is said to be good for bruises and skin problems.

An annual or winter annual of North America, Europe, Asia, and Africa. Self-pollination is the main source of fertilization with only 2–3% of seed production resulting from cross-pollination by insects. The dark brown, finely wrinkled or pitted seeds are dispersed by the sudden splitting of the seed capsule as in *G. maculatum*.

Hydrophyllum virginianum
Virginia Waterleaf*

51

Hydrophyllaceae
Waterleaf Family

*Also John's-cabbage

Flowers white to lavender, in dense terminal clusters; stamens extending from flowers give cluster a fuzzy appearance; leaves alternate, pinnately divided into 5–7 deep lobes, basal pair usually 2–lobed, leaves mottled with lighter green; stems hairy in upper part, to 70 cm. high (28 in.).

May–July

A perennial with a scaly horizontal rhizome. The seed capsule is about 4 mm. (⅙ in.) in diameter and contains only a few relatively large, round, pitted, light or dark brown seeds. These plants have a tendency to grow in dense colonies. The masses of leaves, mottled with irregular areas of light green, are likely to draw more attention than the usually pale flowers.

This is a small genus of 8 species, 4 in the Northeast and 4 in the western states. The eastern species are all somewhat similar woodland species. *H. macrophyllum*, Large-leaved Waterleaf, not illustrated, has a more densely hairy stem than *H. virginianum* and leaves with 7–9 lobes. Its range includes areas south of Pennsylvania and Michigan.

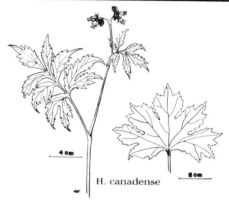

H. canadense

Virginia Waterleaf is found in moist woods and shady woodland borders from Quebec to Manitoba, south to Tennessee and Arkansas.

✖ The young leaves and stem tips, collected before flowering, can be cooked as greens. One or two changes of water during cooking may be necessary to dispel bitterness. *H. virginianum, H. canadense,* and *H. macrophyllum* may be used similarly. In early New York, the first two were collected and cooked and referred to as John's-cabbage.

Hydrophyllum canadense
Broad-leaved Waterleaf

52

Hydrophyllaceae
Waterleaf Family

Flowers similar to H. virginianum; leaves alternate, heart-shaped at base, with 5–9 palmate lobes, extending above flower cluster; stems with few hairs, to 50 cm. high (20 in.).

May–July

A perennial with uniformly green leaves

that somewhat resemble maple leaves. Its seeds are similar to those of *H. virginianum.*

Broad-leaved Waterleaf is found in moist woods from New England to Ontario, south to Georgia and Alabama.

✖ See *H. virginianum.*

Collinsonia canadensis
Stoneroot*

53

*Also Horse Balm, Richweed

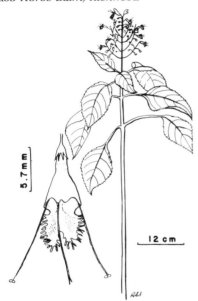

5.7 mm

12 cm

A perennial with a hard woody rhizome, thus the common name Stoneroot. The flowers and bruised leaves have a strong

Flowers pale yellow, in terminal branching clusters, corolla 2-lipped, the lower lip prominent, fringed; stamens 2, long protruding; leaves opposite, several pairs, scattered; stem square, stout, to 1 m. high (40 in.).

July–September

odor of lemon (strong as a horse) which is responsible for the common name Horse Balm. Each of the numerous flowers produces 4 elliptic, shiny, brown seeds.

A related species, *C. verticillata*, not illustrated, is similar but with 4 stamens and only 2 or 3 crowded pairs of leaves. *C. verticillata* occurs only in the southern part of our area, from Virginia south. In some systems of classification it is recognized as *Mitcheliella verticillata*.

Stoneroot is found in moist woods from Quebec and Ontario to Michigan, south to Florida and Mississippi.

℞ The rhizome has diuretic and tonic properties. The leaves have been used as poultices for bruises, wounds, and sores.

Monarda didyma
Oswego Tea*

54

*Also Bee-balm

A perennial with a deep root system. The flowers are visited by honeybees, bumblebees, and butterflies. It is favored by honeybees and is considered an important honey plant, as the name Bee-balm suggests. Each flower produces 4 elliptical, smooth, brown seeds.

M. didyma is very showy when it is in flower, and it is widely cultivated as an ornamental. Through cultivation it has spread

far beyond its natural range in North America and has been introduced into Europe. The name Oswego Tea is said to have been given by John Bartram, an early botanist who found the plant and observed its use for tea at Oswego, New York, then a frontier outpost on Lake Ontario.

A related species, *M. media*, Purple Bergamot, not illustrated, is similar but has flowers that are rose-red or purple-red. It is thought to have originated through the hy-

Flowers scarlet, tubular, 2-lipped, in a dense, crown-like, terminal cluster; leaves opposite, lanceolate, to 15 cm. long (6 in.), sharply toothed, aromatic, small leaves subtending flower cluster reddish; stems square, usually with hairs at nodes, to 1.5 m. high (5 ft.).

July–September

bridization of *M. didyma* and *M. fistulosa*, and has a geographic range and growth habit similar to *M. didyma*.

Oswego Tea is found in moist woods, thickets, along stream banks, and often escaped from cultivation to roadsides and old home sites from Quebec to Michigan, south to West Virginia and Ohio, and along the mountains to Georgia.

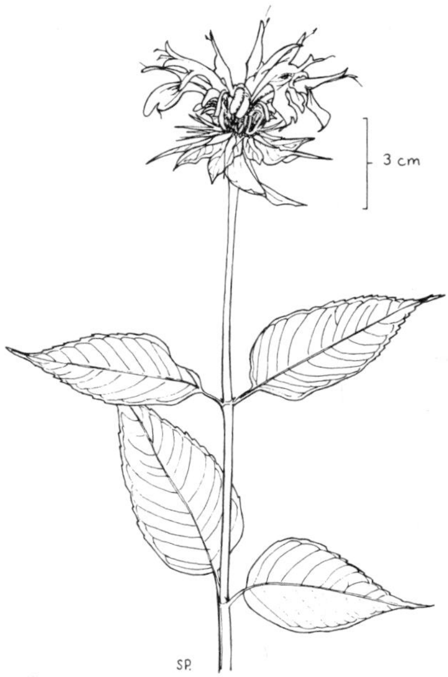

✂ ℞ The dried or fresh leaves can be used to make a fragrant tea. It has been recommended highly as an additive to other teas. A tea made from the plant has been used to relieve nausea and to dispel intestinal gas. The Winnebago and Dakota Indian tribes used the plants of this genus as heart stimulants.

Physostegia virginiana **False Dragonhead*** — 55 — ***Lamiaceae or Labiatae*** **Mint Family**

**Also Obedient Plant*

Flowers pink, showy, funnel-shaped, 2-lipped, upper lip like a hood, lower lip 3-lobed; flowers in a crowded terminal cluster, to 20 cm. long (8 in.); leaves opposite, narrowly lanceolate, with sharp pointed teeth, upper ones greatly reduced; stems square, smooth, usually unbranched, to 1.2 m. high (4 ft.).

June–September

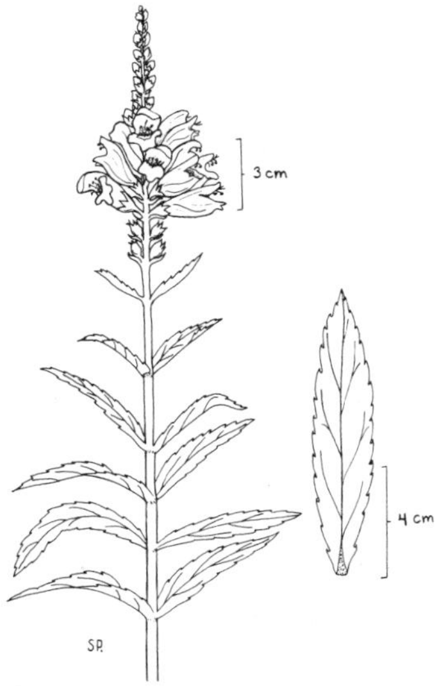

A perennial with a thick stem and basal runners. Each flower produces 4 sharply angled, dull brown seeds. The genus name is derived from Greek words meaning "bladder covering" and refers to the inflated fruiting calyx. If the flowers are pushed to the right or left, they will remain in the new position for a time, thus the name Obedient Plant.

This is a genus native to North America with about 4 species. Because of its large showy flowers, *P. virginiana* is frequently cultivated as an ornamental. In flower gardens it often grows in large clumps that are spectacular at the height of flowering. Two horticultural varieties have been developed, one with white flowers and a large bright pink flowered variety.

False Dragonhead is found in moist woods, woodland borders, along stream banks, and in open spaces from New Brunswick and Quebec to Minnesota, south to Alabama and Texas.

Pycnanthemum virginianum **Virginia Mountain Mint*** — 56 — ***Lamiaceae or Labiatae*** **Mint Family**

**Also Wild Hyssop*

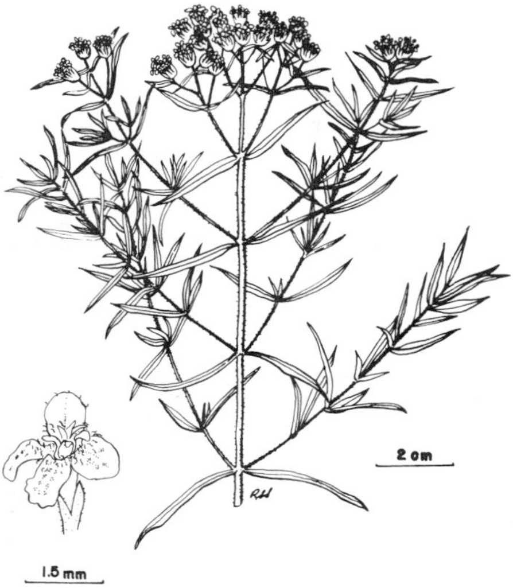

An aromatic perennial with an underground rhizome. It has two types of flowers: those with long protruding stamens and others with short stamens that do not extend beyond lips of the flower. This is an adaptation to prevent self-pollination, and suggests cross-pollination by insects. Each flower produces 4 tiny, finely pitted, dull black seeds.

Flowers white with purple dots, 2-lipped, in dense branched clusters at the ends of stems and branches; leaves opposite, narrowly lanceolate, sessile, entire, rounded at base, numerous; stems square, with hairs on angles, freely branched, to 1 m. high (40 in.).

July–September

This is a North American genus of about 20 species with one on the Pacific Coast and the remainder in the East. The genus name is derived from Greek words meaning "dense flower" and refers to the crowded flower clusters. All species are very similar and sometimes difficult to distinguish from one another. Some are believed to be the result of hybridization. The name Mountain Mint is inappropriate since most species of the genus occur in lowlands.

Virginia Mountain Mint is found in dry woods, thickets, fields and meadows and along stream banks from Maine to North Dakota, south to Georgia and Oklahoma.

✕ The fresh or dried leaves of the Mountain Mint can be used to make a mild aromatic tea.

℞ A hot infusion of the plant is said to be useful for promoting perspiration, and a warm infusion for dispelling intestinal gas.

Scutellaria parvula **Small Skullcap** — 57 — ***Lamiaceae or Labiatae*** **Mint Family**

A perennial with a rhizome constricted at intervals so that it appears like a string of beads. Each flower produces 4 tiny, brown nutlets covered with finger-like projections. Pollination and seed dispersal is similar to that of *S. galericulata*.

Flowers blue, solitary in axils of leaves, 8–10 mm. long (¼ in.), 2-lipped, upper lip notched or entire, lower lip 3-lobed; calyx 2-lipped, a conspicuous cap-shaped protuberance on upper lip; leaves opposite, sessile, oval, usually less than 2.5 cm. long (1 in.); stems square, usually densely hairy, branched from base, to 30 cm. high (1 ft.).

May–July

The specific name is a Latin word meaning "very small" and refers to the low growth habit of this species. In size, growth form, flowers, and rhizome, it is very similar to *S. leonardi* and *S. australis*, and these are sometimes listed as varieties of *S. parvula*.

Small Skullcap is found in dry woods and on moist or dry sand and gravel, especially in limestone areas, from Quebec and Ontario to Minnesota south to Alabama and Texas.

℞ See *S. lateriflora*.

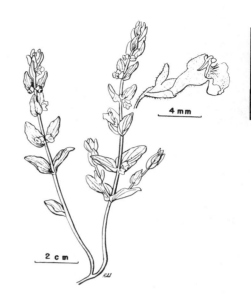

4 mm

2 cm

Teucrium canadense
American Germander*

58

Lamiaceae or Labiatae
Mint Family

*Also Wood Sage

Flowers pink to pale purple, in long terminal clusters, corolla 2-lipped, lower lip prominent, stamens projecting through deep cleft in upper lip; leaves opposite, lanceolate, toothed, to 12.5 cm. long (5 in.); stems square, hairy, usually unbranched, to 1 m. high (40 in.).

July–September

A perennial with a slender creeping rhizome. The long lower lip of the flower provides a convenient landing strip for potential insect pollinators. Each flower produces 4 round, coarsely pitted, brown seeds with scattered white hairs. The fruiting calyx of some members of this genus has a very springy stalk that rebounds when touched by a passing animal, hurling the seeds outward.

There are probably more than 100 species of this genus mostly in the warm regions of the world. The genus name was designated by Linnaeus but it originated with Dioscor-

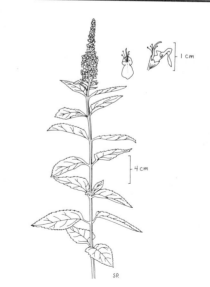

1 cm

4 cm

ides, a botanist of ancient Greece, who is said to have given the name to a plant used medicinally by Teucer, the first king of Troy.

In the sixteenth century, European species of the genus were spread on floors to give rooms a pleasant odor. *T. canadense* is the only species native to the Northeast.

American Germander is found in moist or wet woods, thickets and sandy shores throughout the United States and southern Canada.

℞ *The Dispensatory of the United States of America* does not list the American species of this genus as a medicinal plant, but it does include several European species. Some herbalists are of the opinion that the American species has the same properties as its European relatives. These have been described as useful in the treatment of chronic rheumatism, gout, intermittent fevers, intestinal worms, and constipation.

Allium cernuum
Wild Onion

59

Liliaceae
Lily Family

A perennial with a short rhizome, bearing 1-to-several slender, tapering bulbs. Each cell of the 3-celled seed capsule splits to release 1 or 2 black seeds.

Of the approximately 300 species of this genus, most are found in Asia. There are about 10 species in eastern North America and numerous others in the West. The cultivated forms of the genus include onion, garlic, leeks, and chives.

Flowers white to rose colored, in a terminal umbel, often nodding, individual flowers often bent downward; perianth parts 6, stamens 6, extending beyond perianth; leaves basal, flat, grass-like, several; stems leafless, rising above the leaves, to 70 cm. high (28 in.); whole plant with strong onion odor.

July–August

Wild Onion is found in rocky woods, wooded slopes, and mesic prairies from New York to British Columbia, south to Virginia and Kentucky, and in the mountains to Georgia and Alabama.

🍴 Both the young tender leaves (before flowering) and the bulb are edible. They can be cooked as greens, eaten raw in salads, or used as seasoning. Undesirable flavors in milk and milk products may result if these plants are growing where cows are grazing. Some studies indicate that it is not necessary for the cow to eat the plant for the flavor to be transmitted to the milk. Inhaling the volatile substances from wild members of this genus is sufficient to contaminate the milk.

℞ American Indians rubbed the juices of *A. cernuum* and other members of this genus on their bodies as insect repellents.

Allium tricoccum
Wild Leek*

60

Liliaceae
Lily Family

*Also Ramp

Flowers white, in a terminal globular cluster subtended by 2 bracts; perianth parts 6, about the same length as the 6 stamens; leaves 2 or 3, basal, broad lanceolate, succulent, to 30 cm. long (12 in.); stem leafless, to 50 cm. high (20 in.); all parts of plant with strong onion odor.

June–July

A perennial with bulbs connected by short strands of fibrous roots. Pollination is by bees, and each flower produces 3 shiny black seeds that are exposed when the seed capsule splits. These are often eaten by birds who mistake them for the berries that are produced by some other woodland plants. This deception appears to be an adaptation for seed dispersal.

According to one student of Indian history, an area on the shore of Lake Michigan in which wild leeks grew abundantly was called "Shika'Ko" or "Skunk Place" by the Menomini Indians. By his account, Shika'Ko later became Anglicized to Chicago.

Wild Leek is found in moist shady woods from New Brunswick and Quebec to Minnesota, south to North Carolina, Tennessee and Iowa.

✄ It is said to be the most edible of the wild onions. The leaves and bulbs can be cooked as greens or vegetables, used in salads and for seasoning. Ramp banquets are a spring tradition in southern Appalachia but over-consumption can cause a halitosis problem that breath deodorizers will hardly influence.

For further information on this species see *A. cernuum*.

Clintonia borealis
Clintonia*

61

Liliaceae
Lily Family

*Also Corn Lily, Bluebead Lily

A perennial with a slender branching rhizome. Although the flowers are large, to 2 cm. long (¾ in.), they are relatively inconspicuous because they lack distinctive color and odor. The deep blue berries, each containing several glossy brown seeds, are more colorful than the flowers. The fruits are eaten by chipmunks and other woodland birds and mammals that contribute to seed dispersal. The glossy dark green leaves persist into summer and make this an attractive plant for shaded gardens. The genus was named to honor De Witt Clinton, an early naturalist, statesman, and several times Governor of New York.

The only other species of this genus in the Northeast is *C. umbellulata*, Speckled Wood Lily, not illustrated. It is similar to *C. borealis* but has greenish white flowers, with purple dots, and black berries.

Clintonia is found in moist woods from Newfoundland to Manitoba, south along the mountains to Georgia and Tennessee.

✄ The very young leaves, before they are fully unrolled, can be cooked as greens or used fresh in salads, and have a mild cucumber flavor. This plant is extensively used

Flowers greenish yellow, often nodding, 2-
to-several on a leafless stalk; petals none,
sepals 6, stamens 6; fruit a several seeded,
dark blue berry; leaves basal, elliptical,
pointed, with tapering base, entire, with
parallel venation, to 30 cm. long (12 in.);
flowering stalk to 40 cm. high (16 in.).

May–June

as a potherb, under the name of Cow-tongue,
by country residents in some areas of Maine.
Note: Clintonia occurs in sufficient numbers to be gathered moderately for food only
in the northern part of its range. Food gath-
ering in the southern part of its range runs
the risk of exterminating the plant in those
areas.

Convallaria majalis
Lily-of-the-valley

62

Liliaceae
Lily Family

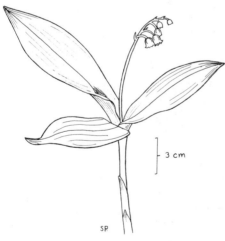

Flowers white, bell-shaped, with 6 lobes,
nodding, in a terminal cluster on a leafless
stalk, fruit a red berry; leaves 2 or 3, basal
or close together, oval, with pointed tips,
parallel veined; flower stalks to 25 cm. high
(10 in.).

May–June

in gardens. The several-seeded fruits should
be planted in autumn. Once they have become established they spread rapidly and
soon become so crowded that they must be
thinned in order to maintain maximum
flowering.

A native species, *C. montana*, occurs in
the mountains of Virginia, West Virginia,
North Carolina, Tennessee, and Georgia. It
is very similar to *C. majalis* but occurs as
scattered plants rather than in colonies.

Lily-of-the-valley was introduced as an
ornamental in shaded gardens but has escaped in many areas of the Northeast to
open woods and shady roadsides.

☠ The rhizome is a heart stimulant and a
diuretic. The plant contains several cardiac
glycosides, the most important being con-
vallatoxin. It has been listed as a poisonous
plant almost as long as published lists of
such plants have been available. There are

A perennial with a slender spreading rhizome, introduced from Europe. Although
the species is mainly self-incompatible, self-
pollination takes place occasionally. Cross-
pollination is effected by bees that apparently collect pollen, since the sweet scented
flowers produce little nectar. Reproduction
is more often by the creeping rhizome than
by seeds.

Typically found around old abandoned
homesites, this plant is fairly easy to initiate

very few documented instances of poisoning, but because of the plant's toxic compounds it should be treated as a dangerous plant. It should not be used in home remedies.

Erythronium americanum — 63
Trout Lily*

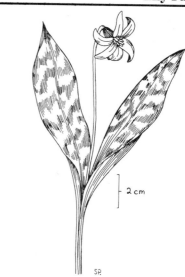

2 cm

Also Dogtooth Violet, Yellow Adder's Tongue, Yellow Fawn Lily

Flowers yellow, lily-like, with 6 backward bending segments borne singly on a leafless stalk; leaves 2, basal, tapering at each end, mottled with brown or purple; flowering stalk and leaves to 25 cm. high (10 in.).

March–May

A perennial arising from a deeply buried bulb. Cross-pollination is effected by early bumblebees, small butterflies and some early flies. The seed capsule splits into 3 sections, releasing numerous wrinkled seeds, each of which has a prominent fleshy appendage that has been interpreted as an adaptation for dispersal by ants.

Seven years are required from seed germination to flowering. The first year after germination, a small bulb and a single leaf are produced. Each year thereafter, until flowering, the leaf and the bulb are larger and the bulb is deeper. At maturity the bulb may be as deep as 37 cm. (15 in.).

These plants grow in colonies consisting of many 1-leaved non-flowering plants interspersed with a few 2-leaved flowering ones. It reproduces vegetatively by multiple horizontal shoots from the bulb. Each shoot develops a new bulb, and thus a new plant, at its tip.

Plants with yellow anthers are often observed growing side-by-side with those having brown-maroon anthers. The genetic relationship of these two types has not been thoroughly investigated, but some botanists have suggested that they may be two separate species. Breeding studies with this species are complicated by the 7-year generation time.

Most of the species of this genus are native to western North America. The genus name is derived from a Greek word meaning red, and refers to a red flowered European species. Another species of the Northeast, *E. albidum*, White Trout Lily, not illustrated, has bluish white flowers and less conspicuously mottled leaves.

Trout Lily is found in moist woods and shady fence rows from Nova Scotia to Minnesota, south to Florida and Alabama.

According to several sources on edible wild plants, the young leaves and the bulbs can be prepared as greens or cooked vegetables. However, a reputable source on medical botany states that the bulbs are known to poison poultry. Several recognized sources on medicinal wild plants list *E. americanum* as an emetic and an emollient.

Collecting quantities of this plant for any use is likely to seriously disrupt the colony in which the collection is made. Its most important value to humans is probably the esthetic value its presence adds to an undisturbed environment.

76

Lilium philadelphicum
Wood Lily*

64

Liliaceae
Lily Family

Also Wild Orange-red Lily

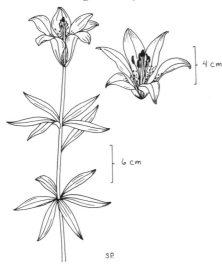

Flowers orange or orange-red, opening upward, in a terminal cluster of 1–3; petals 6, with purple-brown spots on inside; leaves in 2–6 whorls of 4's to 7's, lance-shaped; stems bright green, smooth, to 90 cm. high (3 ft.).

June–August

ored of the wild lilies. A variety of this species that occurs from Ohio westward has alternate leaves except for one whorl subtending the flower or flower cluster. This variety is found mainly in open fields, meadows and along stream banks. In some systems of classification it is listed as *L. umbellatum*.

Wood Lily is found in dry open woods, thickets, clearings, and meadows from Quebec to Wisconsin, south to North Carolina and Kentucky.

✗ The bulbs are edible and are reputed to have been used by American Indians as vegetables. There is little excuse for collecting these plants for food today. The bulb is small, often less than 2.5 cm. in diameter (1 in.), and collecting it destroys the plant. Except for emergencies the natural beauty of a single plant is worth far more than the few calories its bulb may provide.

A perennial with white bulbs bearing thin jointed scales. The flowers are self-incompatible and cross-pollination is necessary for the production of seeds. The 3-celled, 3-sided seed capsule splits into 3 sections, releasing numerous, flat, winged seeds that are adapted for dispersal by wind or water.

This is probably the most beautifully col-

Maianthemum canadense
False Lily-of-the-valley*

65

Liliaceae
Lily Family

Also Canada Mayflower, Beadruby

A perennial with a slender, freely branching rhizome. The 1–2 seeded, translucent red berries may persist into the winter months. They are eaten, to a limited extent, by Ruffed Grouse and small mammals which may serve as agents of dispersal. The name Beadruby refers to the red color of the autumnal fruit.

These plants may grow in dense colonies several meters in diameter, to the complete

exclusion of other plants. New plants often have only one shiny green leaf and no flowers. This species is very similar to Three-leaved Solomon's Seal (*Smilacina trifolia*) but the latter has dark red, unspotted berries and flower parts in 6's.

This genus includes only 3 species; one in the Northeast, one in the Pacific Northwest, and one in Europe and Asia. *M. canadense* has two recognized varieties, one of which occurs in the western part of its range

Flowers white, small, in a dense terminal cluster, to 5 cm. long (2 in.); petals 4, stamens 4; fruit a white berry with dark spots, turning dull red in early autumn; leaves 2 or 3, alternate, shiny green, heart-shaped at base; stems smooth, often zig-zag between leaves, to 20 cm. high (8 in.).

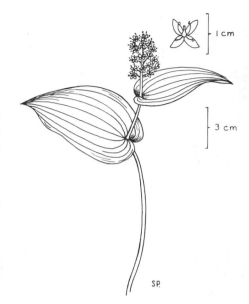

May–June

and flowers about two weeks later than the eastern variety. This species grows well, with some care, in shady wild flower gardens and shrub borders giving a fall and winter display of color to these areas.

False Lily-of-the-valley is found in moist woodlands from Newfoundland to the District of Mackenzie, south to North Carolina and South Dakota.

✖ The mature berries have a bittersweet taste that is considered pleasant by some individuals. However, they are purgative and should be eaten with caution.

Medeola virginiana
Indian Cucumber-root **66** **Liliaceae**
Lily Family

Flowers greenish-yellow, usually drooping to below the upper whorl of leaves, flower segments 6, bent backward, styles and stamens protruding; fruits dark purple-blue berries on erect stalks; leaves in 2 whorls, one near the middle of the stem with 5–9 leaflets, the other at the top, usually with 3 leaflets; stems woolly when young, to 70 cm. high (26 in.).

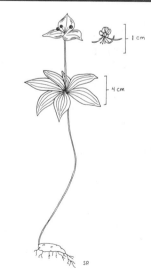

May–June

A perennial with a horizontal tuber-like rhizome. It has been theorized that the woolly stem prevents crawling insects from reaching the flower, leaving the pollen undisturbed for flying insects that visit and effect cross-pollination. The upper whorl of leaves develops some coloration in autumn and this has been credited with attracting birds to the dark blue berries.

This genus is native to northeastern North America and has only one species. It was named for the mythical sorceress Medea, in recognition of its supposed medicinal properties. The species is more often observed growing singly or in groups of a few than in dense colonies. It is a beautiful plant that usually flowers soon after the leaf canopy develops but is more colorful later in the season when the fruits have matured.

Indian Cucumber-root is found in moist woods from Nova Scotia and Quebec to Minnesota, south to Florida, Alabama, and Missouri.

✖ The rhizomes have the taste of cucumber and have been used for salads and pickles. The American Indians are said to have eaten them regularly, hence the name Indian Cucumber-root. However, the rhizome is small, usually 2.5–5 cm. long (1–2 in.), and in order to collect it the plant must be destroyed. Therefore, to avoid local extinction it should not be collected for food.

Polygonatum biflorum — Solomon's Seal — **67** — *Liliaceae* Lily Family

Flowers greenish-yellow, bell-shaped, usually in pairs hanging from leaf axils; fruit a blue-black berry; leaves alternate, smooth, sessile or short stalked, in two rows; stems slender, unbranched and arching, to 90 cm. long (3 ft.).

May–June

in each axillary cluster and hairs on the veins of the undersides of its leaves. The 3 species mentioned above are all woodland plants with extensively overlapping geographic ranges.

This is a genus of the Northern Hemisphere with about 6 species in North America. The name Solomon's Seal, according to one explanation, is derived from the scars left by previous stems on the rhizome. For some early botanist, these resembled the wax seals formerly used on official papers, and he further embellished the concept by associating it with Solomon. Another, and probably older, explanation for the common name dates back to the early Greeks. About 50 A.D. Dioscorides recommended the roots of this genus for sealing wounds and healing broken bones, and those who used this remedy were apparently showing the wisdom of Solomon.

A perennial with a deep, knobby, horizontal rhizome. The flowers are replaced by 3-celled, several-seeded berries which persist until late in the season and give the plants their late summer and fall aspects. These eventually fall or are eaten by forest animals including species of game birds in some parts of the Northeast. All the plants of a colony usually arch in the same direction.

This is a species of variable chromosome number with some forms having multiple sets. These forms, called polyploids, are usually larger, coarser, and have more than 2 flowers in each axillary cluster. They are recognized by most botanists as a separate species, *P. canaliculatum*, not illustrated, which has prominently nerved leaves and may reach a height of 2 m. (6½ ft.).

A closely related species, *P. pubescens*, not illustrated, is similar but has 1 or 2 flowers

Solomon's Seal is found in dry or moist woods from New England and Ontario to Manitoba, south to Florida and Texas.

✖ As an emergency food, the young tender

shoots can be cooked as asparagus. The rhizome was prepared as potatoes by American Indians. This species usually does not occur in sufficient abundance to justify collecting it for food.

☠ The berries are known to cause vomiting and diarrhea if eaten by humans. John Gerarde, a sixteenth-century herbalist, wrote that the crushed rhizome of the green plant was good for black and blue bruises that women may have acquired by stumbling into their husbands' fists.

Smilacina racemosa
False Solomon's Seal*

68

Liliaceae
Lily Family

Also False Spikenard, Solomon's Zigzag
Flowers white, numerous, in a branched
terminal cluster; fruits at first greenish-
yellow, later turning red with brownish
spots; leaves alternate, in 2 rows, short
stalked, pointed, with prominent parallel
veins, to 15 cm. long (6 in.); stems arching,
finely hairy, zigzagging between leaves, to
90 cm. long (3 ft.).

May–July

A perennial with a rough, thick, creeping rhizome. Although the clusters of fragrant flowers may be visited by insects, seeds are produced without fertilization and the new plants that grow from seeds are genetically identical to the parent plant. The aromatic berries, each containing 1–2 roundish, finely wrinkled seeds, persist into autumn.

As the diligent observer will notice, the berries begin to disappear even before they are mature and thus do not persist for long into autumn; they are eaten by the Ruffed Grouse, Gray-cheeked and Olive-backed Thrushes, White-footed Mouse, and other birds and mammals, all of which probably contribute to seed dispersal.

The name False Solomon's Seal refers to the resemblance of this plant to Solomon's Seal (*Polygonatum biflorum*). It differs from Solomon's Seal in having the flowers in terminal clusters, rather than in the axils of leaves. This attractive woodland plant is sometimes planted in moist shady rock gardens, but it should be grown from seeds rather than being transplanted from its native habitant.

False Solomon's Seal is found in moist woods from Nova Scotia to British Columbia, south to Georgia, Mississippi and Arizona.

✖ The young shoots can be used in salads or cooked like asparagus. The rhizomes were used for food by the Ojibwa Indians after soaking them in lye then parboiling. This plant is usually not abundant, however, and should be eaten for food only in emergencies. The berries can be eaten with less damage to the plant, but they may be cathartic and should be eaten sparingly.

80

Streptopus roseus
Rose Twisted Stalk*

69

Liliaceae
Lily Family

Also Rose Mandarin, Rosybells, Liver Berry, Scoot Berry

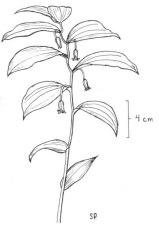

4 cm

sp

A perennial with a branching, root-covered rhizome. The flowers are visited by bumblebees and bee-like flies that may cause pollination. The translucent cherry-red berries may attract birds and small mammals. Each berry contains numerous longitudinally ribbed, white or cream-colored seeds that are likely to be spread by consumers of the berries.

A related species, *S. amplexifolius*, Twisted Stalk or White Mandarin, not illustrated, is similar but has greenish-white flowers, anthers with single lobes, clasping leaves with

Flowers pink or rose-purple, nodding, axillary, bell-shaped, with 6 segments that bend back at tips; stamens 6, anthers with 2 pointed lobes; stigma 3-parted; fruit a red berry; leaves alternate, sessile, not clasping, green on underside, to 9 cm. long (3½ in.); stems often branched, hairy, to 60 cm. high (2 ft.).

May–July

white undersides, and smooth stems. It is more common in the northern part of our range, extending southward in the mountains to North Carolina.

Rose Twisted Stalk is found in moist woods from Newfoundland and Labrador to Minnesota, south to New Jersey and West Virginia, and in the mountains to North Carolina and Tennessee.

✂ The young shoots and the berries of the plants of this genus have been used as cucumber-flavored additives to salads. However, this plant is usually not abundant and should not be collected for food. It is less of a threat to the plant to use only the berries but caution should be observed because they are cathartic and over consumption could be dangerous. Liver Berry and Scoot Berry are colloquial New England names for these plants that refer to the cathartic nature of the berry.

Trillium erectum
Purple or Red Trillium*

70

Liliaceae
Lily Family

Also Stinking Benjamin, Squawroot, Wakerobin, Birthwort

A perennial with a short, thick, brown rhizome. As the name Stinking Benjamin suggests, the flower has an unpleasant odor and may attract carrion flies. However, fertilization is never involved in the production of seeds, and the new plants that grow from them are genetic duplicates of the parent plant.

The purple-brown, 6-angled berry con-

tains many longitudinally wrinkled seeds, each with a soft appendage that attracts ants as agents of dispersal. The berries may be eaten by birds and mammals that also aid in spreading the seeds.

This is one of the most common trilliums in the Northeast. The flower stalk, which is normally erect, is often twisted slightly to one side. There are varieties with green, yellow, and white flowers. One variety in the mountains of Kentucky and Tennessee has

Flower maroon or brown-purple, solitary, on a long stalk, terminal; petals 3, sepals 3, stamens 6; leaves in a single whorl of 3, about as broad as long, sessile, to 20 cm. long (8 in.); plant to 40 cm. high (16 in.).

April–June

flowers that have a pleasant odor.

Purple Trillium is found in moist woods from the Gaspé Peninsula, Quebec, and Ontario south to North Carolina, Georgia, and Tennessee.

✘ As emergency foods the very young unfolding leaves can be used in salads or cooked as greens. However, in their undisturbed woodland setting, these plants are worth more as objects of natural beauty than as food plants. Since picking the leaves may result in the death of the rootstock, this should not be considered a food plant.

℞ A tea made from the roots of trillium species is said to have been used by American Indians to make childbirth easier and to reduce excessive menstrual flow, thus the names Squawroot and Birthwort. The 1851 edition of *The Dispensatory of the United States of America* states of the trilliums, "The complaints in which they are said to have been employed most successfully are the hemorrhages; but they have been used also in cutaneous affections, and externally in obstinate ulcers".

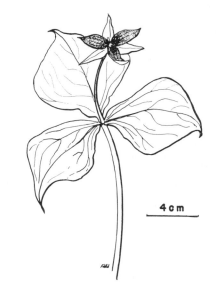

4 cm

Trillium grandiflorum
Large-flowered Trillium*   **71**   *Liliaceae*
Lily Family

*Also White Trillium

Flower white, solitary, on a long stalk, cup-shaped at base, petals 3, arching outward and bending back at tip, sepals 3, stamens 6; fruit a red, 6-angled berry; leaves in a single whorl of 3, broadly oval-shaped, sessile, to 15 cm. long (6 in.); plants to 40 cm. high (16 in.).

April–June

A perennial with a deep, blunt, tuber-like rhizome. The large showy flowers have a pleasant odor and become rose-pink with age. Although the flower may be visited by various insects, seeds are produced without fertilization and the new plants that grow from them are genetic duplicates of the parent plant. The seeds are brown, longitudinally wrinkled, and have soft appendages that attract ants as agents of dispersal.

This is the largest and most showy mem-

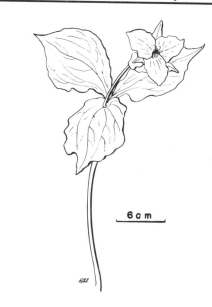

6 cm

ber of the genus. It is also the most variable species with different forms based on leaf variations, flower variations and combinations of leaf and flower variations. It is frequently cultivated in wildflower gardens and grows well in shaded, deep, neutral soils.

Large-flowered Trillium is found in moist woods from Quebec and Ontario to Minnesota, south to Georgia and Arkansas. ✖ ℞ See *T. erectum*.

Trillium sessile Toadshade*	**72**	*Liliaceae* Lily Family

*Also Sessile Trillium

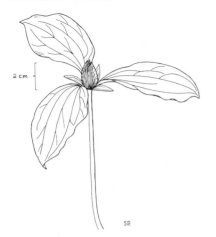

2 cm

SR

Flowers brown-purple, stalkless, petals 3, erect, flower appearing closed, sepals 3, green, erect or partly spreading; fruit a spherical red berry; leaves in a single whorl of 3 subtending the flower, sessile, broadly oval, sometimes mottled with light and dark green areas, to 10 cm. long (4 in.); to 30 cm. high (12 in.).

April–June

contains numerous seeds that, like other eastern species of this genus, are probably adapted for dispersal by ants.

This is one of several species in the Midwest and Northeast that has stalkless flowers. A related species, *T. recurvatum*, Prairie Trillium, not illustrated, is similar but has leaves with petioles and sepals that are widespread or drooping. It occurs in woods from Michigan to Nebraska south to Tennessee and Arkansas.

Toadshade is found in moist woods from western New York to Illinois south to Georgia, Mississippi and Arkansas. ✖ ℞ See *T. erectum*.

A perennial with a short, thick, almost vertically oriented rhizome. The fragrant flowers may be visited by insects, but the seeds are produced without fertilization, and new plants that grow from them are genetic duplicates of the parent plant. The berry

---

Trillium undulatum Painted Trillium	**73**	*Liliaceae* Lily Family

A perennial with a short, thick, brown rhizome. The beautiful but ill-smelling flower may attract carrion flies, but seeds are produced without fertilization. The shiny red berry contains numerous seeds with soft appendages that attract ants as agents of dispersal. The brown, longitudinally wrinkled seeds may also be spread by birds and mammals that eat the berries.

The genus name is derived from the Latin word that means "three" and refers to the occurrence of leaves and flower parts in threes. There are about 25 species of this genus in North America and eastern Asia with the greatest American concentration in the southern Appalachian Mountains. *T. undulatum* is one of the most attractive trilliums in the Northeast. It is easily recog-

Flowers white with red or pink markings at base of each petal, petals 3, wavy margined, sepals 3, green; fruit a 3-lobed, dark red berry; leaves in a single whorl of 3, each with a slender pointed tip and a rounded base, with a short petiole, to 12.5 cm. long (5 in.); plant to 40 cm. high (16 in.).

April–June

nized by the reddish V at the base of each petal, but it has several forms with variations in leaf and petal characteristics.

Painted Trillium is found in moist woods from the Gaspé Peninsula and Quebec to Manitoba, south to New Jersey and Pennsylvania, and along the mountains to Georgia. ✖ ℞ See *T. erectum*.

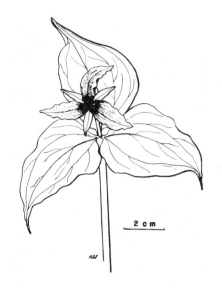

2 cm

Uvularia grandiflora
Large-flowered Bellwort

74

Liliaceae
Lily Family

Flowers yellow, nodding, with 6 segments, smooth inside, to 5 cm. long (2 in.); fruit a 3-lobed or 3-angled capsule; leaves alternate, oval, with fine hairs on underside, surrounding stem at base, to 10 cm. long (4 in.); stems forked in upper half, up to 4 flowers and several leaves on one branch, 4–8 leaves on other, to 60 cm. high (2 ft.).

April–June

A perennial with a short rhizome and fleshy, fibrous roots. The lobes of the seed capsule split exposing the few to several seeds in each cell. Each light brown, finely wrinkled seed has a relatively large, soft appendage that is a food source for ants. The ants carry the seeds to their nest where the appendage is eaten but the hard seed coat is undamaged. Thus the ants not only serve as agents of dispersal, they plant the seeds as well.

The genus name is derived from the Latin word "uvula" and refers to the pendulous nature of the flowers which supposedly resemble the small fleshy lobe that hangs down at the back of the soft palate. *U. grandiflora, U. perfoliata,* and *U. sessilifolia,* Wild

4 cm

Oats, not illustrated, are sometimes cultivated as ornamentals in shady wildflower gardens.

Large-flowered Bellwort prefers limestone soils and is found in moist woods from Quebec and Ontario to Minnesota, south to Georgia, Alabama, and Arkansas. ✗ The young shoots can be prepared like asparagus, and the rhizome and roots are said to be edible but not tasty. However, this beautiful woodland plant is not abundant and should be used for food only in emergencies.

| *Uvularia perfoliata*
Perfoliate Bellwort | **75** | *Liliaceae*
Lily Family |

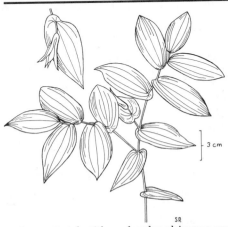

3 cm

A perennial with a slender rhizome and fleshy, fibrous roots. It is self-incompatible, thus cross-pollination by bees and other insects is essential for seed production. The type and numbers of seeds and the method of seed dispersal are probably similar to *U. grandiflora*.

Flowers yellow, nodding, with 6 segments, rough, orange-pebbled on inside, to 3.5 cm. long (1¼ in.), fragrant; fruit a capsule with 2 outer ridges and 2 peaks; leaves alternate, oval, smooth on underside, surrounding stem at base, to 9 cm. long (3½ in.); stems forked in upper half, several leaves and flowers on one branch, only leaves on other, to 45 cm. high (18 in.).

May–June

This is a genus of only 5 species, all native to eastern North America. Besides the two that are illustrated, two others occur in the Northeast: *U. pudica*, Mountain Bellwort, and *U. sessiliflora*, Sessile Bellwort. Both have leaves that do not surround the stem. *U. floridana* is similar to the latter two but the flower is subtended by a leaf-like bract.

Perfoliate Bellwort is found in moist acid woods from Quebec and Ontario south to Florida and Louisiana.

✗ See *U. grandiflora*.

| *Phoradendron flavescens*
American Mistletoe | **76** | *Loranthaceae*
Mistletoe Family |

A shrubby, evergreen perennial which is partially parasitic on several deciduous tree species. Staminate and pistillate flowers are on separate plants, making cross-pollination a necessity for fertilization. The berries are especially favored by the Cedar Waxwing and the Eastern Bluebird, but they are eaten by several other species of songbirds, all of which spread the seeds in their droppings.

These plants carry on photosynthesis and are thus dependent on their hosts for only water and mineral nutrients. They are probably best known for their use as Christmas decorations. Although they are not cultivated, they are harvested widely for the floristic trade. This species has been adopted as the state flower by Oklahoma.

This is a large genus mainly of tropical and subtropical America with one species in the Northeast, two in Florida, and several

Flowers greenish, small, on axillary branches to 5 cm. long (2 in.); fruit a white 1-seeded berry; leaves opposite, greenish-yellow, rather thick, oval, to 5 cm. long (2 in.); stems stout, brittle, bushy-branched, to 40 cm. long (16 in.).

September–November

in the Southwest. The origin of the practice of kissing under the mistletoe is obscure. It is probably related to the medieval ceremonial use of a related European species that was thought to ward off evil spirits. The kissing tradition today is often influenced by other kinds of spirits.

American Mistletoe is found on numerous species of deciduous trees from New Jersey, Pennsylvania, and Indiana, south to Florida and Texas.

🔍 American Indians of the Pacific Coast are reported to have used a tea made from the leaves of a western species, *P. villosum*, not

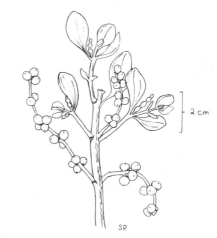

illustrated, to cause abortion. The berries of *P. flavescens* are poisonous, and several deaths of children have been attributed to eating them. Contact with the plant may cause dermatitis in sensitive individuals.

Menispermum canadense
Moonseed*

77

Menispermaceae
Moonseed Family

*Also Yellow Parilla
Flowers greenish white, in stalked axillary clusters, unisexual; sepals 4–8, petals 6–8, sepals longer than petals, stamens 12–24, pistils 2–4; fruit a black 1-seeded berry; leaves alternate, long-stalked, usually with 5–7 shallow palmate lobes, to 20 cm. wide (8 in.); stems climbing, smooth, to 3.6 m. long (12 ft.).

June–July

A perennial woody twining vine with a substantial rhizome. Pistillate and staminate flowers are on different plants, thus cross-pollination is a necessity for seed production. The mature fruits resemble bunches of wild grapes with each berry containing a single crescent-shaped seed. The fruits and seeds are known to be eaten by some species of birds.

This is a genus of only 2 species, the other one native to eastern Asia. A tropical member of this family, *Chrondodendron tomentosum*, not illustrated, is the source of curare,

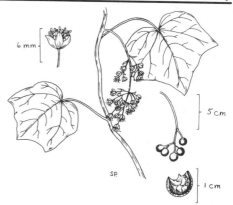

used as an arrow poison by South American Indians. It causes death by paralysis of the respiratory muscles. Curare is an important muscle relaxant used medically in certain types of surgery.

Moonseed is found in moist woods and along stream banks and fence rows from Quebec to Manitoba, south to Georgia, Alabama and Arkansas.

🌿 In herbology this species is described as a tonic, a laxative and a diuretic. However, the rhizome and the berries contain several toxic alkaloids, one of which has curare-like action. Eating the clusters of berries, mistaking them for wild grapes, is believed to have caused the deaths of several children in Ohio and Pennsylvania. This plant should not be used in home remedies.

| *Circaea quadrisulcata*
Enchanter's Nightshade | **78** | *Onagraceae*
Evening Primrose Family |

Flowers white, small, numerous, in elongate terminal or axillary clusters; petals 2, deeply 2-lobed, sepals 2, stamens 2; fruit covered with bristles, on stalks inclined downward; leaves opposite, oval, toothed, long stalked; stems stiff, to 60 cm. high (2 ft.).

June–August

A perennial, with a thin rhizome and thread-like runners, of North America and Asia. Plants growing in moist protected areas are pollinated by pollen eating Hover Flies (*Syrphios*), while those in dryer, more breezy locations are pollinated by small Halictus bees. The bristly 2-seeded fruits attach readily to the coats of passing animals.

There are 3 similar woodland species of this genus in eastern North America. *C. alpina*, not illustrated, ranges from Newfoundland to Alaska, south to the northern part of our area. *C. canadensis*, not illustrated, occurs in North America and Europe, does not reproduce sexually, and may be a hybrid of *C. alpina* and *C. quadrisulcata*.

This genus was named for Circe, a beautiful enchantress in Greek mythology who had the power to turn men into beasts. The name was originally applied to a poisonous plant of the ancient world but was later transferred to the present non-poisonous genus.

| *Aplectrum hyemale*
Puttyroot* | **79** | *Orchidaceae*
Orchid Family |

*Also Adam-and-Eve

Early in the season the flowering stalk arises from an underground tuber. The tuber later produces a slender rhizome that develops a second tuber at its tip. In late summer or autumn a single over-wintering leaf develops from the second tuber. The leaf withers in early spring and the flowering stalk grows from near its base. The name Adam-and-Eve is derived from the pair of tubers. The seed capsule splits into 3 sections, releasing a great many dust-like seeds that are dispersed by air currents.

This is a North American genus with a

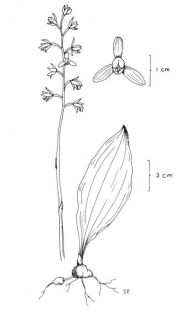

Flowers brownish or purplish, lip white with purple spots, to 16 mm. long (⅔ in.); in a cluster of 7–15 at tip of a leafless stalk; leaf 1, basal, elliptical, tapering at both ends, to 15 cm. long (6 in.); flowering stalk to 40 cm. high (16 in.).

May–June

single species. The genus name is derived from Greek words that mean "without a spur" and refer to the flower. The specific name is a Latin word for "winter" and alludes to the evergreen leaf. The tubers may be up to 2.5 cm. (1 in.) in diameter and have a sticky-gummy internal consistency. This may be responsible for the name Puttyroot.

Puttyroot is found in moist woods from Quebec to Saskatchewan, south to Georgia and Arkansas.

✗ ℞ The tuber is edible as a cooked vegetable and has been used in home remedies to treat bronchial problems. However, this plant should not be collected for other than survival emergencies. Like most of the other native orchids it was never abundant, but in recent years it has become increasingly rare.

Cypripedium acaule
Moccasin Flower*

80

Orchidaceae
Orchid Family

*Also Stemless Lady's Slipper, Pink Lady's Slipper, Nerve-root
Flowers pink, solitary, on leafless stalks, lip with deeply colored veins; leaves 2, basal, with prominent parallel veins; flowering stem to 40 cm. high (16 in.).

May–June

A perennial with thick fibrous roots. Pollination is by bees which may be lured to the flower by color or odor and require at least two experiences to learn that it offers no food pollen or nectar. Each plant produces about 30,000 tiny, hair-like seeds which are dispersed by wind. The seeds contain no stored food; thus, probably less than 1 in 10,000 falls in a suitable place for germination.

The genus name is derived from Greek words that mean "Aphrodite's Shoe", or from the Latinized form which means "Venus's shoe." Aphrodite, called Venus by the Romans, was the goddess of love and beauty in Greek mythology.

The flower soon withers, but the leaves may persist throughout the year. This is the most common Lady's Slipper. It is illegal to pick or injure this plant in New York and some other states of the Northeast.

Moccasin Flower is found in the acid soils of woods and bogs from Newfoundland to Alberta, south to Georgia and Alabama. ℞ See *C. calceolus*.

Cypripedium calceolus **Yellow Lady's Slipper**	**81**	*Orchidaceae* **Orchid Family**

4 cm

SP

A perennial of North America, Europe and Asia. Plants growing in the wet substrate of bogs, swamps and wet woods are not as tall and have flowers less than 2.5 cm. long (1 in.), with purplish lateral petals. At one time

Flowers yellow, often with purple veins, usually solitary, lateral petals brownish, twisted; leaves alternate, bases forming sheaths around stem; stems erect, to 60 cm. high (2 ft.).

May–July

these were recognized as a distinct species, *C. parviflorum*. Plants growing in moist or dry upland woods are taller, with flowers larger than 2.5 cm. long (1 in.). These have been called *C. pubescens*.

In North America, Yellow Lady's Slipper ranges from Newfoundland to northern British Columbia, south to Georgia and Alabama.

℞ The roots of this species and others in this genus were at one time considered to be effective in the treatment of nerve disorders, neuralgia, and insomnia. They were used by American Indians to treat nervous diseases, subdue pain, and promote sleep. Fortunately, there are synthetic drugs available today that are more effective for these uses than Lady's Slipper roots. It is illegal to pick any member of this genus in New York and some other states of the Northeast.

Epipactis latifolia **Helleborine***	**82**	*Orchidaceae* **Orchid Family**

*Also Weed Orchid

A perennial with a creeping rootstock, introduced from Europe. This genus includes species that show a gradation from cross-pollination by insects to self-pollination to the formation of seeds without fertilization. *E. latifolia* is chiefly self-pollinated, and it produces great numbers of dust-like seeds that are dispersed by air currents.

This is a small genus of about 10 hardy

Flowers purple or green-purple, in an elongated terminal cluster, perianth parts 6, lower lip heart-shaped, forming a sac at its base; leaves alternate, oval, clasping the stem; stems unbranched, to 60 cm. high (2 ft.).

July–September

species of the North Temperate Zone. Only one other species occurs in North America, *E. gigantea*, Giant Helleborine, not illustrated, a native of the Pacific coast states and the Northwest. The genus name is a Greek word meaning "to coagulate" and refers to a milk-curdling property that may have been associated with an earlier group of plants for which this name was used.

Helleborine is classified by some botanists as *E. helleborine*. It is found in woods, woodland borders, and thickets, and along shady roadsides from Quebec and Ontario, south to New Jersey, Pennsylvania, and Missouri. It appears to be spreading.

Woodlands

Goodyera pubescens
Downy Rattlesnake Plantain

83

Orchidaceae
Orchid Family

Flowers greenish white, in a dense cylindrical, terminal cluster; leaves basal, oval, dark green, with a network of white veins; flowering stem leafless, densely covered with hairs, bearing several small bracts, to 50 cm. high (20 in.).

July–August

A perennial with a short thick rhizome, stout fibrous roots, and a rosette of evergreen leaves. Pollination is effected by bumblebees. As with other members of this family, the numerous tiny, irregular, dustlike seeds are dispersed by wind. It is a very attractive plant throughout the year with its bluish-green leaves and conspicuous white netted veins.

The genus includes 4 woodland species in eastern North America. It was named in honor of John Goodyer, a seventeenth-century British botanist who assisted in the preparation of a book on medicinal plants that was well known in its day as Gerard's *Herbal*. Older names for this species are *Epipactis pubescens* and *Peramium pubescens*.

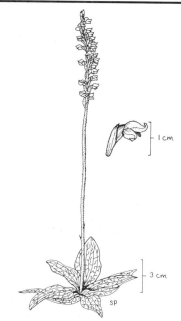

There is always the temptation among uninformed plant lovers to collect this spe-

90

cies for indoor terrariums and wild plant gardens. However, it is usually not abundant and in some areas it is protected by law. It can be appreciated most when it grows wild in its natural habitat.

Downy Rattlesnake Plantain is found in moist or dry woods from Newfoundland to Ontario, south to Florida and Alabama.

℞ The fresh leaves have been used as a poultice for skin irritations, bruises, and insect bites. An alternate method has been to soak the leaves in milk then apply as a poultice.

Habenaria psycodes
Purple Fringed Orchid*

84

Orchidaceae
Orchid Family

Also Soldier's-plume

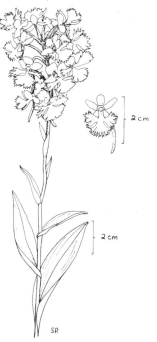
2 cm
2 cm
SP.

Flowers rose-purple to lavender, in a dense, cylindrical, terminal cluster; lower lip of flower 3-lobed, each lobe finely toothed, appearing fringed; leaves alternate, sheathing the stem, lower ones lanceolate to oval, decreasing in size upward becoming very narrow; stems thick, to 90 cm. high (3 ft.).

June–August

seeds do not fall on a suitable medium for germination, and this beautiful plant is never abundant. When one is fortunate enough to find a plant growing in its natural habitat, it should not be disturbed.

This is a large complex genus and sometimes hybridization between species creates problems in classification. When they occur in the same area _H. psycodes_ interbreeds with _H. lacera_, Ragged Fringed Orchid, not illustrated, and produces hybrids with characteristics intermediate between the two. In some systems of classification one variety of _H. psycodes_ is listed as a species, _H. fimbriata_.

Purple Fringed Orchid is found in moist open woods and wet meadows and along the margins of wooded swamps from Newfoundland to Manitoba, south to New Jersey and Pennsylvania and along the mountains to Tennessee and Georgia.

A perennial with a dense cluster of fleshy roots at the base of the stem. The very showy and fragrant flowers are cross-pollinated mainly by moths, and the numerous tiny seeds are dispersed by wind. Most of the

Orchis spectabilis
Showy Orchis

85

Woodlands

Flowers 2-lipped, lower lip prolonged into a long spur, white; upper lip a hood formed by the fusion of 2 lateral petals and 3 sepals, pink to purple; flowers in a short terminal cluster of 5–6, each subtended by a leaf-like bract; leaves 2, basal, oval, to 20 cm. long (8 in.); stems 4–5 angled, fleshy, to 25 cm. high (10 in.).

April–June

A perennial with a short rhizome and thick fleshy roots. Bumblebees visiting the flowers acquire sticky balls of pollen which they transfer to the stigma of the next flower visited. The several angled seed capsule splits into 3 sections releasing a great number of wind-borne seeds, very few of which germinate. (See *Cypripedium acule.*)

This beautiful native orchid is the most widely distributed member of the genus in the Northeast. The rhizome may lie dormant underground for several years then produce small clusters of plants. However, it is never abundant and should not be picked. It deserves the same legal protection that other native orchids have in some areas.

The genus includes about 100 species in the Northern Hemisphere. The genus name is derived from a Greek word that means "testicle" and it refers to the shape of the rhizome in some species. *O. spectabilis* and

O. rotundifolia, not illustrated, are the only species that occur in the Northeast. *O. rotundifolia*, Small Round-leaved Orchis, ranges from Greenland to the Yukon, south to New York and Michigan.

Showy Orchis is found in rich, deciduous woods, often in limestone areas from New Brunswick to Minnesota, south to Georgia, Alabama, and Arkansas.

Conopholis americana
Squawroot*

86

*Also Cancer Root

Flowers yellowish, densely crowded on upper half, or more, of stem, each turned downward, overlapping; leaves alternate, non-green, fleshy, reduced to scales, overlapping and completely covering stem, tan or light brown; stems unbranched, stout and blunt, to 20 cm. high (8 in.).

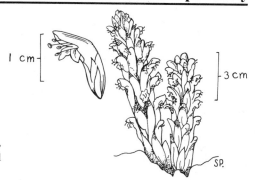

May–July

A parasite on the roots of several trees, often favoring oak and hemlock. The seed capsule splits into 2 sections, releasing numerous, tiny, irregular, glossy brown seeds. These are no more than 1 mm. (½₅ in.) in any dimension and may be spread by air currents. These plants are often overlooked in woodlands because their brownish color blends with fallen leaves which may partly or completely cover them.

This genus includes only 1 species in the Northeast with 3 others in the southwestern states, Mexico, and Central America. The genus name is Greek for "cone scale" and refers to the flowering stem, which somewhat resembles a weathered White Pine cone.

Squawroot is found in rich, moist woods, often in dense clusters, ranging from Nova Scotia to Wisconsin, south to Florida and Alabama.

Epifagus virginiana
Beech-drops

87

Orobanchaceae
Broom-rape Family

6 cm

An annual or perennial root parasite usually found growing under beech trees. The upper flowers are completely developed and are visited by bees but are sterile. The uncolored, smaller, lower flowers never open and produce seeds either by self-pollination or by asexual means. The seed capsule opens by splitting across the top to release numerous tiny, dust-like seeds.

Upper flowers whitish with purple stripes, lower ones smaller, not opening, flowers in axils of scales; leaves reduced to scales, alternate, non-green; stems light brown to purplish, freely branched, to 50 cm. high (20 in.).

August–October

This is the only species of a genus that is native to North America. This family has only 3 genera in eastern North America, with a total of 7 species, all of which are parasitic on the roots of other plants.

E. virginiana is the most common representative of the family in the Northeast. It is a characteristic plant of the autumnal herbaceous flora in the climax beech-maple-hemlock forest. The dead stems commonly persist throughout the winter and into the next growing season. There is very little difference in the appearance of dead and live stems. Damage to host plants seems to be insignificant.

Beech-drops is often abundant under beech trees from Nova Scotia and Quebec to Wisconsin, south to Florida and Louisiana. ℞ The plant is reputed to have astringent properties that are useful in the external treatment of bruises, cuts, and skin irritations.

Orobanche uniflora
One-flowered Cancer-root*

88

Orobanchaceae
Broom-rape Family

*Also Broom-rape

Flowers pale lavender, 5-lobed, tubular, tube curved, solitary at tip of leafless stalk; stems leafless, without chlorophyll, hairy, to 25 cm. high (10 in.).

April–June

A perennial with a short underground stem. This attractive, fragrant-flowered little plant is parasitic on the roots of several species. Fertilization is most often the result of cross-pollination by small bees of the genus *Halictus* and bumblebees. Persistently enclosed by the withering corolla, the seed capsule opens at maturity by two valves releasing great numbers of dust-like seeds that are dispersed by air currents.

This is a large genus of worldwide distribution. The genus name is derived from Greek words meaning literally "vetch strangler," apparently an allusion to their parasitic habit. Several species are parasitic on the roots of cultivated plants, but they are usually not serious problems in North America. There are five species in the Northeast including two that are natives of Europe and Asia. *O. uniflora* is the best known and the most widely distributed.

A related species, *O. minor*, Clover Broomrape, not illustrated, has a terminal cluster of yellowish flowers, scale-like leaves, and reaches a height of 45 cm. (18 in.). This native of Europe is a parasite on the roots of clover and tobacco along the Atlantic coast from New Jersey to North Carolina.

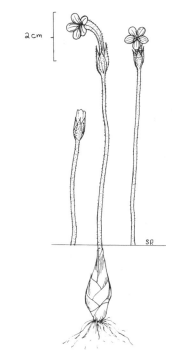

One-flowered Cancer-root is found in moist woods and damp thickets from Newfoundland and Quebec to British Columbia, south to Florida, Texas, and California.

✹ The tender underground parts of the stems of *O. minor* are said to be edible and can be prepared in the same way as asparagus.

Oxalis acetosella
Common Wood Sorrel*

89

Oxalidaceae
Wood Sorrel Family

*Also Wood Shamrock

A perennial with a creeping, scaly, horizontal rhizome. The flowers are visited and possibly pollinated by small bees and syrphus flies. Late in the flowering season, or after, flowers are produced near the surface, or underground, that never open and are self-pollinated. Fertile seeds are produced

by both types of flowers. The seed capsule splits into 5 sections, releasing usually numerous dark brown, longitudinally ribbed seeds.

This is a very fragile, light-sensitive species that typically, after sunset, has closed flowers and drooping, folded leaves. The similarity of its leaves to those of some of

94

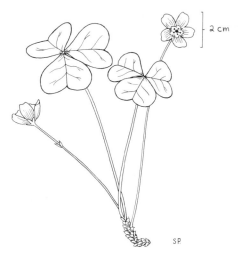

the clovers is the source of the name Wood Shamrock. This plant is also widespread in Europe and Asia, and some botanists recognize the American representatives as a separate species, *O. montana*.

Common Wood Sorrel is found in moist woods from Newfoundland and Quebec to

Flowers pink or white, with deeper pink lines, to 18 mm. wide (¾ in.), solitary, on long leafless stalks; sepals 5, petals 5, notched, stamens 10, pistil 1; leaves basal, long stalked, with 3 heart-shaped leaflets; flower stalks usually extending above leaves, to 15 cm. high (6 in.).

May–July

Saskatchewan, south to Pennsylvania and Ohio, and along the mountains to North Carolina and Tennessee.

✖ The leaves of the plants in this genus are tart trailside nibbles. They can be added in small quantities to salads or steeped in hot water for a sour drink. The genus name is derived from a Greek word that means "sour." The taste of the leaves is the result of the presence of oxalic acid, which is toxic in large amounts. Thus, these plants should be consumed in moderation.

℞ In herbology it is described as a diuretic and a spring tonic. A solution made from the seeds has been used in Europe for the treatment of spermatorhea (the emission of semen without stimulation).

Sanguinaria canadensis
Bloodroot*

90

Papaveraceae
Poppy Family

*Also Red Puccoon

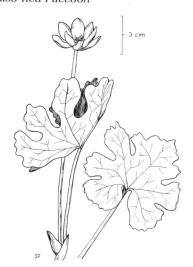

Flowers white, solitary, on leafless stalks, to 5 cm. wide (2 in.), petals usually 8, stamens numerous; one basal leaf, with 5–9 deep palmate lobes, heart-shaped at base; flowering stalk to 15 cm. high (6 in.), to 30 cm. high (1 ft.) in fruit.

March–May

A perennial arising from a thick horizontal rhizome with red-orange juice. Self-pollination is avoided by the maturation of the anthers after the stigma has shrivelled. The bees and bee-like flies that visit the flower are attracted by the pollen since the plant does not produce nectar. The egg-shaped, wrinkled, glossy brown seeds are dispersed by ants which are attracted by an edible appendage on the end of each seed.

In an early stage of development the leaf is wrapped around the emerging flower which eventually rises above the leaf. The flower opens only in full sunlight and the petals disappear in 1–2 days. The leaf continues to grow after flowering and may attain a width of 20 cm. (8 in.).

Bloodroot is found in moist woods from Nova Scotia to Manitoba, south to Florida, Alabama and Oklahoma.

℞ The red-orange juice is the active ingredient in the plant's medicinal uses. Upon hearing from Indian traders that American Indians used the juice for the treatment of cancerous diseases, J. W. Fell in 1857 used an extract of the plant to make a paste which

he claimed was effective in the treatment of breast cancer.

Taken internally in small doses an extract of the plant is said to promote discharge of mucus from the respiratory tract, but an overdose can be fatal. According to *The Dispensatory of the United States of America* (1851), "Four persons lost their lives at Bellevue Hospital, New York, in consequence of drinking largely a tincture of bloodroot, which they mistook for ardent spirit." A colonial physician in 1807 observed that the most prominent effect of Bloodroot, even when taken in moderate doses, was to induce vomiting.

Phryma leptostachya
Lopseed
91
Phrymaceae
Lopseed Family

Flowers pale purple to pink, 2-lipped, in pairs, along the leafless tips of stems and branches, bending downward and hugging the stem in fruit; leaves opposite, oval, petioled, toothed, each pair attached at right angles to the pair above and below; stems slender, 4 angled, unbranched or with few branches, smooth, to 90 cm. high (3 ft.).

June–August

A perennial with bright green leaves, of North America and eastern Asia. Each flower produces a single seed which is enclosed in the persistent calyx. At maturity the calyx, with its enclosed seed, breaks free forcefully when touched and may travel for a considerable distance. The long slender lobes of the calyx are slightly curved or hooked at the tip and may become attached to the coats of passing animals.

This is a family with a single genus and only three species. Besides the one that occurs in both North America and eastern Asia, two others have been identified in eastern Asia. The name Lopseed refers to the way the calyx lops or hangs down, and this is a

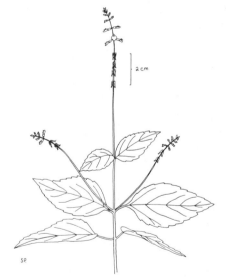

characteristic that makes the plant easy to identify.

Lopseed is found in moist woods from New Brunswick and Quebec to Manitoba, south to Florida and Oklahoma.

Phlox divaricata
Wild Blue Phlox*

92

Polemoniaceae
Phlox Family

*Also Wild Sweet William

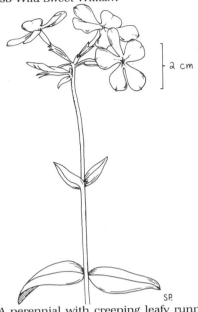

2 cm

S.P.

Flowers pale blue-purple, with long tubes and 5 spreading, often shallowly notched lobes, in branched terminal clusters; pistil and stamens enclosed in flower tube and not visible; leaves opposite, sessile, the widest part below the middle, to 5 cm. long (2 in.), in scattered pairs; stems often prostrate at base, sparsely hairy, to 20 cm. high (8 in.).

April–June

A perennial with creeping leafy runners that root at intervals. Pollination is commonly by moths or butterflies, and hybridization with other species of the genus occasionally takes place. The seed capsule splits from the top into 3 sections, releasing 3–12 finely wrinkled, light green or brown seeds.

This species has two distinct geographical varieties. In the western part of its range, it is characterized by rounded petal lobes, while in the eastern part the petal lobes are notched. The ranges of these varieties usually do not overlap. The name Wild Sweet William is also applied to P. maculata, not illustrated, a species that flowers in June and July and has numerous red-purple flowers.

This is a genus of about 50 species of North America and at least one in Siberia. Many species are cultivated as the phlox and moss-pink of flower gardens. Hybridization between P. devaricata and P. paniculata, not illustrated, has produced a many-flowered species with several forms cultivated as Elizabeth Phlox, Amanda Phlox, Louise Phlox, etc.

Wild Blue Phlox is found in moist woods from Quebec to Minnesota, south to Florida and Texas.

Polemonium reptans
Greek Valerian*

93

Polemoniaceae
Phlox Family

*Also Sweatroot, Abscess Root

A perennial with a creeping horizontal rhizome. The oval seed capsule splits into 3 sections, releasing 15–36 seeds. This attractive plant is often cultivated as an ornamental in shady wild flower gardens. It can be grown from seeds which germinate in 3–4 weeks at temperatures of 68–86°F. In-

itial plantings should be acquired from commercial greenhouses.

This genus is much more abundant in western North America than in the Northeast. Of the two native species in eastern North America, P. reptans has the widest geographic distribution. A closely related species, P. van-bruntiae, Jacob's Ladder, not

Flowers blue, in few-flowered clusters at the ends of stems and branches; flowers bell-shaped, nodding, 5-lobed, stamens not protruding; leaves alternate, pinnately compound, with 11–17 leaflets; stems smooth, slender, often weak and sprawling, branched, to 40 cm. long (16 in.).

April–June

illustrated, is very similar but has stamens that extend beyond the lobes of the carolla. It is found in swamps and bogs in mountainous areas from New England and New York south to Maryland and West Virginia.

Greek Valerian is found in moist woods from New York to Minnesota, south to Georgia, Mississippi, and Oklahoma.

℞ The rhizome is reported to have astringent properties and to cause sweating, thus the names Sweatroot and Abscess Root. It has been recommended for diarrhea, snake and insect bites, inflammations, and lung problems. *P. caeruleum*, not illustrated, a native of Europe and Asia cultivated in North America as an ornamental, has similar properties.

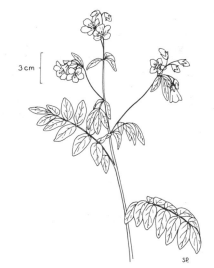

Polygala paucifolia
Fringed Polygala*

94

Polygalaceae
Milkwort Family

*Also Flowering Wintergreen, Bird-on-the-wing, Gaywings

Flowers rose-purple, tubular, with a fringe at the tip, flanked by 2 spreading, petal-like sepals; flowers 1–4, solitary, in leaf axils at tip of stem; leaves alternate, lower ones scale-like, scattered, upper ones oval, crowded at tip of stem; stems to 15 cm. high (6 in.).

May–June

An evergreen perennial with a slender, shallow rhizome and rooting surface runners. Cross-pollination is effected by honeybees and small bees of the genera *Halictus* and *Andrenidae*. The weight of the bee depresses the lower lobe of the flower, causing the style and the stamens to burst out and come into contact with the bee's body. If the weight of the visitor is insufficient, the stigma will not be exposed. During summer and autumn, small flowers that never open and are self-pollinated are produced on short underground branches.

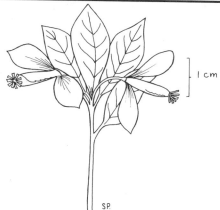

The roundish 2-celled seed capsule splits into two sections, releasing 2 black, hairy, egg-shaped seeds. Each seed has a soft appendage at one end which serves to attract ants as agents of dispersal. The seeds produced by both aerial and subterranean flowers are fertile.

The genus name is derived from Greek words that mean "much milk." This refers to an early belief that if eaten by cows, some species of this genus would increase the flow of milk. It is a large genus of worldwide distribution with about 16 species in northeastern North America.

Fringed Polygala is found in moist woodlands from New Brunswick and Quebec to Manitoba, south to Virginia and West Virginia, and in the mountains to Georgia and Tennessee.

℞ The roots are bitter and have been used to cause sweating in the treatment of fevers and to promote the discharge of mucus from the respiratory tract. A related species, *P. senega*, Seneca Snakeroot, not illustrated, was valued by the Seneca Indians as a treatment for snakebite.

Polygonum scandens
False Buckwheat
95
Polygonaceae
Smartweed Family

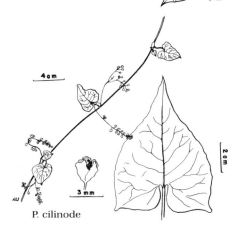

P. cilinode

Flowers greenish-white, in interrupted clusters, on long axillary stalks; fruit a 3-angled, shiny black nutlet, enclosed by a strongly 3-winged calyx; leaves alternate, pointed, deeply heart-shaped at base, to 14 cm. long (5½ in.); stems angled, twining or trailing, to 5 m. long (16½ ft.).

August–October

A perennial with a substantial rhizome, native to North America and Europe. Each flower produces a single, triangular, glossy black seed tightly enclosed by a broadly 3-winged calyx adapting it for dispersal by wind. The seeds are eaten by upland game birds which may contribute to seed dispersal. It sometimes becomes a troublesome weed in vegetable gardens and along ditches.

A related species, *P. cilinode*, Fringed Bindweed, see illustration, is similar but has white flowers in branched clusters; the seed enclosed in a 3-angled but not strongly winged calyx; leaves usually with some sharp angles in the basal lobes; stem with a ring of bristles at each node. It is a perennial of dry woods and thickets from southeastern Canada to Pennsylvania and in the mountains to North Carolina and Tennessee.

Another related species, *P. convolvulus*, Black Bindweed, not illustrated, is similar to *P. cilinode* and *P. scandens* but has green flowers; dull black seeds; calyx 3-angled but not winged; nodes of stem smooth rather than bristly. It is an annual of open areas, native of Europe and widespread in the Northeast. These three species are easy to distinguish from other members of the genus in eastern North America by their trailing or twining growth habit.

False Buckwheat is found in moist woods and thickets, along fence rows and road-sides, and in waste areas from New Brunswick and Quebec to Manitoba, south to Florida and Texas.

✕ The seeds of all three of the species mentioned above are said to have been used by American Indians to make a meal similar to buckwheat flour.

Polygonum virginianum
Jumpseed*

96

Polygonaceae
Smartweed Family

*Also Virginia Knotweed

Flowers greenish-white or tinged with pink, tiny, scattered on terminal or axillary stalks, to 50 cm. long (20 in.); fruit a lens-shaped, glossy brown nutlet with 2 persistent, hooked styles; leaves alternate, oval to lanceolate, usually hairy on the underside, to 15 cm. long (6 in.); stems erect, nodes with a ring of bristles, to 1.2 m. high (4 ft.).

July–October

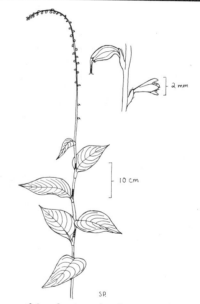

A perennial with a knotty, somewhat woody rhizome, of eastern North America and India. The name Jumpseed refers to a peculiar method of seed dispersal exhibited by this plant. The slightest pressure on the hardened styles causes a rebound that may fling the seed outward for a distance of up to 3 m. (10 ft.). As an additional possibility for seed dispersal, the persistent hooks on the tips of the styles may become entangled in the coat of a passing animal.

This species is easy to distinguish from other Polygonums by the long, slender, interrupted flowering stalks and the 2 persistent hooks on each seed. It is a relatively uniform species with little intraspecific variation, although a form with reddish flowers is occasionally observed. The differences between this plant and other members of the genus are so distinct that it has been placed by some botanists in a separate genus with the name *Tovara virginiana*.

Jumpseed is found in moist woods, thickets, and lake margins from Quebec to Minnesota, south to Florida and Texas.

Claytonia virginica
Spring Beauty

97

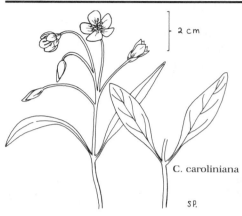

2 cm

C. caroliniana

S.P.

Flowers white to pink, in terminal clusters of 2-several, petals 5, with darker pink lines; leaves opposite, single pair near middle of stem, narrow, to 15 cm. long (6 in.); stems often decumbent, succulent, smooth, delicate, to 30 cm. long (12 in.).

March–May

A perennial with a deep bulb-like rootstock. Pollination is by bumblebees and bee flies that get both pollen and nectar from the flowers. The anthers often mature before the stigma, increasing the probability of cross-fertilization. The seed capsule splits into 3 sections which roll inward from the top, releasing 6 shiny black seeds. Each seed has a soft appendage which has been interpreted as an adaptation for dispersal by ants.

There are two closely related and very similar species of this genus in the Northeast. They have a brief period of flowering and vegetative growth, then usually disappear by early summer. After several months of dormancy the bulbs resume activity during warm periods in January and February.

Claytonia caroliniana, Carolina Spring Beauty, is very similar to *C. virginiana*, except its leaves have distinct petioles and are broader. It is found in rich woods from Newfoundland to Saskatchewan, south to southern New England, and along the mountains to North Carolina and Tennessee.

The genus was named to honor John Clayton, an amateur botanist in Colonial Virginia, who collected extensively and sent specimens to England. These were used to prepare the first scientific publication on the flora of Virginia.

C. virginica is found in moist woods from Nova Scotia to Saskatchewan, south to Georgia, Louisiana, and Texas. It is often found in more open areas in the southern part of its range.

✖ The leaves and stems can be cooked as greens, and the bulbs can be prepared in the same manner as potatoes. However, these plants should be collected for food only in emergencies or where they are superabundant. Their value as food plants is far less than the value of the beauty they impart to spring woodlands.

Dodecatheon meadia
Shooting Star*

98

*Also American Cowslip

A perennial with a basal rosette of leaves and a fibrous root system. For a given flower, the stigmatic surface becomes receptive to pollen before the anthers mature and cross-pollination is accomplished mainly by bees. The reddish-brown, cone-shaped seed capsule opens by 5 short terminal teeth, releasing numerous seeds.

This is a genus of about 30 species native to North America, mostly in the western states. *D. meadia* has several varieties and forms distinguished by variations in the shape of the leaves, the red color at the base of the leaves, and the length of the anthers. It was called "Prairie Pointers" by the early settlers and was much more abundant before its habitat was converted to croplands.

Flowers pink, nodding, in a cluster of 6–30, at the tip of a long leafless stalk; flower divisions 5, strongly bent backward, stamens 5, anthers fused into a cone-shaped protrusion; leaves basal, narrowly elliptical, reddish at base, to 15 cm. long (6 in.); flowering stalk to 50 cm. high (20 in.).

April–June

A related species, *D. radicatum*, Jeweled Shooting Star, not illustrated, is similar but has reddish-purple flowers, in a cluster of 2–11, on a stalk to 30 cm. high (1 ft.), and leaves that do not have a reddish base. It flowers earlier than *D. meadia* and its range does not extend south of Missouri. White-flowered forms occur in both species.

Shooting Star is found in moist or dry woods, meadows, and prairies from Pennsylvania to Wisconsin, south to Georgia and Texas.

Lysimachia quadrifolia
Whorled Loosestrife*

99

Primulaceae
Primrose Family

*Also Four-leaved Loosestrife
Flowers yellow, solitary, on long slender stalks, from the axils of upper leaves; petals 5, streaked or dotted with red; leaves in whorls of 4's or 5's, lanceolate, spreading at right angles to the stem, to 10 cm. long (4 in.); stems usually unbranched, to 75 cm. high (2½ ft.).

June–August

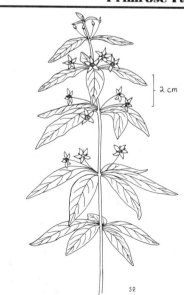

A perennial of delicate beauty with its light green leaves and pale golden yellow flowers. The pistil extends far beyond the stamens so that self-pollination is rare. Insects that visit the flowers for the purpose of collecting pollen are bumblebees, honeybees, and bees of the genus *Macropis*. The seed capsule splits from the top into several segments, releasing numerous angular, faintly pitted, dark brown seeds. These are so small that they are readily windborne.

This is a genus of about 13 species in the Northeast, 10 of which are native to the region. All species are similar in having yellow, star-shaped flowers, often with dark lines

102

or dots, and opposite or whorled leaves. *L. quadrifolia* is easy to distinguish from related species by its leaves in whorls of 4's and 5's and its small flowers.

Whorled Loosestrife is found in moist or dry open woods and thickets from New Brunswick to Wisconsin, south to Georgia and Alabama.

Trientalis borealis
Starflower*

100

Primulaceae
Primrose Family

*Also Chickweed Wintergreen

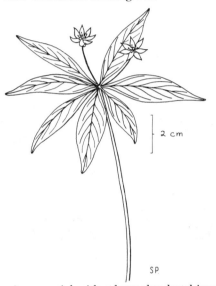

2 cm

S.P.

A perennial with a long slender rhizome and sometimes surface runners. On a given plant the stigma matures before pollen is released assuring cross-pollination which is

Flowers white, usually 2, to 15 mm. wide (⅔ in.), on slender terminal stalks, usually with 7 pointed petals; leaves in a single terminal whorl of 5–9, shiny green, lanceolate, with tapering point, to 10 cm. long (4 in.); stems to 25 cm. high (10 in.).

May–June

usually effected by bee-like flies. The seed capsule opens from the top by 5 teeth, releasing numerous irregular, black, finely reticulate seeds.

The genus name is derived from a Latin word meaning "a third" and may refer to the length of the flower stalk which is sometimes about one-third the length of the plant. There are only 3 species of the genus, and *T. borealis* is the only representative in the Northeast. It is the only member of the Primrose Family that occurs above the timberline in the mountains of New England. It is listed as *T. americana* in older systems of classification.

Starflower is found in moist woods and bogs from Newfoundland and Labrador to Alberta, south to Virginia and Illinois.

Aconitum uncinatum
Wild Monkshood

101

Ranunculaceae
Crowfoot Family

A perennial with a short tuberous rhizome, often with a few small secondary tubers. On a given plant the stamens mature before the pistils, thus cross-pollination, probably by bumblebees, is assured. Each flower usually produces 3 seedpods that split lengthwise, releasing numerous light

to dark brown, angular, winged seeds. The presence of a flattened surface, or wing, on a seed is usually an indication of dispersal by wind.

This is a genus of 80–100 species occurring in the temperate regions of North America, Europe and Asia. Of the three spe-

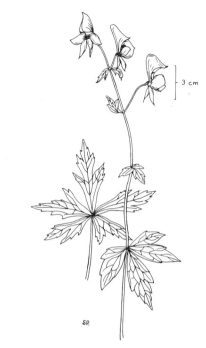

Flowers blue, occurring singly on stalks, usually clustered at tip of stem; petals 2–5, small and inconspicuous; sepals petal-like, upper one helmet-shaped, other 4 narrower, unequal; leaves alternate, with 3–5 deeply cut palmate lobes, lower ones long stalked, upper ones shorter stalked or sessile; stems weak, often leaning on other plants, to 1.2 m. high (4 ft.).

August–October

cies native to northeastern North America, *A. uncinatum* has the greatest geographic distribution. A closely related species, *A. noveboracense*, Northern Monkshood, not illustrated, is very similar but the hood-shaped upper sepal is extended backward into a short blunt spur. It is very rare and since 1978 has been protected under the national Endangered Species Act.

Wild Monkshood is found in moist woods from New Jersey and Pennsylvania to Indiana, south to Georgia and Alabama.

☠ All the species of this genus contain the alkaloid aconitine and others, and are poisonous to humans and livestock. All parts of the plant contain the toxic compounds, but they are concentrated in the tuberous rhizomes and the seeds. The leaves are most toxic just before flowering. Death to both humans and domestic animals, often from respiratory paralysis, usually occurs within a few hours after ingestion. In ancient times a solution made from these plants was given to condemned criminals as a form of execution.

Actaea alba
White Baneberry*

102

Ranunculaceae
Crowfoot Family

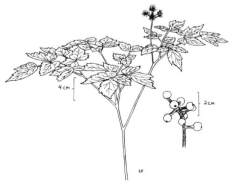

*Also Doll's Eyes, White Cohosh
Flowers white, to 6 mm. wide (¼ in.), in a dense cylindrical cluster on a long stalk; sepals and petals shedding early, stamens numerous, giving the flower an almost globular appearance, pistil 1, with a very large stigma; fruit a white berry on a thick red stalk, the persistent stigma forming a prominent black dot; leaves alternate, 2 or 3 times divided, leaflets to 10 cm. long (4 in.); stems erect to 75 cm. high (30 in.).

May–June

A perennial with a well developed rhizome. The flowers do not produce nectar but pollination is effected by small bees that visit the flowers for pollen. The white berries on the bright red stalks are more impressive than the flowers and are the source of the name Doll's Eyes. Each berry contains several dark brown, angular seeds that may be dispersed by the Ruffed Grouse and other birds that eat the berries.

This species is recognized by some botanists as *A. pachypoda*. The fruiting period is July to October, and occasionally a red fruited form is observed. A related species, *A. rubra*, Red baneberry, not illustrated, is very similar but has red berries. The latter has a white fruited form but is easily distinguished from *A. alba* by the slender fruit stalks.

White Baneberry is found in rich moist woods from Quebec to Minnesota, south to Georgia and Louisiana.

☠ All members of this genus are considered poisonous, and all parts of the plant contain the poison. The rhizome is a very violent purgative and emetic. Ingestion of only a few of the berries may cause increased heart rate, irritation of the intestinal tract, and dizziness. Although there are no records of loss of life in the United States, death of children has been reported in Europe as a result of eating the berries of *A. spicata*, European Baneberry, not illustrated.

Anemone quinquefolia
Wood Anemone*

103

Ranunculaceae
Crowfoot Family

*Also Wind Flower

2 cm

S.P.

A perennial with a slender, yellowish, horizontal rhizome. The flowers are usually closed at night and during cloudy damp weather. The anthers mature over a 7–14 day period and the sepals are shed immediately after the stamens. Cross-pollination

Flowers white, solitary, terminal, to 2.5 cm. wide (1 in.); petals none, petal-like sepals usually 5, stamens numerous, pistils numerous; leaves with 3–5 palmate divisions, basal leaves long-stalked; stem leaves in a single whorl of 3 below the long-stalked flower; stems slender, smooth, to 20 cm. high (8 in.).

April–June

is accomplished by bees and bee-like flies. Each flower produces 6–15 brown, densely stiff-hairy, seed-like fruits.

The genus name is derived from a Semitic name for Adonis of Greek mythology, whose blood is said to have given the color to a crimson flowered *Anemone* of the Orient. This and a few other species are cultivated as showy garden ornamentals. *A. quinquifolia* is sometimes cultivated, but to avoid local extinction it should not be disturbed in its native habitat. A western variety has large blue flowers on short stalks.

Wood Anemone is found in open woods, woods borders, and thickets from Quebec and Ontario to Manitoba, south to West Virginia, Kentucky, and Illinois, and along the mountains to Georgia.

Anemonella thalictroides
Rue Anemone*

104

Also Woods Potato
Flowers white to pink, to 2.5 cm. wide (1
in.), on slender stalks, in a terminal cluster
of 2 or 3; petals none, petal-like sepals 5–
10, stamens numerous, pistils several;
basal leaves long-stalked, twice ternately
divided, stem leaves in a whorl subtending
flower cluster, leaves and leaflets with 3
shallow, blunt lobes; stems smooth, slen-
der, to 20 cm. high (8 in.).

April–June

A delicate perennial with a cluster of spindle-shaped tubers. The flowers are visited by numerous early bees and bee-like flies that effect pollination. Each flower produces 8–12 smooth, tan to brown, spindle-shaped, prominently 8-ribbed, seed-like fruits.

This is a genus with only one species,limited to eastern North America. The flowers resemble those of *Anemone* (Wood Anemone) and the leaves are similar to those of *Thalictrum* (Meadow Rue). Both the Latin and the common names allude to this similarity. It is easily distinguished from these plants since Wood Anemone has a single terminal flower and Meadow Rue is usually several feet high.

It is sometimes cultivated as an ornamental in shady rock gardens. Green flowered forms are occasionally observed in nature, and rose colored and double flowered forms are available from some greenhouses. It is usually not abundant and thus should not be collected in its natural habitat. Commercial handlers who collect it in the wild should not be encouraged.

Rue Anemone is found in moist or dry woods from Maine and Ontario to Minnesota, south to Florida, Mississippi, and Arkansas.

The tubers are high in starch and can be eaten as an emergency food. However, they are small and collecting them destroys the plant. In order to avoid local extinction, it should not be considered as a food plant.

Aquilegia canadensis
Wild Columbine

105

A perennial with a short persistent stem base and a branched horizontal rhizome with dense fibrous roots. Nectar collects in each of the hollow spurs and is extracted by the long beak of the Ruby-throated Hummingbird, which is an effective pollinator. Each flower usually produces 5 spindle- shaped seedpods that split longitudinally along the inner side, releasing numerous glossy black seeds.

Of the 15 species of this genus in North America all but 2 are in the West. There are 3 species groups within the genus, and it has been hypothesized that the main bar-

3 cm

SP.

Flowers red with yellow centers, to 5 cm. long (2 in.), nodding, petals 5, each extending back in a long hollow spur; stamens numerous, extending beyond petals; sepals 5, petal-like; leaves alternate, 2 or 3 times divided into 3's, to 15 cm. wide (6 in.), upper ones smaller, usually with a few large basal leaves; stems slender, usually branched, to 60 cm. high (2 ft.).

April–July

rier to inbreeding among them is that each group has a different type of pollinator. The *A. canadensis* complex is normally pollinated by hummingbirds, another group by bumblebees, and the third by hawkmoths.

A related species, *A. vulgaris*, Garden or European Columbine, not illustrated, has blue-purple or white flowers, stamens that do not protrude, spurs that are curved at their tips, and flowers that are about as long as wide. This species is in the bumblebee-pollinated group but is partially self-polli-

nated. It is a cultivated perennial introduced from Europe but has escaped widely and grows wild along roadsides and in the borders of woods, especially in the northern part of our area. A blue-flowered native species, *A. caerulea*, Rocky Mountain Columbine, not illustrated, is the state flower of Colorado.

Wild Columbine is found in dry rocky woods, on shaded cliffs and ledges, and in moist ravines from Newfoundland and Quebec to Wisconsin, south to Florida and Texas. ☠ In herbology the European Columbine is described as astringent, diuretic, and diaphoretic. It has been used for sore mouths and throats, for diarrhea, and to promote perspiration. Medical botanists warn that all plants of the genus are probably poisonous. Consumption of the seeds is reported to have been fatal to children. Wild Columbine is a beautiful woodland species that should be protected.

Cimicifuga racemosa
Black Snakeroot*
106
Ranunculaceae
Crowfoot Family

*Also Black Cohosh, Bugbane, Rattletop

A perennial with a thick rough rhizome and fibrous roots. The flowers have an unpleasant odor and are visited by green flesh flies that probably effect pollination. Each flower produces several brown, finely wrinkled or pitted, 3-angled seeds with slight wings on the angles. The dry seeds inside the numerous seedpods give the name Rattletop to the plant.

The genus name is derived from Latin words that mean "to drive away bugs," ap-

parently referring to the offensive odor of the flowers, thus the name Bugbane. A related species, *C. americana*, American or Mountain Bugbane, not illustrated, has flowers that do not have an unpleasant odor, has 2 or more pistils, and is somewhat smaller, to 1.8 m. high (6 ft.). It occurs in mountainous areas from Pennsylvania to North Carolina and Tennessee.

Black Snakeroot is found in moist or dry woods from New England and Ontario, south to Georgia and Missouri.

Flowers white, in one or more long terminal clusters, to 15 cm. long (6 in.); sepals 4 or 5, shedding early, petals none, stamens numerous giving flower a fuzzy appearance, pistil 1; leaves alternate, 1 or 2 times pinnately compound, leaflets with coarse sharp teeth; stems slender, unbranched below flower clusters, to 2.4 m. high (8 ft.).

June–August

℞ There are conflicting reports regarding the medicinal properties of the rhizome of this plant. In *The Dispensatory of the United States of America* (1851), it is described as a mild tonic that stimulates secretions of the skin, kidneys, and mucous membranes. In large doses it produced dizziness, impaired vision, nausea and vomiting, and a reduction in circulation.

In modern books on medical botany it is listed as having hypoglycemia-inducing properties and thus may have potential use in the treatment of diabetes.

Coptis trifolia
Goldthread*

107

Ranunculaceae
Crowfoot Family

**Also Canker-root*

Flowers white, solitary, on slender leafless stalks; petals none, petal-like sepals 4–7, stamens numerous; leaves basal, ternately compound, glossy dark green; flower stalks and leaves to 15 cm. high (6 in.).

May–July

An evergreen perennial with a bright yellow, branching rhizome. Five to seven golden yellow, sterile stamens function as nectaries, and pollination is most often effected by fungus gnats and a species of small beetle. The flower produces 3–7 seedpods, each with several small, egg-shaped shiny black seeds. The foliage is eaten by the Ruffed Grouse.

This beautiful little plant, recognized as *C. groenlandica* by some botanists, is the only representative of the genus in the Northeast. The common name Goldthread refers to the slender, creeping, yellow rhizome. It grows well and is an attractive plant for indoor terrariums and moist, shady, wild plant gardens.

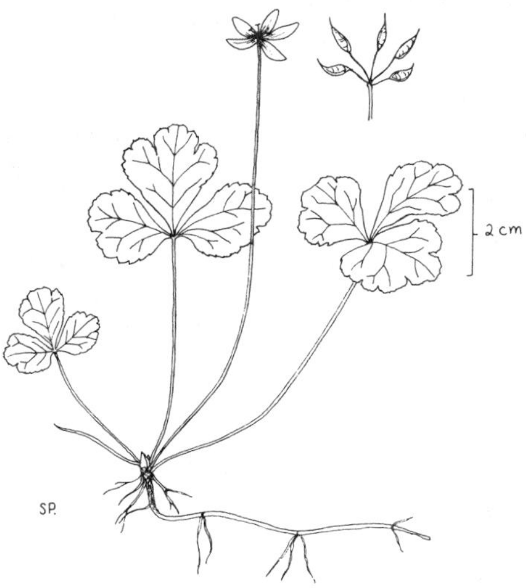

Goldthread is found in damp mossy woods, wooded swamps, and bogs from

108

Greenland to Alaska, south to New Jersey and along the mountains to North Carolina and Tennessee. One variety of this plant extends into eastern Asia.

℞ An extract of the rhizomes is a bitter non-astringent tonic recommended for upset stomach and ulcers of the mouth and throat. When mixed with equal parts of Goldenseal (*Hydrastis canadensis*) it is said to be a cure for alcoholism.

The rhizome has been used in some countries as a source of dye.

Delphinium tricorne
Dwarf Larkspur

108

Ranunculaceae
Crowfoot Family

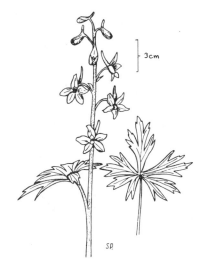

Flowers blue, or occasionally white, with slender spurs, in a few-flowered, loose, terminal cluster; pistils 3, stamens numerous; stem leaves few, alternate, most of leaves basal, deeply palmately lobed with 5–7 segments; stems unbranched, succulent, to 75 cm. high (30 in.).

April–May

mately 150 species in the North Temperate Zone, 4 are native to the Northeast, and two others occur as introduced ornamentals that have escaped from cultivation. The native species are more common south and west of New York State.

Dwarf Larkspur is found in moist woods from Pennsylvania to Minnesota, south to Georgia, Alabama, and Oklahoma.

☠ The species of this genus contain the toxic alkaloids Ajacine, Delphinine, and Delphineidine, which are most concentrated in the seeds and young plants. If ingested by humans they may cause nervous disorders, upset stomach and death if the plant is eaten in large quantities. In southern Europe and the Near East a tincture made from the seeds of 2 species is used as an insecticide for human head lice. Handling of plants of this genus may cause dermatitis in sensitive individuals.

In the western states larkspur poisoning of livestock is a serious problem, and losses of cattle are exceeded only by poisoning from the locoweeds (*Astragalus spp.*). There is some evidence that sheep are less susceptible to larkspur poisoning than cattle.

A perennial with a cluster of thick tuberous roots. The long spur of the flower serves as a repository for nectar which invites, as pollinators, long-tongued insects such as bumblebees and butterflies. Since the flowers do not close at night, they may also be visited by moths. Each pistil develops into a seedpod containing several smooth 3-angled seeds.

Many color varieties and double flowered ornamental forms of this genus have been developed by hybridization. Traditionally the annual forms are referred to by gardeners as larkspurs, while the perennials are designated as delphiniums. Of the approxi-

Hepatica acutiloba
Sharp-lobed Hepatica*

109

Ranunculaceae
Crowfoot Family

*Also Liverleaf

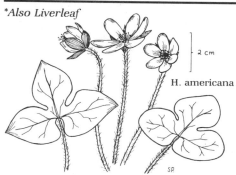

H. americana

A perennial with a greatly branched fibrous root system. The leaves persist through the winter, and the very early flowers appear before the new leaves have unfolded. Pollination is by early flies and bees, and each pistil becomes a green or brown, hairy, seed-like fruit. A short persistent style may serve as an attractant for dispersal by ants.

A closely related species, *H. americana*, Round-lobed Hepatica, illustrated, is very similar but has leaves with rounded lobes and usually occurs in dry woods with an acid or neutral substrate. Since *H. acutiloba* prefers moist woods in limestone areas, the

Flowers bluish-lavender, pink or white, to 2.5 cm. wide (1 in.), solitary on leafless, hairy stalks; petals none, petal-like sepals 6–12, stamens numerous, pistils numerous; leaves basal, with 3 pointed lobes, stalks hairy, to 10 cm. wide (4 in.); flower stalks to 15 cm. high (6 in.).

March–May

two species usually do not occur together. When they do, hybridization produces plants with intermediate leaf characteristics. Both species range from Nova Scotia and Quebec to Minnesota, south to Georgia and Alabama.

℞ The genus and common names for these species are derived from a Greek word that means "liver," which the 3-lobed leaves are supposed to resemble. Consequently, according to the "Doctrine of Signatures" Hepatica is good for liver disorders. In herbology it is recommended for chronic hepatitis, gallstones, and other liver ailments as well as urinary conditions, kidney problems and hemorrhaging. There is no justification in modern medicine for these uses.

Hydrastic canadensis
Golden Seal*

110

Ranunculaceae
Crowfoot Family

*Also Orangeroot, Yellow Puccoon

A perennial with a thick, knobby, yellow rhizome. In fruit the pistils are fused into a cluster of 5–12 dark red berries, each containing 2 smooth, shiny black, keeled seeds. Instead of a basal leaf, sometimes a second, smaller flowering stem with 2 alternate leaves is produced. The common names all refer to the rhizome, which was used as a source of yellow dye by American Indians.

This species has been extensively collected for use in herbal remedies. As with Ginseng (*Panax quinquefolium*), the result has been its near extermination in many areas. If it is to survive as a member of the native

flora this plant will probably need protection. It has been listed as commercially vulnerable and threatened by the Smithsonian Institute.

Golden Seal is found in rich moist woods from Vermont to Minnesota, south to Georgia and Arkansas. For its herbal uses it has been widely planted outside of its range.

℞ The rhizome has tonic properties and has been used for inflammation of mucous membranes, hemorrhoids and for disorders of the stomach and liver. Midwestern Indians and early settlers used it to make a solution for inflamed eyes. The Cherokee Indians mixed the powdered rhizome with

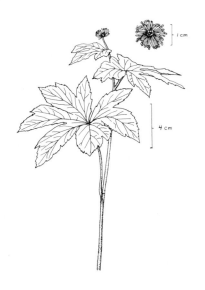

Flowers greenish-white, to 12 mm. wide (½ in.), solitary, terminal; sepals 3, shedding early, petals none, stamens numerous and prominent, pistils numerous; usually a single long-stalked basal leaf and 2 alternate stem leaves, heart-shaped at base, with 5–7 lobes, to 25 cm. wide (10 in.) at maturity; stems hairy, to 37.5 cm. high (15 in.).

April–May

bear grease to make an insect repellent.

All of these uses can be more effectively served today by a visit to a physician or a pharmacy. If this plant is to avoid endangerment and possible extinction it should not be collected for home remedies.

Ranunculus abortivus
Small-flowered Crowfoot*

111

Ranunculaceae
Crowfoot Family

*Also Kidneyleaf Crowfoot

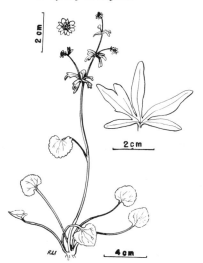

Flowers yellow, few to many, solitary, on long terminal stalks; petals 5, drooping, shorter than sepals; basal leaves long-stalked, somewhat kidney-shaped or lobed, with rounded teeth; stem leaves alternate, sessile, with 3–5 narrow divisions; stems smooth, branched, succulent, to 60 cm. high (2 ft.).

April–June

tities by several species of birds and small mammals which are potential agents of dispersal.

This plant sometimes becomes a weed pest in fields and gardens. It can be eradicated by clean cultivation and close mowing before seeds are produced. The species is highly variable with several overlapping geographic varieties based mainly on leaf characteristics. Field studies have suggested that one variety may be a perennial.

Small-flowered Crowfoot is found in moist or dry woods, on shady streambanks, and in low moist meadows and fields from Labrador to Alaska, south to Florida, Texas, and Washington.

☠ For toxic properties see *R. acris*.

An annual or biennial with thread-like fibrous roots. The fruiting receptacle is globular and includes up to 50 flattened, yellowish-brown, smooth or slightly roughened nutlets. The seeds and vegetation of plants in this genus are eaten in small quan-

Thalictrum dioicum
Early Meadow Rue*

112

*Also Quicksilver-weed

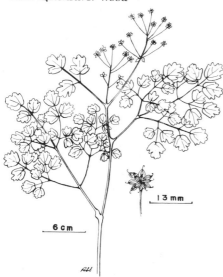

3 mm

6 cm

Flowers whitish or greenish, unisexual, male and female flowers on different plants, staminate flowers with long drooping stamens and yellow anthers; leaves alternate, with long petioles, the lower ones more than once ternately compound, leaflets often with 3 or more shallow lobes; stems usually smooth, to 70 cm. high (28 in.).

April–May

A perennial with a persistent stem base and fibrous root system. Flowering occurs just before the full development of the deciduous forest canopy of leaves. The freely exposed anthers are typical of wind pollinated plants, and cross-pollination is assured by staminate and pistillate flowers on different plants. Each pistillate flower produces a cluster of several brown to black, longitudinally ridged seeds.

This is a genus of over 100 species, worldwide in distribution but concentrated in the North Temperate Zone. The approximately 12 species in the Northeast are all native to North America and are recognizable by their ternately divided leaves. Although an attractive plant in woods and fields, the genus is represented by only a few cultivated ornamental varieties.

Early Meadow Rue is found in moist woods from Quebec to Manitoba, south to Georgia, Alabama, and Missouri.

Ceanothus americanus
New Jersey Tea*

113

*Also Red-root

A shrubby perennial with a deeply anchored, red taproot. It is insect-pollinated and each of the tiny fragrant flowers produces 3 glossy, brown, elliptical seeds. In the western states this is a very important wildlife food genus. The seeds are eaten by upland game birds, and the leaves and stems are grazed or browsed by numerous species of large and small mammals. The eastern species are browsed by the White-tailed Deer.

This is a shrubby genus native to North America with most species occurring west of the Rocky Mountains. Of the approximately 50 species, over 40 are endemic to California and 3 are native to the Northeast. *C. ovatus*, Smaller Red-root, not illustrated, is an eastern species similar to *C. americanus* but has flower clusters at the ends of leafy branches, and narrower leaves. It is most common in the western part of our range.

New Jersey Tea is found in dry open

SP.

4 cm

Flowers white, tiny, in dense cylindrical or oval clusters, on long axillary stalks, upper ones progressively shorter; leaves alternate, oval, sharp pointed, with 3 prominent veins, toothed, to 10 cm. long (4 in.); stems brown-green, usually branched, to 1.2 m. high (4 ft.).

June–July

woods, pine barrens, and clearings from Quebec to Manitoba, south to Florida and Texas.

✖ The dried leaves can be used as a substitute for tea. It was widely used during the Revolutionary War and has been rated variously by wild food enthusiasts from "excellent" to "pretty good" to "indifferent."

℞ In herbology the bark of the root is listed as an astringent and an expectorant. It has been used for mouth and throat irritations, chronic bronchitis, whooping cough, and asthma.

Agrimonia gryposepala
Tall Agrimony

114

Rosaceae
Rose Family

Flowers yellow, to 8 mm. wide (⅓ in.), in long slender clusters at the ends of stems and branches; petals 5, stamens 5–15, pistils 2; seed cluster with a ring of hooked bristles; leaves alternate, pinnately compound, with 5–9 toothed major leaflets interspersed with tiny leaflets; stems hairy, sturdy, to 1.5 m. high (5 ft.).

July–August

A perennial with a short thick rhizome and long fibrous roots. The bristly fruits readily become attached to the coats of passing animals for dispersal. The plant is covered with resinous glands and the stem is somewhat aromatic when crushed. It may become a weed in cultivated fields but usually yields to cultivation.

This is a genus with 7 very similar species in the Northeast, one a native of Europe and Asia. *A. gryposepala* is the most abundant

1 cm

2 cm

4 cm

6 cm

RW

species. The specific name means "with hooked sepals." A related species, *A. pubescens*, not illustrated, has tuberous roots and dense hairs but no glands along the flowering part of the stem. It occurs in open woods from Ontario to Georgia.

Tall Agrimony is found in thickets, shady woodland borders, and neglected fields from Nova Scotia and Quebec to North Dakota, south to North Carolina and Missouri and less abundantly along the Pacific coast.

℞ *A. gryposepala* and *A. eupatoria*, an introduced weed of open fields, are both listed as medicinal herbs. They have been used for coughs resulting from colds, to make a gargle for irritation of the mouth and throat, and to treat jaundice, diarrhea, and bedwetting. *Caution*: Contact with *A. eupatoria* and subsequent exposure to sunlight may cause an allergic reaction in sensitive individuals.

Dalibarda repens **Dewdrop***	**115**	*Rosaceae* **Rose Family**

**Also False Violet, Robin-run-away*
Flowers white, solitary, on long stalks; petals 5, stamens numerous, with long, slender filaments; leaves basal, roundish, heart-shaped at base, petioles covered with soft hairs; flowering stalk and leaves to 25 cm. high (6 in.).

June–September

A perennial with a slender creeping stem. Flowers are of 2 types: a few infertile ones with white petals, and numerous fertile ones with no petals. The fertile flowers have shorter stalks, frequently bending toward the ground. Each fertile flower produces 2 finely wrinkled, hairy, brown seeds.

This genus is native to North America and includes only one species. It was named in honor of Thomas Dalibard, a French horticulturist and contemporary of Linnaeus. This is a hardy species that is sometimes planted and grows well in shady rock gardens.

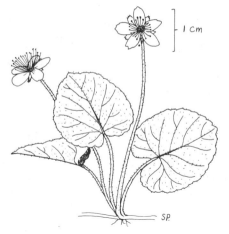

Dewdrop is found in moist woods from Nova Scotia and Quebec to Minnesota, south to New Jersey, West Virginia, and Ohio, and southward in the mountains to North Carolina.

Geum canadense **White Avens**	**116**	*Rosaceae* **Rose Family**

A perennial with a well developed rhizome. Self-pollination is rare and the flower is visited by several potential insect pollinators. After pollination the tissues of the style harden and separate at the loop leaving a stiff hook that attaches to the fur of

passing animals for dispersal.

This genus includes about 60 species that are distributed throughout the northern hemisphere, mostly in the temperate zone. The 9 species that are native to the Northeast usually occur in different habitats or have different flowering dates. However, when their ranges overlap several of them hybridize freely. For example, experimental crosses between *G. canadense* and *G. macrophyllum*, not illustrated, resulted in vigorous hybrids. In crosses between other species the offspring were healthy vegetatively but were sterile.

Some species of the genus may become weeds in meadows or cultivated fields but usually are not serious pests and are easily controlled by cultivation. The plants of this

Flowers white, petals 5, about same size as sepals, stamens numerous; pistils numerous, styles persistent, each with a loop above the middle, interspersed with long stiff hairs; after the petals fall the fruit becomes bristly and bur-like; leaves alternate, upper ones usually simple, lower ones compound, with 3 sharply toothed terminal leaflets, 2–4 smaller leaflets; stems slender, branched above middle, usually hairy, to 1 m. high (40 in.).

May–August

genus are not of great importance as wildlife food, but the leaves are browsed by the Ruffed Grouse. In the higher elevations of the Rocky Mountains the leaves are stored as hay for winter consumption by the Pika, a small guinea pig-like mammal.

White Avens is found in moist or dry woods, woodland borders, and shady roadsides from Nova Scotia to Minnesota, south to Georgia and Texas.

✖ One species, *G. rivale*, Water or Purple Avens, not illustrated, is sometimes referred to as Chocolate Root or Indian Chocolate. The rhizome can be used to make a beverage that, in the opinion of some, tastes somewhat like hot chocolate. According to another wild food biographer, the beverage will be enjoyed more if it is not expected to taste like chocolate.

℞ Several species of this genus are listed as medicinal plants. The rhizomes have astringent properties and are recommended in the treatment of diarrhea, sorethroat, fevers, and upset stomach.

Rubus odoratus
Purple-flowering Raspberry*

117

Rosaceae
Rose Family

*Also Thimbleberry

A shrubby perennial with fragrant Wild Rose-like flowers. It is self-incompatible, and cross-pollination is by insects. This is a very important wildlife food genus, the fruits comprising up to 25% of the summer and

fall diets of the Ring-necked Pheasant, Catbird, Yellow-breasted Chat, Pine Grossbeak, and Summer Tanager. They are also eaten by bears, squirrels, rabbits, skunks, chipmunks, raccoons, and other animals, all of which contribute to seed dispersal.

Flowers rose-purple, showy, to 5 cm. wide (2 in.), in open branched clusters, stamens numerous, petals 5; leaves alternate, with 5 sharp lobes, heart-shaped at base, to 20 cm. wide (8 in.), stalks covered with sticky hairs; stems freely branched, thornless, upper parts covered with red or brown sticky hairs, to 1.5 m. high (6 ft.).

June–September

A closely related species *R. parviflorus,* Thimbleberry, not illustrated, has similar but smaller leaves and white flowers. It is more common in western North America and the western part of our range. A very small species, *R. chamaemorus,* Cloudberry or Baked-apple Berry, not illustrated, has similar leaves and white flowers but does not exceed 30 cm. in height (1 ft.). Its range extends from the Arctic to the mountains of New England. The other species of this genus in the Northeast have thorny stems and compound leaves.

Purple-flowering Raspberry is found in moist woods, woodland margins, and shady roadsides from Nova Scotia and Quebec to

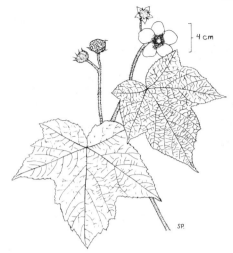

northern Michigan, south to Georgia and Tennessee.

✄ The fruits of *R. odoratus* are edible but rather tasteless when compared with other raspberries. The fruits of Cloudberry or Baked-apple Berry are regularly gathered for food in Newfoundland and the Arctic. The soft ripe berries are said to be delicious.

**Galium aparine
Cleavers***

118

**Rubiaceae
Madder Family**

Also Bedstraw, Goosegrass
Flowers white, on branched axillary stalks, tiny, 4-lobed, stamens 4; leaves mostly in whorls of 8's, to 8 cm. long (3 in.); stems 4-angled, weak, sprawling on other plants, with backward angled spines, to 1.2 m. long (4 ft.).

May–July

An annual of North America, Europe, and Asia. At maturity the 2-lobed ovary splits into 2 round, gray-brown seeds covered with hooked bristles. These are spread by adhering to the coats of passing animals. It sometimes becomes established as a weed in gardens, meadows, and bare areas but is easy to control by clean cultivation or mowing before seeds are formed.

The Galiums are a distinctive group of plants, usually easy to recognize by their square stems, whorls of leaves, and very

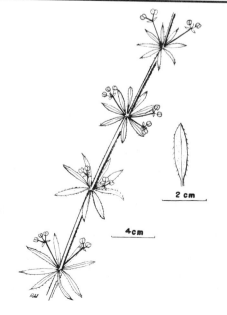

Woodlands

small flowers. There are about 28 species in northeastern North America, at least 6 of which have been introduced from Europe and Asia. The various species are very similar and often difficult to identify. A mature fruit is sometimes necessary since identification may be based on the nature of the bristles.

Cleavers is found in rich woods, woodland borders, thickets, moist meadows, fence rows, and waste areas from Newfoundland to British Columbia, south throughout the United States.

✖ The young tender shoots can be cooked and eaten as greens or added to salads. No information is available as to its nutritive value, but as early as the sixteenth century this plant was described as a food for those who did not wish to become fat.

The dried and roasted seeds have long been known to be an excellent substitute for coffee. The probable reason for this is that Cleavers belongs to the same plant family as the commercial coffee plant. Its seeds may contain caffein and caffeol, the oil that gives coffee its aromatic flavor.

℞ This plant is described in herbology as a diuretic, a tonic and a diaphoretic. It has been recommended for irritated mucous membranes, kidney stones, and other problems of the urinary tract. The fresh leaves, crushed or made into a salve, have been used for sunburn and skin problems.

In addition to the commercial coffee plant genus, *Coffea*, this family also includes the genus *Cinchona*, which is the source of commercial quinine, and the fragrant, attractive ornamental shrub *Gardenia*.

Galium triflorum
Sweet-scented Bedstraw*

119

Rubiaceae
Madder Family

*Also Fragrant Bedstraw

4 cm

Flowers greenish-white, in stalked 3-flowered, terminal and axillary clusters: corolla 4-lobed, stamens 4, sepals none; leaves in whorls of 6's, bright green, to 6 cm. long (2½ in.); stems weak, reclining on other vegetation, usually smooth, to 1 m. long (40 in.).

June–August

A perennial of North America, Europe, and Asia. Each flower produces 2 densely bristly seeds that are dispersed by attaching to the coats of animals. The common names of this plant refer to its pleasant odor after drying.

Sweet-scented Bedstraw is found in moist woods from Newfoundland to Alaska, south to Florida and Mexico.

☠ The seeds have been shown to contain a blood anticoagulant known as coumarin. This substance may enter the bloodstream through oral ingestion of the plant, and it causes a reduction in the production of a blood protein essential for the clotting of blood. For additional information see *Melilotus alba*.

For other information on the genus and family see *G. aparine* and *G. verum*.

Mitchella repens
Partridge Berry*

120

Rubiaceae
Madder Family

*Also Two-eyed Berry, Running Box, Twin-
berry, Checkerberry, Squaw Vine
Flowers white or pink, funnel-shaped, 4-
lobed, in pairs at the ends of branches,
fragrant; fruit a red berry formed by fusion
of the ovaries of 2 flowers; leaves opposite,
small, roundish, often with greenish-white
veins; stems creepiing, branched, forming
mats, to 30 cm. long (1 ft.).*

2 cm

May–July

An evergreen perennial rooting at the
nodes of a trailing stem. The flowers are of
2 types; in some plants they have long styles
and short stamens, in others they have short
styles and long stamens. They are self-incompatible, and cross-pollination is effected by bees and small butterflies. Each
pair of flowers produces a single red berry
that contains 8 gray, finely wrinkled seeds.

This is a beautiful woodland plant with
its dark green leaves and pale or whitish
veins. It is an attractive ornamental for indoor terrariums and shady rock gardens.
However, it should be collected in moderation and only where it is abundant.

The colorful fruit persists throughout the
winter and, while it is not a major source
of food for any species of wildlife, it is eaten
by the Ruffed Grouse, Wild Turkey, Red Fox,
Eastern Skunk, and White-footed Mouse.
These animals undoubtedly serve as agents
of seed dispersal.

In addition to the American species, this
genus includes one other that is a native of
Japan. A rare white-berried form of *M. repens* with fused flowers is occasionally observed. The genus was named in honor of
John Mitchell, a Virginia physician and botanist who corresponded with Linnaeus and
described many American plants in botanical publications.

Partridge Berry is found in dry or moist
woods from Newfoundland to Minnesota,
south to Florida and Texas.

✄ The berries are edible as a survival food
but are dry, tasteless, and full of seeds.

℞ The plant has been used as an astringent, a diuretic, and a tonic. The women in
several American Indian tribes used a tea
made from the plant during the last month
of pregnancy to make childbirth easier, thus
the name Squaw Vine. The Menominee Indians used the tea as a treatment for
insomnia.

Heuchera americana
Alumroot*

121

Saxifragaceae
Saxifrage Family

Also Rock Geranium

A perennial with a knotty horizontal rhizome. The 2-beaked seed capsule splits between the beaks into 2 sections, releasing
numerous tiny, black, spiny seeds. These
are small enough to become windborne in
a moderate breeze, but they may also be
dispersed by becoming attached to the coats
of passing animals.

This is a North American genus with 9
species native to the Northeast. The genus
was named in honor of Johann Heucher, an
eighteenth-century German physician who
wrote several books on medical botany. *H.
americana* has 4 recognized varieties, the 2
within our range distinguished by the relative lengths of their petals and sepals. A
related species, *H. sanguinea*, Coral Bells,

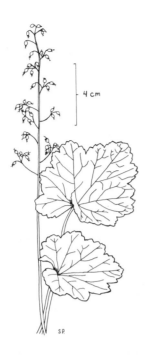

4 cm

Flowers green or reddish, on short branches from a long, usually leafless stalk; corolla bell-shaped, nodding, with stamens protruding, anthers orange, petals 5, calyx lobes 5; leaves mostly basal, long-stalked, heart-shaped at base, with 5–9 shallow lobes; flowering stalk slightly hairy, to 90 cm. high (3 ft.).

April–June

not illustrated, is a popular flower garden ornamental with bright red flowers and several horticultural varieties.

Alumroot is found in dry woods and rocky slopes, especially in limestone areas, from southern New England and Ontario to Michigan, south to Georgia, Alabama, and Oklahoma.

℞ The rhizome has astringent properties and has been used in the treatment of diarrhea. It was used by American Indians as a poultice for wounds and skin abrasions and for persistent sores.

Mitella diphylla Mitrewort*	**122**	*Saxifragaceae* Saxifrage Family

*Also Bishop's Cap, Coolwort

A perennial with a thick rhizome and no runners. The 2-beaked seed capsule opens from the top by 2 valves to release a few ridged, glossy, black seeds. The plant is eaten by the Ruffed Grouse, providing 2–5% of its summer diet in some areas.

This is a very delicate plant that is hardly noticeable when growing alone but provides an attractive display when it occurs in woodland colonies. Close examination is necessary if one wishes to observe the symmetry and beauty of the very small blossoms. Its flowering time is mid-to late spring, usually before the leaf canopy of the trees is completely developed.

It is a genus of about 12 species of the temperate zone in North America and eastern Asia. The genus name is the Latin diminutive for "cap" and refers to the young seed capsule which supposedly resembles a bishop's cap or miter. *M. diphylla* is by far the most common and widespread of the 2 species that occur in the Northeast.

Flowers white, small, on very short stalks, in a terminal cluster to 15 cm. long (6 in.); each petal finely dissected, giving the flower the appearance of a 5-pointed snowflake; stem leaves 2, opposite, small, sessile, 3-lobed, near middle of stem; basal leaves larger, long stalked, heart-shaped at base, 3–5 lobed; stems hairy, to 40 cm. high (16 in.).

April–June

A related species, *M. nuda*, Naked Mitrewort, not illustrated, is similar but smaller, has slender runners, yellowish-green flowers and a leafless flowering stem. It is a species with a northern distribution extending no farther south than Pennsylvania and northern Ohio. Plants with intermediate characteristics, recognized as *M. intermedia* by some botanists, are occasionally observed and are believed to be hybrids of *M. diphylla* and *M. nuda*.

Mitrewort is found in moist woods from Quebec to Minnesota, south to North Carolina, Alabama, and Missouri.

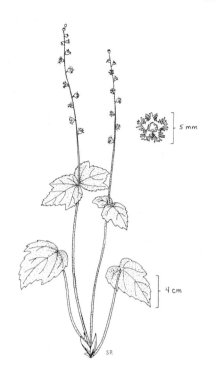

Woodlands

Saxifraga virginiensis
Early Saxifrage

123

Saxifragaceae
Saxifrage Family

Flowers white, in branching clusters on a leafless stalk; petals 5, stamens 10, bright yellow; leaves basal, oval, toothed, often purple on underside; stem with sticky hairs, flowering when 10 cm. high (4 in.) but later reaching a height of 40 cm. (16 in.); as stem elongates flower clusters become scattered.

April–June

A perennial with a thick cylindrical rhizome and an over-wintering basal rosette of leaves. The flowers produce nectar and are visited by early bees and butterflies which probably effect pollination. In some species of the genus self-pollination is avoided by the maturation and shedding of pollen before the stigma is receptive. The 2-beaked seed capsule splits into two sections, releasing numerous seeds that are small enough to become windborne in a slight breeze.

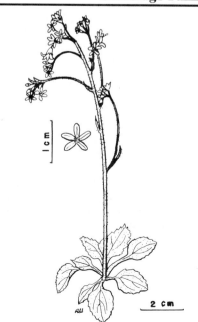

This is a genus of about 100 species in North America with over a dozen in the Northeast. *S. virginiensis* is a popular wildflower of rocky woodlands, and it grows well in shaded rock gardens. It is the most widespread species of the genus in the Northeast. The cultivated window plant known as Strawberry Geranium, a native of Asia, is a member of this genus.

Early Saxifrage is found in moist or dry woods, rock ledges, gorges, and cliffs from New Brunswick to Manitoba, south to Georgia and Oklahoma.

✄ The young leaves of several members of this genus are edible in salads or as cooked greens. The leaves of Early Saxifrage are small and the plant is not abundant, so even if it is found to be edible this plant is not a good candidate for a wild plant meal.

Tiarella cordifolia **Foamflower***	**124**	*Saxifragaceae* **Saxifrage Family**

*Also False Miterwort, Coolwort

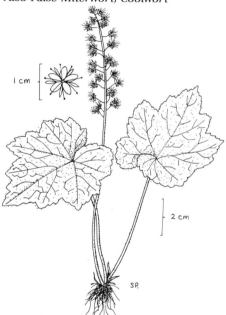

1 cm

2 cm

SP.

Flowers white, in a long terminal cluster, on a leafless stalk; petals 5, entire, stamens 10, long and slender, giving the flower cluster a fuzzy appearance; leaves basal, long stalked, hairy, with several shallow lobes, heart-shaped at base, to 15 cm. long (6 in.); flowering stem to 30 cm. high (12 in.).

April–June

comprise up to 5% of the autumn diet of the Ruffed Grouse in some areas.

The genus name is derived from the unequal seedpods, which form a structure that in some early botanist's imagination resembled a tiara. *T. cordifolia* is the only species in the Northeast, but the genus includes 5 others in North America. A variety without runners occurs from Virginia southward and is recognized as *T. wherryi* by some botanists. This species is sometimes available from commercial greenhouses and can be planted as an attractive addition to wildflower gardens.

Foamflower is found in moist woods from Nova Scotia to Minnesota, south to Georgia and Alabama.

℞ It is listed in herbology as a tonic and a diuretic. It has been used for kidney stones, liver problems, and congestion of the lungs. *Note*: This attractive wildflower should not be collected in its natural habitat for home remedies or for flower gardens.

A perennial with a scaly horizontal rhizome and seasonal runners. The flowers are visited by small bees, syrphus flies, and butterflies that may effect pollination. There are 2 unequal seed capsules that split along inside seams, releasing several shiny black, finely pitted seeds. The plant is known to

Gerardia tenuifolia
Slender Gerardia

125

Scrophulariaceae
Figwort Family

Flowers light purple, funnel-shaped, with 5 unequal lobes, on long stalks from upper axils, numerous; leaves opposite, very narrow; stems smooth, freely branched, to 60 cm. high (2 ft.).

August–October

An annual with a fibrous root system. Each flower lasts only one day and flowering may continue until the first frost. The roundish seed capsule splits into 2 sections to release numerous tiny, gray to brown, irregular, pitted seeds. These are so small that they may be carried for considerable distances by wind.

This is a somewhat variable species with 3 geographic varieties based on diameter of the seed capsule and width of leaves. However, the differences between these varieties are measured in millimeters, and thus are not readily discernible. *G. tenuifolia* is the most abundant species of the genus in the Northeast.

The genus was named for John Gerarde, a well known sixteenth-century herbalist. There are about 13 very similar species, mostly annuals, native to northeastern North America. In an earlier system of classifica-

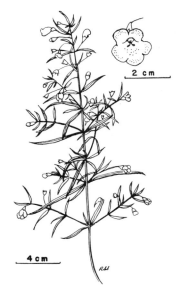

tion most of these were placed in the genus *Agalinis*.

Slender Gerardia is found in dry woods, thickets, prairies, and moist shores from Maine to Michigan, south to Florida and Texas.

Melampyrum lineare
Cowwheat

126

Scrophulariageae
Figwort Family

An annual with a small, tubular, 2-lipped corolla, the lower lip 3-lobed. The flowers are regularly visited by several species of butterflies and bees. The flattened, pointed seed capsule usually contains 4 brown seeds. Each of these has a prominent appendage at one end, and they are harvested by ants. This is an effective adaptation for dispersal since, after spreading and planting the seeds, the ants consume only the soft appendages. In some European species the mechanism is made even more efficient by the production of post-floral nectar to attract ants to the seeds.

This is a somewhat variable species with 4 recognized intergrading geographic varieties. One variety occurs above the timberline in the Northeast, and another is restricted to the Atlantic coastal plain. Individual plants are rather inconspicuous, but they sometimes grow in dense colonies in dry sandy soil of pine forests.

There is only one species of the genus in North America; about 15 others occur in western Europe. The genus name is derived from Greek and means "black wheat," referring to the color of the seeds of some species. The name Cowwheat probably dates

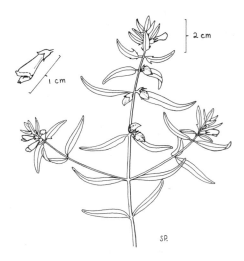

Flowers whitish with yellow tips, solitary on short stalks, in axils of leaves at the ends of stems and branches; leaves opposite, lower ones narrowly lanceolate and toothless, upper ones with a few sharp teeth near the base; stems slender, branched, to 40 cm. high (16 in.).

June–August

to the Middle Ages and was originally applied to a European species, *M. pratense,* not illustrated.

Cowwheat is found in dry or moist woods from Newfoundland to British Columbia, south to Virginia and Indiana, and along the mountains to Georgia.

Pedicularis canadensis
Wood Betony*

127

Scrophulariaceae
Figwort Family

*Also Lousewort

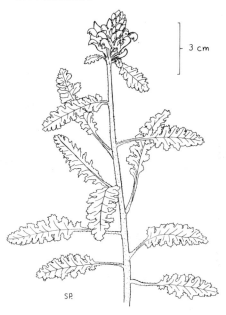

Flowers yellow, red or purple, to 2 cm. long (¾ in.), in a short, dense, terminal cluster; corolla 2-lipped, upper lip hoodlike over the shorter 3-lobed lower lip; calyx not lobed or toothed; leaves alternate, stalked, deeply pinnately divided, hairy, to 12.5 cm. long (5 in.); stems erect, hairy, usually several in a cluster, to 40 cm. high (16 in.).

April–June

fected by robust visitors such as bumblebees. The seed capsule splits along the upper side, releasing several blunt, cone-shaped, finely wrinkled, brown seeds. In some European species the seeds are adapted for dispersal by ants, but this does not seem to be so for the species in eastern America.

The genus name is derived from a Latin word that means "louse," (thus the name Lousewort) and refers to an early superstition in Europe that cattle and sheep grazing in a pasture where this plant grew would become infested with lice.

A related species, *P. lanceolata,* Swamp Lousewort, not illustrated, is similar to *P. canadensis* but has pale yellow flowers, op-

A perennial arising from a crown of roots. Insect visitors must force open the flower to reach the nectar; thus pollination is ef-

posite leaves, and a calyx with 2 lobes. It occurs in wet areas from Massachusetts and Ontario to Manitoba, south to North Carolina and Missouri.

Wood Betony is found in woods, woodland borders, clearings, and prairies from Quebec to Manitoba, south to Florida and Texas.

Penstemon calycosus
Long-sepaled Beardtongue
128
Scrophulariaceae
Figwort Family

Flowers pale purple externally, white inside, in terminal and stalked axillary clusters; flowers 2-lipped, upper lip 2-lobed, lower lip 3-lobed; fertile stamens 4, sterile stamen 1, with tuft of hair at its tip; leaves opposite, sessile, lanceolate, sharply toothed, to 15 cm. long (6 in.); stems often hairy, hairs turned downward, to 1 m. high (40 cm.).

May–July

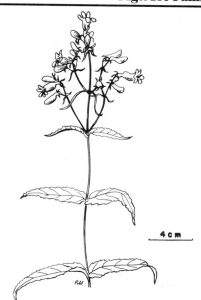

4 cm

A perennial with a rosette of stalked basal leaves. Cross-pollination is probably effected by bees or butterflies as in many other species of this genus. The conical seed capsule splits into two sections, releasing numerous gray, irregularly angled seeds. The name *calycosus* refers to the large calyx which is often 8 mm. long (⅓ in.).

This is a genus of about 300 species with one in northeastern Asia, and most of the remainder in Mexico and western North America. It is a complex genus with several species that appear to have originated by hybridization. There are about 16 species in the Northeast, several of which are so similar that they are difficult to distinguish.

All species of this genus are similar in having a tubular corolla with 2 upper and 3 lower lobes, 5 stamens (4 fertile and 1 sterile), and opposite leaves without petioles. The sterile stamen has a tuft of hairs at its tip which is the source of the name Beardtongue. About 48 species are cultivated as ornamentals in North America.

Long-sepaled Beardtongue is found in moist or dry woods, in meadows, and along roadsides, especially in limestone areas, from Maine to Michigan, south to Georgia and Alabama.

Laportea canadensis
Wood Nettle
129
Urticeae
Nettle Family

A coarse perennial with a fibrous root system. The pistillate flowers are in long branching clusters that are longer than the petioles of the upper leaves. Staminate flowers are in smaller clusters in lower leaf axils, and pollination is by insects. Each pistillate flower produces a single flattened, faintly wrinkled, slightly winged, black seed.

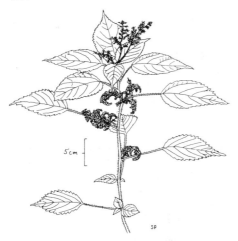

5 cm

SP

Flowers greenish, small, unisexual, in dense branching terminal and axillary clusters; leaves alternate, egg-shaped, with long petioles, coarsely toothed; stems unbranched, often seeming to zig-zag from leaf to leaf, covered with stinging hairs, to 1 m. high (40 in.).

July–August

leases a histamine-like substance that causes a very unpleasant, burning sensation that, once it has been experienced, is hard to forget.

Wood Nettle is found in moist woods and along stream banks from Nova Scotia to Manitoba, south to Georgia and Oklahoma. ✖ The young plants, picked early in the season when they are 30 cm. high (1 ft.) or less, can be cooked as greens. Although gloves may be necessary to collect them, they are highly reputed as a vegetable dish and can be used in any recipe that calls for spinach. The cooking water can be used to make soup.

This is mainly a tropical genus with only one species native to northern North America. Plants of this genus occurring in the tropics are woody and tree-like. The Wood Nettles are not poisonous, but the sharp brittle hairs of the stems and leaves may penetrate soft skin when touched. This re-

---

Pilea pumila
Clearweed*

130

Urticaceae
Nettle Family

*Also Richweed, Coolwort

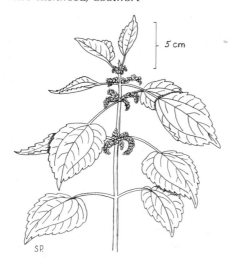

5 cm

SP.

Flowers greenish, inconspicuous, unisexual, in dense axillary clusters; leaves opposite, long petioled, rounded at base, with 3 prominent veins, coarsely toothed, to 15 cm. long (6 in.); stems smooth, often prostrate at base, smooth, soft, translucent, to 50 cm. high (20 in.).

July–October

An annual without stinging hairs, with a juicy appearing stem and a fibrous root system. Male and female flowers sometimes occur on the same plant and sometimes on separate plants. This is an adaptation that decreases the possibility of self-pollination. Each female flower produces a green, often purple marked, narrowly winged, seed-like fruit. In some species, the fruit is ejected forcefully when it is mature.

This is a genus with several hundred species, mostly of tropical distribution. Only two are native to the Northeast; three others, which go by the names Artillery Plant, Creeping Charley, and Panamigo, are cultivated for hanging baskets and borders. The other native species, *P. fontana*, not illustrated, is very similar to *P. pumila* but has dull black seeds. It is most common in swampy areas in the western part of our range.

Clearweed is found in moist shady bottomlands and woods, often occurring in colonies, from Quebec to Minnesota, south to Florida, Louisiana, and Oklahoma. ✖ It can be cooked as a bland potherb. It requires only brief cooking and can be used to dilute the strong flavor of other greens. Some of the tropical species are also used as potherbs.

| *Cubelium concolor*
Green Violet | **131** | *Violaceae*
Violet Family |

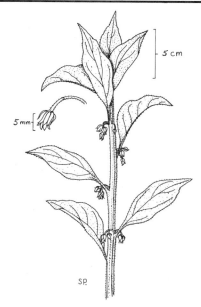

Flowers greenish-white, hanging from middle and upper leaf axils, solitary or in few-flowered clusters; flowers about 6 mm. long (¼ in.), sepals 5, narrow, petals 5, stamens 5, united by anthers in tube around pistil; leaves alternate; elliptical, tapering at both ends, to 15 cm. long (6 in.); stems hairy, usually unbranched, to 90 cm. high (3 ft.).

April–June

A perennial with a dense cluster of fibrous roots. The seed capsule splits into 3 sections, releasing several nearly globular, smooth, light brown seeds. The seeds have soft appendages that may result in their dispersal by ants. This plant does not closely resemble the other violets, but the lowest petal is slightly spurred and the stamens and seed capsules are characteristic of the Violet family.

This is a native genus of only one species. It has been suggested that the genus was named, for reasons that are unclear, for Cybele, the goddess of the growth of natural things in Greek mythology. The specific name means "one color" and refers to the uniformly greenish sepals and petals. Some botanists recognize this species as *Hybanthus concolor*.

Green Violet is found in moist woods and ravines, especially in limestone areas, from southern New England and Ontario to Wisconsin, south to Georgia, Mississippi, and Arkansas.

Viola canadensis
Canada Violet*

132

Also Tall White Violet

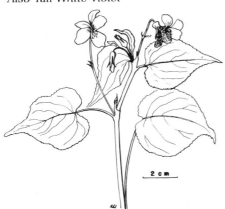

2 cm

Flowers white, on axillary stalks; corolla with yellow throat, back of petals tinged with purple; basal leaves several, long stalked, stem leaves alternate, closer together near top of stem, heart-shaped, to 10 cm. long (4 in.); stems smooth or slightly hairy, to 40 cm. high (16 in.).

May–July

katchewan, south to South Carolina, Alabama, and Arizona.

⚔ According to some wild food experts all of the species of this genus are edible. If some are more tasty than others it remains for the forager to determine by trial. The young leaves can be added to salads or cooked as greens. Some species have a high mucilage content and can be added to soups and stews as thickeners. The flowers can be candied or used to make a pleasant flavored jelly.

℞ The violets have a long history of medicinal use but their use as foods indicate that the medicinal influence of the leaves and flowers is mild at best. The leaves and the rhizomes are said to be emetic, expectorant, and laxative. A syrup made from the flowers has been used for consumption, coughs, and whooping cough. Some tribes of American Indians used a tea made from the plant as a treatment for heart pains. *Note*: W. H. Lewis and M. P. F. Elvin-Lewis state that a large dose of the rhizomes and seeds of *V. odorata* may cause severe gastroenteritis, nervousness, and respiratory and circulatory depression.

A perennial with a thick semi-woody rhizome. Nectar is produced by basal outgrowths of two stamens and collects in the short spur at the base of the corolla. This attracts bees and butterflies that probably effect cross-pollination. After the flowering season, other flowers are produced throughout the summer that do not open and are self-pollinated.

Both of these flower types bear numerous glossy, light brown, fertile seeds. Each seed has an appendage at the base that contributes to dispersal by ants. The ants carry the seeds to their nest where the edible parts are eaten. The rest of the seed, protected by a hard coat, is undamaged.

Canada Violet is probably the most widely distributed North American violet. It is found in moist woods from New Brunswick to Sas-

Viola pubescens
Downy Yellow Violet

133

A perennial with a thick woody rhizome and heavy fibrous roots. for pollination ecology and methods of seed dispersal, see *V. canadensis*.

This species is easy to distinguish from

the other yellow violets by its hairy stems and leaves. A related species, *V. eriocarpa*, Smooth Yellow Violet, not illustrated, is very similar but has non-hairy stems and leaves, leaves with 25–30 teeth and 1–3 basal leaves.

*Flowers yellow with purple veins, on slen-
der, hairy, axillary stalks, at same level or
slightly above leaves; leaves alternate, usually broader than long, with 30–45 teeth,
densely soft-haired on underside, often
with a single, long-stemmed basal leaf;
stems soft-hairy, to 40 cm. high (16 in.).*

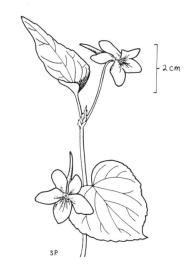

May–June

Another yellow flowered species, *V. rotundifolia*, Round-leaved Yellow Violet, not illustrated, is distinguished from the above
species by the absence of a stem, the leaves
and flower stalks arising directly from a horizontal rhizome.

Downy Yellow Violet is found in moist or
dry woods from Nova Scotia and Quebec to
Minnesota, south to Georgia, Mississippi,
and Oklahoma.

✗ ℞ See *V. canadensis*.

Viola renifolia		
Kidney-leaved Violet	**134**	*Violaceae* **Violet Family**

*Flowers white, with purple lines near base
of 3 lower petals, to 12 mm. wide (½ in.),
on leafless stalks; leaves all basal, kidney-
shaped, hairy at least on lower surface, leaf
stalks hairy; stemless, to 10 cm. high (4
in.).*

May–June

A perennial with a thick scaly rhizome
and no surface runners. For pollination
ecology and methods of seed dispersal see
V. canadensis.

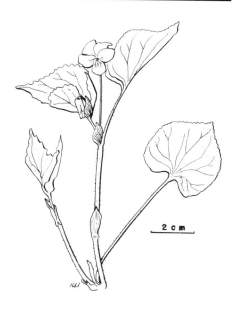

A related species, *V. blanda*, Sweet White
Violet, not illustrated, is very similar but is
more fragrant, has leafy runners, pointed
leaves, and reddish leaf and flower stalks. It
occurs in deciduous woods from New England and Quebec to Minnesota, south along
the mountains to Georgia.

Kidney-leaved Violet is found in cool,
moist woods from Newfoundland to Alaska,
south to Pennsylvania, Michigan, and
Colorado.

✗ ℞ See *V. canadensis*.

128

Viola rostrata
Long-spurred Violet

135

Violaceae
Violet Family

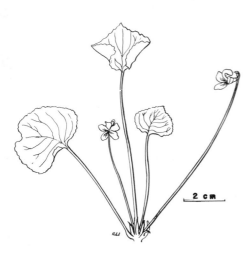

2 cm

Flowers pale lilac, with darker lines on 3 lower petals, on long stalks usually extending above leaves; corolla with long spur often curving upward at tip, to 16 mm. long (⅔ in.), petals without hairs at base; leaves alternate, heart-shaped at base, pointed, stipules toothed; stems smooth, often sprawling, to 20 cm. high (8 in.).

April–June

taining a row of seeds that are ejected forcibly as the section drys and shrinks. As an additional means of dispersal the seeds have soft appendages that attract ants.

A closely related species, *V. conspersa*, American Dog Violet, not illustrated, is very similar but has side petals with bearded bases and a spur to 6 mm. long (¼ in.). It occurs throughout the Northeast, often in the same type of habitats as *V. rostrata* with which it sometimes hybridizes.

Long-spurred Violet is found in moist woods, especially in limestone areas, from Quebec to Michigan and Wisconsin, south to Georgia and Alabama.

✗ ℞ See *V. canadensis*.

A perennial with a branched, woody rhizome. Flowers are of two types: those that open and are cross-pollinated by insects, and others that do not open and are self-pollinated. The egg-shaped seed capsule splits into 3 pod-like sections, each con-

Plants of Open Fields, Roadsides, and Waste Spaces

The early stages of succession are characterized by plants that have high light requirements. They have high growth rates and produce large numbers of seeds and thus are able to establish themselves quickly on bare soil. The first plants to arrive are usually those that complete their life cycles in one year and spend the winter months as seeds. Following these annuals, but not in a rigid order, are biennials and perennials in successive years.

Biennials are plants that take two years to complete their life cycles. During the first year they usually grow vigorously, storing food in roots or underground stems. During the second year they flower, produce seeds, and die. Perennials are plants that live for more than two years. They spend the winter months not only as seeds but also as living roots or stems or both. As succession progresses, the number of perennials becomes greater so that in a climax forest all the trees and shrubs and many of the herbaceous plants are characterized by this growth habit.

Plants of early successional stages are called opportunists because they will invade and become established on any bare area that becomes available. Many of them produce great numbers of airborne seeds that become widespread in surrounding areas. Even in a forest, far from an open field, an opening caused by fire or windfall will be colonized by some open field species. An open field plowed for cultivation, or a fallow field after harvest, provides an ideal environment for the spread of opportunistic species. In order to control this spread, the farmer must cultivate more frequently and, unfortunately, he too often resorts to the use of herbicides, some of which contain small amounts of very toxic substances.

Early successional species may also be observed along roadways, rail and power line rights-of-way, and waste places such as garbage dumps, land fills, and excavation sites. A surprisingly large number of these species may be observed along roadsides of secondary roads.

Tree and shrub seedlings may germinate along with opportunistic species. However, since they have slower growth rates and may not produce flowers and seeds for ten years or more after germination, their influence on the community may be delayed. Some of the recent studies on succession have suggested that the total time from abandoned field to climax forest may be on the order of 150-200 years. On almost any trip into the countryside one can observe several stages in the successional sequence.

Introduced Species

From the earliest settlement of the Northeast, there has been a steady increase in the number of exotic or introduced species of plants. Some of these were deliberately introduced as cultivated plants, then escaped, but most have been introduced accidentally through commerce and world travel. Amos Eaton, Professor of Botany at Yale University, wrote a book on the flora of North America in 1840 in which he identified 350 introduced species. The Seventh Edition of *Gray's Manual of Botany* in 1908 lists 666 introduced species and the Eighth Edition in 1950 lists 1098. These have been introduced from all over the world, but most have come from Europe and Asia.

Most of the introduced species produce numerous seeds and have high growth rates and high light requirements. Since these are characteristics of opportunistic species, most exotic species are found in early successional communities. The climax forest is a more highly integrated community and much less receptive to plants that have not evolved there.

It follows, then, that open fields, roadsides, and waste places provide environmental conditions that are most favorable to exotic species. Of the five habitats described in this book, the greatest proportion of introduced species occur in the following section. Each introduced species is so designated in the description of its ecology.

The following species are sometimes observed in dry, open, sunny areas, but they also occur in other types of habitats. These plants are described and illustrated in the section that describes the habitat where they are believed to be most common.
Amaryllidaceae, the Amaryllis Family
 Hypoxis hirsuta, Star Grass
Apiaceae, the Parsley Family
 Cicuta maculata, Water Hemlock
Apocynaceae, the Dogbane Family
 Apocynum androsaemifolium, Spreading Dogbane
Asclepiadaceae, the Milkweed Family
 Asclepias incarnata, Swamp Milkweed
Asteraceae, the Aster Family
 Antennaria plantaginifolia, Plantain-leaved Pussytoes
 Aster linariifolius, Stiff Aster
 Mikania scandens, Climbing Hempweed
 Rudbeckia laciniata, Green-headed Coneflower
 Rudbeckia triloba, Thin-leaved Coneflower

Senecio aureus, Golden Ragwort
Silphium perfoliatum, Cup Plant
Berberidaceae, the Barberry Family
   Podophyllum peltatum, Mayapple
Boraginaceae, the Borage Family
   Mertensia virginica, Virginia Bluebells
   Myosotis scorpioides, True Forget-me-not
Brassicaceae, the Mustard Family
   Cardamine bulbosa, Spring Cress
   Cardamine pratensis, Cuckoo Flower
Caesalpiniaceae, the Caesalpinia Family
   Cassia fasciculata, Partridge-pea
Campanulaceae, the Harebell Family
   Campanula aparinoides, Marsh Bellflower
Caprifoliaceae, the Honeysuckle Family
   Triosteum perfoliatum, Wild Coffee
Caryophyllaceae, the Pink Family
   Arenaria lateriflora, Grove Sandwort
   Arenaria serpyllifolia, Thyme-leaved Sandwort
Chenopodiaceae, the Goosefoot Family
   Chenopodium hybridum, Maple-leaved Goosefoot
Commelinaceae, the Spiderwort Family
   Tradescantia virginiana, Spiderwort
Convolvulaceae, the Morning-glory Family
   Cuscuta gronovii, Dodder
Crassulaceae, the Orpine Family
   Penthorum sedoides, Ditch Stonecrop
Fabaceae, the Bean Family
   Desmodium canadense, Showy Tick Trefoil
   Lespedeza repens, Creeping Bush Clover
   Lupinus perennis, Wild Lupine
Geraniaceae, the Geranium Family
   Geranium maculatum, Wild Geranium
Iridaceae, the Iris Family
   Iris versicolor, Blue Flag
Lamiaceae, the Mint Family
   Lycopus americanus, Water Horehound
   Monarda didyma, Oswego Tea
   Physostegia virginiana, Obedient Plant
   Satureja vulgaris, Wild Basil
   Scutellaria parvula, Small Skullcap
   Stachys hispida, Hedge-nettle
   Teucrium canadense, American Germander
   Trichostema dichotomum, Blue Curls
Liliaceae, the Lily Family
   Allium cernuum, Nodding Wild Onion
   Lilium philadelphicum, Wood Lily
   Smilacina stellata, Star-flowered Solomon's Seal
Lobeliaceae, the Lobelia Family
   Lobelia cardinalis, Cardinal Flower
Menispermaceae, the Moonseed Family
   Menispermum canadense, Moonseed
Orchidaceae, the Orchid Family

Epipactis helleborine, Helleborine
Habenaria psycodes, Small Purple Fringed Orchis
Spiranthes cernua, Nodding Ladies' Tresses
Polygonaceae, the Smartweed Family
Polygonum cilinode, Fringed Bindweed
Primulaceae, the Primrose Family
Dodecatheon meadia, Shooting Star
Lysimachia ciliata, Fringed Loosestrife
Lysimachia terrestris, Swamp Candles
Ranunculaceae, the Crowfoot Family
Ranunculus abortivus, Small Flowered Crowfoot
Thalictrum polygamum, Tall Meadow Rue
Rhamnaceae, the Buckthorn Family
Ceanothus americanus, New Jersey Tea
Rosaceae, the Rose Family
Agrimonia gryposepala, Agrimony
Saxifragaceae, the Saxifrage Family
Parnassia glauca, Grass of Parnassus
Scrophulariaceae, the Figwort Family
Chelone glabra, Turtlehead
Gerardia tenuifolia, Slender Gerardia
Pedicularis canadensis, Lousewort
Penstemon calycosus, Beard-tongue

Mollugo verticillata **Carpetweed*** 136 ***Aizoaceae*** **Carpetweed Family**

**Also Indian Chickweed, Whorled Chickweed*

Flowers greenish, on long axillary stalks, 2–5 from each node, petals 5; leaves usually in whorls of 5–6, narrow, to 3 cm. long (1½ in.); stems smooth, freely branched, forming carpet-like mats, to 50 cm. in diameter (20 in.).

June–October

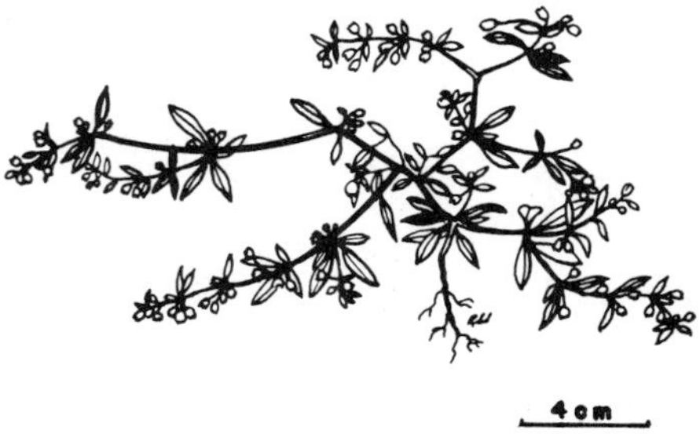

An annual with a deep, scarcely branched taproot, introduced from tropical America. The oval seed capsule opens by 3 valves to release numerous tiny, kidney-shaped, glossy reddish-brown, windblown seeds. It sometimes becomes established as a weed in cultivated areas and can be controlled by clean cultivation until late in the growing season.

This is not an early spring plant, germination usually occurring later in the season when conditions are more like those of its tropical native habitat. However, its late start is compensated for by a very rapid rate of growth in summer and fall when it may become a nuisance in lawns and gardens.

The genus is mostly a tropical one of about 15 species. The name of the genus is derived from *Galium mollugo* and is an older name for that species. *M. verticillata* is similar to *G. mollugo* in that both species have whorled leaves. *M. verticillata* is widespread in Africa.

In North America, Carpetweed is found on sandy shores, sandy fields and gardens, roadsides, and waste areas from New Brunswick to Washington, south to Florida, Mexico, and California.

✕ The plant may be cooked and eaten as a potherb.

Amaranthus retroflexus **Redroot Pigweed*** 137 ***Amaranthaceae*** **Amaranth Family**

**Also Green Amaranth, Wild Beet*

An annual with a shallow, red taproot, introduced from tropical America. Pollination is by wind, but it does not shed large quantities of pollen and is not a hayfever plant in the Northeast. However, it is reported to be a hayfever agent in Arizona, Oklahoma, California, and Oregon.

The plants of this genus are important food plants for many species of birds, including waterfowl, upland gamebirds and songbirds, small mammals, and, in the West, some of the hoofed browsers. Feeding experiments have demonstrated that the seeds of *A. retroflexus* remain viable after passing through the digestive tracts of horses, cattle, sheep, and swine. These very durable seeds have remained germinable after storage for 40 years in the soil.

This is one of the commonest weeds of gardens, cultivated fields, and waste spaces throughout southern Canada, the United States, and Mexico.

✕ Young plants may be cooked as very mild

Flowers greenish, inconspicuous, in dense, elongate, terminal and axillary clusters; staminate flowers with 5 stamens; flowers subtended by 3 stiff, spiny bracts which are longer than the sepals; leaves alternate, long petioled, to 15 cm. long (6 in.); stems hairy, freely branched, to 2 m. high (6½ ft.).

August–October

greens or used fresh in salads. The seeds can be ground into a nutritious flour. The southwestern Indians cultivated this plant for the seeds which were roasted and eaten, or ground into flour.

℞ An infusion made from the dried leaves of several species of this genus is said to be useful for mouth and throat irritations and for diarrhea. The cultivated Cockscomb, *A. hypochondriacus*, not illustrated, has been used to treat excessive menstrual bleeding, diarrhea, and dysentery.

Leucojum aestivum Snowflake	**138**	*Amaryllidaceae* Amaryllis Family

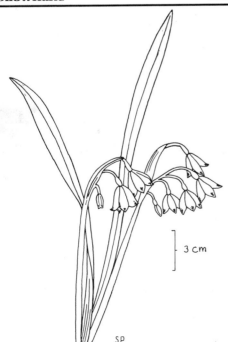

Flowers white with green edges, nodding, funnel or cup-shaped, with 6 equal segments, in a terminal cluster of 2-several on a long leafless stalk; leaves basal, grass-like, to 45 cm. long (18 in.); flowering stem smooth, to 60 cm. high (2 ft.).

April–June

A perennial with a small, white, onion-like bulb, introduced from central Europe. The anthers split longitudinally and present pollen to pollinators at lower temperatures on sunny than on cloudy spring days. The inverted egg-shaped seed capsule opens by 3 valves to release numerous round seeds.

The genus name is derived from Greek words that mean "white violet." The genus includes about 9 species, all natives of Europe and the Mediterranean region. Three species have been introduced into North America as ornamentals, but only *L. aestivum* has widely escaped cultivation. A closely related species, *L. autumnale*, not illustrated, is similar but flowers in the fall and has blossoms edged with red.

Snowflake is often locally abundant on alluvial plains and roadsides and in meadows and open moist woods from Nova Scotia to Ontario, south to Virginia.

℞ In parts of Europe this plant is used to treat a nerve disorder known as myasthenia gravis. This condition is characterized by rapid fatigue and excessive weakness of muscles.

Conium maculatum
Poison Hemlock

139

Apiaceae or Umbelliferae
Parsley Family

Flowers white, tiny, numerous, in branched, flat-topped, terminal clusters; leaves alternate, 3–4 times pinnately compound, with stalk enlarged at base, sheathing the stem, to 40 cm. long (16 in.); stems branched, purple spotted, to 1.8 m. high (6 ft.).

June–August

A biennial with a long white taproot, introduced from Europe. Each flower produces a small, brown, wavy-ribbed, two-seeded fruit. These are sometimes dispersed to meadows and fields where the plant is an unwanted weed. Recommended methods of control include cutting the first year taproot below the basal rosette of leaves and mowing the second year growth before seeds are formed.

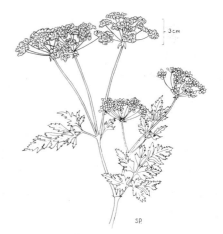

This is a genus of 2 species, with only one in North America. It resembles Wild Carrot (*Daucus carota*), a member of the same family, but can be distinguished by the absence of hairs on its stems and leaves. Two related species, *Cicuta maculata* (Water Hemlock) and *Sium suave* (Water Parsnip), are somewhat similar but can be distinguished by their less dissected leaves and distinct, toothed leaflets. The finely dissected leaves of *Conium maculatum* are fern-like, and it has been called "Water Fern."

Poison Hemlock is found along roadsides and in fields, pastures, meadows, and waste areas throughout most of the United States and southern Canada.

☠ The toxic properties of this species has been known for thousands of years. It was used in ancient Greece as a means of execution for condemned criminals. An extract of the leaves was administered to Socrates, who was condemned for corrupting the youth of the state. It causes death by paralysis of the respiratory system.

All parts of the plant are poisonous, containing an alkaloid called coniine, but this substance is most concentrated in the young leaves. Fatal poisoning of cattle, sheep, horses, swine, goats, and fowl have been reported. Poisoning and often death in humans has resulted from mistaking the young leaves for parsley, the seeds for anise, and the roots for parsnips.

A recent discovery of interest is that there is a structural similarity between the alkaloid coniine and the venom of the much publicized fire ants of the southern states.

136

Daucus carota
Wild Carrot*

140

Also Queen Ann's Lace, Bird's Nest, Devil's Plague

A biennial with a thick taproot, introduced from Europe. Fertilization is usually the result of cross-pollination effected by the numerous insects that visit the flowers. These include flies, butterflies, bees, and moths that apparently are attracted by the strong odor of the plant. The cultivated carrot is a variety of this species.

The bristly seeds are efficiently adapted for dispersal by adhering to passing animals. They are relatively long-lived, as demonstrated by germination studies which indicated 71% viability after dry storage for

Flowers white, tiny, in a cluster of flat-topped umbels, often with 1 central purple flower; 2-seeded fruit covered with short bristles; leaves alternate, pinnately dissected into fine segments; stems hairy, branched, to 1 m. high (40 in.).

May–October

5 years. Although very abundant, this species is of limited value as food for wildlife. The seeds are eaten by a few species of birds and small mammals. The root also is eaten by small mammals.

Admired by many as a beautiful wildflower, it is not so highly regarded by farmers, as the common name Devil's Plague indicates. It is a host for carrot blight, a disease that attacks the cultivated carrot, and it has been declared a noxious weed in the seed laws of ten, mostly eastern, states.

A native species, *D. pusillus*, not illustrated, is similar but is smaller and less hairy and is an annual. Its range is from South Carolina to Florida and Texas, and along the Pacific Coast from California to British Columbia.

Wild Carrot (*D. carota*) is found along roadsides and in fields, meadows, and pastures throughout southern Canada and the United States.

✖ The first year root can be collected in the fall or early spring and cooked as a vegetable. It is smaller, tougher, and white, but the flavor resembles that of cultivated carrots.
℞ A tea made from the whole plant is said to be a stimulant and a diuretic and to be useful in dispelling intestinal gas.

Heracleum lanatum
Cow Parsnip*

141

Also Masterwort

A perennial with a large thick root, of North America and Siberia. The flowers are shallow and massed in flattish clusters. The nectar is easily accessible and not hidden

in deep corolla tubes. These conditions are especially favorable to short-tongued flies and beetles, although the flower cluster is visited by a variety of potential insect pollinators. The large, flattened, heart-shaped,

Flowers white, in a flat-topped terminal cluster to 20 cm. wide (8 in.); fruits narrowed at base, ribs extending to middle or less; leaves alternate, compound, with 3 leaflets, each to 60 cm. long (2 ft.), base of petiole enlarged, forming a sheath around stem, leaflets heart-shaped at base; stems hairy, thick, hollow, ridged, to 3 m. high (10 ft.); plant with strong odor.

June–August

seeds have a broad marginal wing that aids in dispersal by wind.

The large size of the plants in this genus undoubtedly led to their being named in honor of Herakles, the strong man of Greek mythology. The amount of vegetative growth that takes place in one growing season is impressive and not typical of herbaceous plants. The genus includes about 60 species in the northern hemisphere. Three species occur in the Northeast but only *H. lanatum* is native to the region. *H. lanatum* is classified as *H. maximum* by some botanists.

A related species, *H. sphondylium*, Eltrot, not illustrated, is very similar in size and growth habitat but has leaves that are once pinnately compound. It is a native of Europe and Asia, and ranges in North America from southern New England and New York to southeastern Newfoundland. Although it is reputed to have food uses in Europe and Asia, it is considered a weed pest in Newfoundland.

Cow Parsnip is found on the rich alluvial soil of flood plains, open thickets, stream

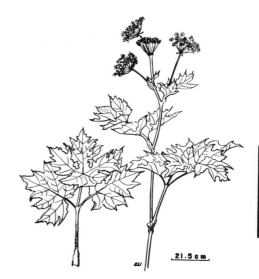

21.5 cm

banks and shorelines from Labrador to Alaska, south to Georgia and Arizona.

✖ The young stems and leaf stalks, when peeled and cooked in at least two changes of water, are similar to stewed celery. The young full grown roots can be cooked in a variety of ways and have a strong parsnip flavor. The dried seeds have been used as a seasoning. Some American Indians are said to have added small sections of the dried hollow stem base to other foods as a substitute for salt.

℞ The roots and seeds have been used in remedies for epilepsy, asthma, upset stomach and cramps. Applied externally they are reputed to be good for sores and wounds.

Heracleum mantegazzianum — Giant Hogweed — 142 — Apiaceae or Umbelliferae — Parsley Family

A perennial with a very large tuberous root, introduced from Europe. Numerous seeds are produced each growing season that are spread by wind, water, and gravity. It has the potential for becoming a serious weed pest in agricultural land that is not cultivated regularly. Once it has become established control measures include clipping the flower clusters before seeds are

formed and repeated mowing of the shoots to starve the roots. There is a chemical called glyphosate, available commercially, that will kill the plant, but it is expensive and it also kills surrounding plants. It should be used with caution and as a last resort.

This plant is attractive to many people as a garden curiosity because of its size. It may be the largest herbaceous plant in the

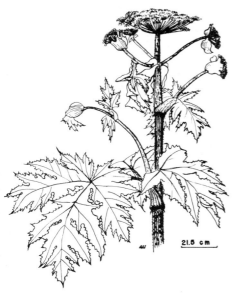

21.5 cm

Flowers white, in a flat-topped terminal cluster to 75 cm. wide (30 in.); fruits rounded at base, ribs extending well past middle; leaves alternate, compound, with 3 greatly dissected leaflets, base of petiole forming a sheath around stem, with tuft of coarse white hairs where leaf attaches to stem, leaf to 1.5 m. wide (5 ft.); stems covered with coarse hairs, with numerous purple streaks and blotches, to 4.2 m. high (14 ft.).

June–July

ditches and waste areas. It has been reported in the provinces of Ontario and British Columbia, and in over a dozen counties in New York State.

☠ Some individuals are very sensitive to the juices of Giant Hogweed. When these persons come into contact with sap from the stem or leaves of this plant and the skin is subsequently exposed to sunlight, painful blisters and/or red discolorations appear within a few days. After the blisters disappear, purplish scars may remain for a long period of time. Susceptible individuals who have been exposed to the juices of this plant and then experienced the ensuing photodermatitis have reported that it is considerably worse than poison ivy.

Northeast. The exact date of its introduction into North America is unknown, but it was cultivated in Rochester, New York, as early as 1917. Introduced as an unusual ornamental, it has escaped from many gardens and seems to be spreading.

Giant Hogweed is found in rich moist soil of alluvial deposits, stream banks, roadside

---

**Pastinaca sativa
Wild Parsnip**

143

**Apiaceae or Umbelliferae
Parsley Family**

A biennial with a first year rosette of leaves and a fleshy taproot, introduced from Europe. During the second year of growth, a thick flowering stem develops, and fertilization is the result of both self-pollination and cross-pollination by insects. Each of the numerous flowers produces a single seed that is elliptical, flattened with longitudinal ridges on each side, and has a narrow mar-

ginal wing that aids in dispersal by wind.

This species has been cultivated for more than 2000 years in Europe. It was introduced into North America as the cultivated garden parsnip but has escaped widely. European botanists have classified the cultivated and escaped plants as different varieties. The escaped plant has a more slender taproot and is a common weed.

*Flowers yellow, small, numerous, in flat-
topped clusters at the ends of stems and
branches; leaves alternate, pinnately compound, leaflets 5–15, oval, sharply toothed
or lobed, upper ones sheathing the stem;
stems strongly grooved, smooth, to 1.5 m.
high (5 ft.).*

May–October

Wild Parsnip is found in old fields and
meadows, along roadsides, and in waste
areas throughout northern United States and
southern Canada.

✄ The first year root is edible and can be
prepared in the same manner as the cultivated parsnip. Caution should be exer-
cised not to confuse this plant with the
poisonous Water Hemlock (*Cicuta*), which
has a root with the odor of parsnip.

Caution: Contact with the wet leaves of
Wild Parsnip may cause severe dermatitis
in sensitive individuals.

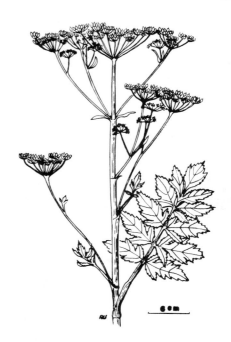

Open Fields

Zizia aurea
Golden Alexanders

144

Apiaceae or Umbelliferae
Parsley Family

*Flowers yellow, numerous, in branched ter-
minal clusters, center flower of each cluster usually sessile; leaves alternate, upper
ones divided into 3's, lower ones twice
compound into 3's, leaflets lanceolate,
finely toothed; stems usually branched, to
75 cm. high (30 in.).*

May–June

A perennial with a dense cluster of coarse
roots. Fertilization is the result of self-pollination and cross-pollination by insects. In
field studies of this species, bagged flower
clusters produced almost as many seeds as
unbagged clusters. The seeds are oblong and
flattened with several lighter colored, longitudinal ridges.

This is a genus of 3 species, all native to
eastern North America. *Z. aurea* is probably
the most common one in the Northeast. A

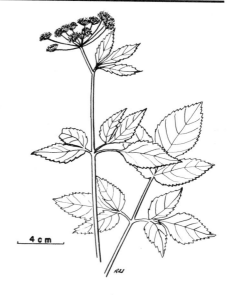

related species, *Z. aptera*, Heart-leaved Alexanders, not illustrated, has basal leaves that are simple, long stalked, and broadly heart-shaped at the base. It occurs in open woods and prairies and is most common west of the Mississippi River.

Golden Alexanders is found along shady roadsides and river banks and in moist fields and open woods from Maine and Quebec to Saskatchewan, south to Florida and Texas.

Apocynum androsaemifolium **Spreading Dogbane**	**145**	*Apocynaceae* **Dogbane Family**

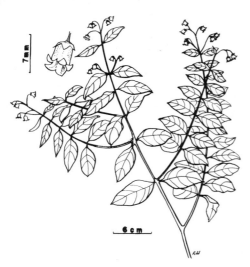

Flowers pink with deeply colored stripes inside, to 8 mm. wide (⅓ in.), in terminal and axillary clusters; corolla bell-shaped, nodding, with 5 lobes, bent backward; leaves opposite, stalked, oval, hairy on underside, to 10 cm. long (4 in.); stems freely branched, smooth, to 1.2 m. high (4 ft.); plants with milky juice.

June–August

Spreading Dogbane is found in old fields, meadows, thickets, and the edges of woods and along roadsides from Newfoundland to British Columbia, south to Georgia, Texas, and Arizona.

☠ Spreading Dogbane and Indian Hemp (*A. cannabinum*) are both described as poisonous plants and they appear to have similar medicinal properties. The medicinal part is the root, described in herbology as an emetic, diuretic, and cathartic. According to *The Dispensatory of the United States of America* (1851) the most beneficial use of Indian Hemp is in the treatment of dropsy (edema) and to dispel serous fluid in the abdomen. Both species contain substances that have a digitalis-like effect on the heart. Indians of the plains boiled the unripe pods in water and used it as heart medicine.

Both species are poisonous to livestock if eaten dried or green. Although animals usually find these plants distasteful, there have been instances of poisoning in cattle, sheep, and horses. As little as 15 to 30 grams of the green leaves of Indian hemp will cause death of a horse or cow.

These species should not be experimented with in home remedies.

A perennial with a creeping horizontal rhizome. Pollination is mainly by bees and butterflies, including frequent visits by the Monarch Butterfly. An almost constant inhabitant of the plant is the metallic red and green Dogbane Beetle. Each flower produces 2 slender seedpods, about 10 cm. long (4 in.), that split along one side to release numerous tiny, reddish-brown seeds with tufts of silky white hairs.

There are 7 species of this genus in North America with 3 native to the Northeast. The occurrence of intermediate forms suggests that hybridization takes place among all 3 species. The hybrid forms of *A. androsaemifolium* are traditionally classified as *A. medium*. Western plants of the former are characterized by the absence of hairs on their leaves.

Apocynum cannabinum
Indian Hemp
146
Apocynaceae
Dogbane Family

Flowers greenish-white, to 6 mm. long (¼ in.), in clusters at ends of stems and axillary branches, branches often higher than stem tip; corolla with 5 erect, pointed lobes; leaves opposite, with distinct stalks, often with pointed tips, rounded or tapering at base, to 12.5 cm. long (5 in.); stems erect, branched in upper part, to 1 m. high (40 in.).

June–September

A perennial with a long horizontal rhizome. Pollination is by insects, and this species hybridizes with other members of the genus in the Northeast. The fruit consists of 2 long slender pods, to 20 cm. long (8 in.), that split along one side releasing numerous brown seeds with tufts of silky white hairs.

The wind-borne seeds sometimes invade crop fields where deep plowing and clean cultivation are necessary control measures for dense infestations. This is a highly variable species with several varieties in different parts of its range. A related species, *A. sibiricum*, Clasping-leaved Dogbane, not illustrated, is very similar but has leaves that are rounded at the base and are stalkless. It occurs in the same types of habitats as *A. cannabinum*, with which it sometimes hybridizes.

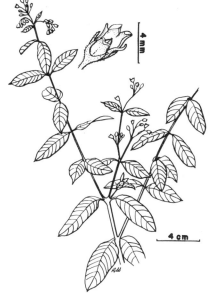

The tough fibers of the stem were used by the American Indians to make ropes and thread; thus the name Indian Hemp.

Indian Hemp is found in abandoned fields, open spaces, thickets, borders of woods, and waste areas throughout most of the United States and in Canada from Quebec to Alberta. ℞ See *A. androsaemifolium*.

Vinca minor
Periwinkle*
147
Apocynaceae
Dogbane Family

*Also Myrtle

A semi-shrubby, perennial evergreen introduced from Europe. Pollination is probably by bees and butterflies, and since it is rarely self-compatible cross-pollination is usually necessary for seed production. Possibly as a result of the requirement of cross-pollination, few seeds are formed and re-

production is mainly by surface runners.

This is a small genus native to Europe and Asia. Three species have been introduced into eastern North America but only *V. minor* is widespread. A related species *V. major*, Large Periwinkle, not illustrated, has flowering stems to 50 cm. high (20 in.) and occurs from northern Virginia to Georgia.

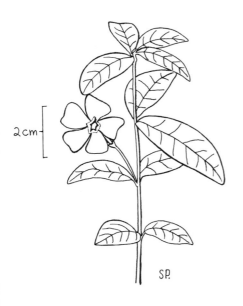

Flowers blue, on axillary stalks, only 1 for a given pair of leaves; corolla funnel-shaped, with 5 spreading lobes; leaves opposite, shiny green, leathery, entire, to 5 cm. long (2 in.); stems creeping, rooting at nodes, forming dense mats, to 1 m. long (40 in.).

April–June

Periwinkle was introduced in colonial times as a ground cover, but it has become thoroughly naturalized. Its tendency to form dense mats makes it a good soil stabilizer on slopes that are too shady for other types of vegetation. It is found along roadsides and shady banks, in cemeteries, around old home sites, and in open woods throughout the Northeast.

℞ In herbology it is identified as an astringent and a sedative, a remedy for diarrhea, excessive menstruation, and hemorrhaging.

2cm

SP.

Asclepias syriaca
Common Milkweed*

148

Asclepiadaceae
Milkweed Family

*Also Silkweed

0.7 m.

4 cm

6 cm

Flowers rose to brownish-purple, fragrant, in numerous rounded, terminal, and upper axillary umbels; pods erect, pointed, gray-green, with warty surfaces; leaves opposite, oblong or oval, gray-downy on under side; stems erect, unbranched, hairy, to 1.5 m. high (5 ft.); plants with milky juice.

June–August

terflies. As a result of the specialized structure of the flower, visiting insects sometimes get their legs caught and, being unable to free themselves, die there or escape minus a leg. This limits cross-pollination, with the result that most of the numerous flowers do not produce mature seedpods.

A tuft of silky hairs on each of the numerous brown, flat, winged seeds facilitates dispersal by wind. The plant is host for the larvae of Monarch Butterflies, which feed upon the leaves.

Common Milkweed is found in open fields, pastures, and waste spaces from New Brunswick to Saskatchewan, south to Georgia and Kansas.

A perennial with a long creeping rhizome. The flowers are self-incompatible, and cross-pollination is effected by bees, flies, and but-

The young shoots and pods can be cooked as greens with at least one change of water. The young pods can be cooked and canned like other vegetables and are said to be comparable with okra. The eighteenth-century French Canadians are reported to have boiled the flowers to make a palatable brown sugar.

Although several of the milkweeds are nourishing food when cooked, it should be noted that, eaten raw, they are known to cause poisoning in livestock, and they are probably toxic to humans.

℞ In herbal medicine, the dried rhizome has been recommended as an agent to induce vomiting, a diuretic, and a treatment for intestinal worms. Indians of Quebec used it to induce temporary sterility in women.

Children in some parts of the country, in the early 1900's, used the hardened latex for chewing gum.

During World War II the milky juice was experimented with as a possible raw material for synthetic rubber, and the fluff from the seeds was used as stuffing for life jackets.

Open Fields

Asclepias tuberosa
Butterfly-weed*

149

Asclepiadaceae
Milkweed Family

Also Pleurisy-root, Chigger-flower
Flowers yellow to orange-red, usually numerous in terminal or axillary umbels; leaves alternate, sometimes opposite on upper branches, hairy, lanceolate; stems hairy, usually branching toward the top, juice not milky, to 60 cm. high (2 ft.).

June–September

A perennial with a well developed rootstock. (See *A. incarnata*.)

Butterfly-weed is found in open fields, pastures, and waste spaces, especially in sandy areas, from Ontario to Minnesota, south to Florida and Texas.

℞ This plant has a long history of medicinal uses by American Indians and early American physicians. The medicinal substance was prepared by boiling the dried root in water. This solution was used to promote perspiration and to cause the discharge of mucus from the respiratory tract.

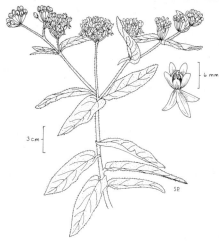

In heavier doses, it caused vomiting and diarrhea. Some tribes of American Indians used the macerated root for wounds, bruises, and sores. (See also *A. incarnata*.)

Achillea millefolium Yarrow*

150

*Also Milfoil

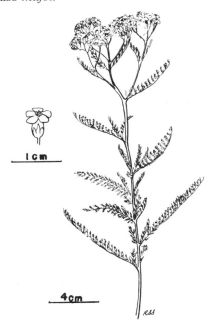

1 cm

4 cm

RSS

Flower heads with 4–6 white or pink rays; heads numerous, in a flat or round-topped terminal cluster; leaves alternate, finely pinnately dissected, sessile, aromatic; stems unbranched below flower clusters, hairy, to 1 m. high (40 in.).

June–September

A perennial with a substantial underground rhizome, having both native and introduced varieties. The plants are self-incompatible and cross-pollination may be effected by bees or short-tongued flies. Each flower head produces numerous light brown, flattened, finely ridged, seed-like fruits with narrow marginal wings that may aid in dispersal by wind. It is sometimes a weed in meadows and pastures where, if eaten by cattle, it may cause an unpleasant flavor in milk.

A native subspecies common on western rangelands is recognized by some botanists as a distinct species, *A. lanulosa*. Varieties of this subspecies are the most common ones in our area. These plants are eaten by a few species of birds and mammals that may contribute to dispersal. Feeding experiments have demonstrated that the seeds can pass undamaged through the intestinal tracts of horses, cows, swine, and sheep.

Yarrow is found in open fields and lawns, along roadsides, and in waste areas throughout most of the United States and southern Canada.

✄ The dried leaves can be steeped in hot water to make a bitter aromatic tea.

℞ The genus name is derived from Achilles, who, according to legend, used the plant to treat the wounds of his soldiers. Some tribes of American Indians used the pulverized plant for wounds and burns. In herbology it is recommended for nosebleeds, bleeding hemorrhoids, and excessive menstrual bleeding. Yarrow tea is said to be a good remedy for colds.

Ambrosia artemisiifolia Common Ragweed

151

An annual with an extensive root system. Pollination is by wind and a great quantity of airborne pollen is produced. This plant is probably the most important cause of hayfever in eastern North America. Each plant bears numerous seeds that may remain viable for 5 years or more. It is sometimes a troublesome weed in cultivated areas, where it competes aggressively for the available water. For the production of a pound of dry matter this plant requires 3 times more water than is needed by corn.

In contrast to its adverse affect on many humans, this is an important wildlife food genus. The most important species are Common Ragweed and Western Ragweed,

Flower heads green, unisexual, inconspi-cuous; pistillate heads in axils of leaves, staminate heads above, in long terminal clusters, to 15 cm. long (6 in.); lower leaves opposite, upper ones alternate, twice pin-nately dissected, to 10 cm. long (4 in.); stems branched, to 2 m. high (6½ ft.).

August–October

A. psilostachya, not illustrated. These con-tribute to the diets of many species of birds, including waterfowl, shore birds, upland game birds, and songbirds. The seeds and foliage are eaten also by several species of small mammals and the White-tailed Deer. Since some seeds may pass through their digestive tract undamaged, these animals serve as agents of dispersal.

A related species, *A. trifida*, Giant Rag-weed, illustrated, has similar flowers, op-posite leaves with 3 deep palmate lobes and a rough-hairy stem to 5.4 m. high (18 ft.). This species and Common Ragweed are the cause of most late summer hayfever in the Northeast. Its seeds are large and tough-coated; thus it is of little value as food for birds. It occurs on moist floodplains and rich bottomlands from Quebec to British

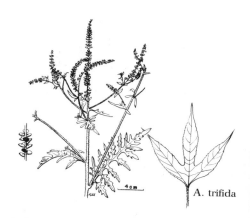

A. trifida

Columbia, south to Florida, Texas, and Arizona.

Common Ragweed is found in pastures, recently abandoned fields, and cultivated lands, along roadsides, and in waste spaces throughout most of the United States and southern Canada.

℞ In herbology these plants are reputed to be stimulating, astringent, and antiseptic. They have been used to treat intestinal dis-turbances, and in Mexico, to treat intestinal worms and fever.

Open Fields

Anaphalis margaritacea
Pearly Everlasting

152

Asteraceae or Compositae
Aster Family

A perennial with a short rhizome, of North America, Europe, and northern Asia. Pistil-late and staminate flowers are usually on different plants, and cross-pollination may be effected by the numerous insects, es-pecially the moths and butterflies, that visit the flower heads. Each pistillate head pro-duces numerous tiny, cylindrical, light brown, seed-like fruits that are dispersed by wind with the aid of a parachute of hairs.

This is a genus of the Northern Hemi-sphere with only one species in eastern North America. It is a highly variable species with regard to number of leaves, number of flower heads, and degree of hairiness. As the showiest species of the genus it is some-times dyed and sold in dried flower ar-rangements. It is sometimes planted as an ornamental in sunny wild flower rock gardens.

146

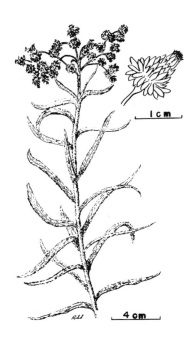

Bracts of flower heads pearly white and petal-like, flower heads numerous, to 12 mm. wide (½ in.), in a branched terminal cluster; leaves alternate, woolly-white on underside, narrow, to 13 cm. long (5.2 in.); stems erect, woolly-white, to 90 cm. high (3 ft.).

July–September

Pearly Everlasting is found in dry meadows and pastures, roadsides, and waste areas from Nova Scotia and Quebec to British Columbia, south to North Carolina and Missouri.

℞ A solution made from the plant in water is a mild astringent that was used by Indians and early American colonists for hemorrhoids. Some American Indians treated paralysis by burning the flower heads and having the patient inhale the smoke. A tea made from the plant has been used for inflammation of the mucous membranes of the nose and throat. The leaves have been used as a nicotine-free substitute for tobacco.

Antennaria plantaginifolia
Plantain-leaved Everlasting*

153

Asteraceae or Compositae
Aster Family

*Also Pussy-toes, Ladies Tobacco
Male and female flower heads on different plants, pistillate heads pink to red, staminate heads white; basal leaves in a rosette, oval, 3–5 nerved, cauline leaves alternate, smaller, lanceolate, all leaves silvery white beneath; stems white woolly, supporting several flower heads, to 40 cm. high (16 in.).*

April–June

A perennial with a well developed root and surface runners. Seeds are produced, at least part of the time, without fertilization. Each tiny nutlet is brown with white resinous dots and has a dense tuft of hairs which facilitates dispersal by wind.

When this plant invades pastureland it is normally not eaten by livestock, and when overgrazing removes competing species it spreads. Thus, in fields where it abounds, plantain-leaved Everlasting may be an indicator of overgrazing.

The 5 species of this complex genus in the Northeast are similar in being white-woolly, dioecious perennials with alternate cauline leaves and a rosette of basal leaves.

Plantain-leaved Everlasting is found in dry fields, pastures, and open woods from Quebec to Minnesota, south to Florida and Texas.

Arctium minus
Common Burdock

154

Asteraceae or Compositae
Aster Family

Flower heads lavender, to 2.5 cm. wide (1 in.), sessile or on short stalks, in axillary clusters; bracts of flower heads are inward curving spines, becoming prickly burs in fruit; leaves alternate, heart-shaped, woolly on underside, with hollow stalks, lower ones to 50 cm. long (20 in.); stems freely branched, to 1.5 m. high (5 ft.).

July–October

A biennial, introduced from Europe, with a large fleshy taproot and a dense rosette of leaves in the first year of growth. After the flowering season in the second year, each flower head becomes a many-seeded bur. These persist throughout the winter and may be dispersed by becoming attached to the clothing of a field biologist or some other passing animal.

This is a small genus of about 5 species, all natives of Europe or Asia. The 3 species that have been introduced into North America are common weeds of open spaces, and each will hybridize with the other two. *A. lappa*, Great Burdock, not illustrated, has larger flower heads, to 5 cm. wide (2 in.) on long stalks, leaves with solid rather than hollow stems, and may reach a height of 2.7 m. (9 ft.).

Common Burdock is found in rich soil of neglected farmlands, along roadsides, and in waste areas from Newfoundland to British Columbia, south to Georgia, Oklahoma, and northern California.

✖ The very young leaves can be added fresh to salads or cooked as greens in several changes of water. After the first year of growth, the tough, bitter, outer layer of the root can be peeled off and the inner portion boiled in at least two changes of water. In

4 cm

the second year of growth, before the flowers appear, the leaf stalks and the young plant stem can be prepared in the same way. All of the burdocks have been widely used for food in many parts of Europe, and one variety is cultivated in Japan. This seems an excellent way to use the otherwise troublesome weed.

℞ In herbology the burdocks are described as tonics, diuretics, and diaphoretics. *The Dispensatory of the United States of America* (1851) states of the root of *A. lappa* that it "has been recommended in gouty, scorbutic, venereal, rheumatic, scrofulous, leprous and nephritic affections." The fresh bruised leaves have been used as a poultice for bruises, swellings, and poison ivy.

According to mythology, these plants were named as her own by Venus, the Roman

148

goddess of natural productivity and later of love and beauty. By applying the seeds and leaves to her navel, a pregnant woman could, supposedly, prevent an abortion.

Artemisia vulgaris
Common Mugwort

155

Asteraceae or Compositae
Aster Family

4.5mm

4 cm

Flower heads greenish, erect, crowded in a branched terminal cluster; leaves alternate, pinnately dissected, the segments further toothed or divided, usually with 1 or 2 pairs of stipule-like lobes at base of petiole, white wooly on underside; stems smooth below flower heads, to 1.5 m. high (5 ft.).

July–October

to brown, slightly ribbed seeds are very small and may be spread by wind.

This is a large genus with about 40 species in North America. Most of these occur in the western states and include several species of sagebrush.

Common Mugwort is found in pastures, fields, and waste spaces, typically in limestone areas, from Newfoundland and Ontario to Minnesota, south to Georgia, and along the Pacific Coast from Oregon to southern California.

✄ The leaves are sometimes used as seasoning when bitter aromatic herbs are required. The dried leaves have been used in Great Britain as a substitute for tea. It is said to have been used to flavor beer before the introduction of hops. The common name Mugwort may have originated with the latter use.

An aromatic perennial with a short branching rhizome introduced from Europe. Pollination is by wind, and it may be an important source of hayfever pollen in some parts of North America. The dark gray

Aster linariifolius
Stiff Aster

156

Asteraceae or Compositae
Aster Family

A perennial with a persistent, short, semi-woody base and a dense tuft of fibrous roots. It is self-incompatible and cross-pollination is effected by insects. Each flower head produces 25–50 seed-like fruits with 4–7 ribs and a parachute of hairs that aids in dispersal by wind.

The genus name is derived from a Greek word meaning "a star" and refers to the flower heads with their radiating ray flow-

Flower heads with blue or violet rays, to 2.5 cm. wide (1 in.), solitary or a few on terminal stalks; leaves alternate, numerous, very narrow, rough, stiff, with a prominent midrib; stems erect, stiff, hairy, to 60 cm. high (2 ft.).

August–October

ers. The specific name alludes to the similarity of the leaves to those of *Linaria vulgaris* (Butter-and-eggs). Since it has leaves with margins that are rough to the touch, another appropriate name for this species is Sand-paper Starwort. A form with white ray flowers is occasionally observed.

Stiff Aster is found in dry open areas and open woods, especially in sandy or rocky soil, from New Brunswick and Quebec to Minnesota, south to Florida and Texas.

2 cm

Aster novae-angliae
New England Aster

157

Asteraceae or Compositae
Aster Family

Ray flowers purple-violet, flower heads numerous, showy; leaves alternate, lanceolate, clasping; stems stiff-hairy, branching to 2.1 m. high (7 ft.).

August–October

A perennial with a short thick rhizome. Cross-pollination is essential for seed production and most likely effected by butterflies and bees. The longitudinally ribbed, hairy, seed-like fruits are equipped with parachutes of hairs that facilitate dispersal by wind.

This is one of the most colorful flowers of late summer and fall and is sometimes grown as an ornamental in flower gardens. It does not survive clean cultivation and is thus not a problem weed of crop fields.

The wild asters are a complex group of approximately 250 species of North America, South America, and Eurasia, but mostly in the northern hemisphere. North America may have the greatest number of native spe-

2 cm

6 cm

cies, and the genus has its greatest concentration and complexity in the eastern United

States. As food for wildlife it is of relatively little importance, but the seeds and leaves are eaten occasionally by a few upland game birds and mammals. New England Aster is found in moist areas along roadsides and in abandoned fields and meadows ranging from Quebec to Saskatchewan south to South Carolina, Alabama, and Colorado.

℞ Aster flowers have been used in pot-pourris as room air fresheners. Several species of *Aster* in our area are recognized as selenium absorbers and are potentially toxic for grazing animals, but these are more of a threat in western states which have soils high in selenium.

Aster simplex
White Field Aster*

158

Asteraceae or Compositae
Composite Family

*Also Panicled Aster

4 cm

A perennial with a long creeping rhizome. The ray flowers are pistillate and more numerous than the disk flowers, which have both stamens and pistils. Each flower head produces numerous light brown, flattened, hairy seed-like fruits with white parachutes of hairs that aid in dispersal by wind. In

Flower heads with white rays, to 2 cm. wide (¾ in.), numerous, on leafy branches; leaves alternate, narrowly lanceolate, slightly toothed or entire, smooth, largest ones to 15 cm. long (6 in.); stems thick, freely branched, smooth at least on lower part, to 1.5 m. high (5 ft.).

August–October

fields that have become heavily infested with this species, the planting of clean cultivated row crops for two years will usually result in its eradication.

It spreads by a creeping rhizome and sometimes forms dense colonies. There are several varieties of this plant that differ in width and height of the flower heads and in the ratio of length to width of the leaves. Forms with lavender or blue ray flowers are occasionally observed. In older systems of classification, it was listed as *A. paniculata*.

White Field Aster is found in abandoned fields, pastures, and neglected gardens and often on recently plowed land from Nova Scotia and Quebec to Saskatchewan, south to North Carolina, Missouri, and Texas.

Bidens bipinnata
Spanish Needles

159

Flower heads yellow, numerous, long and slender, solitary, at the ends of long stalks; seed-like fruit brown or black, 4-angled, to 18 mm. long (¾ in.), with 3 or 4 short barbs on upper end; leaves opposite, 2 or 3 times pinnately divided, stalked, to 20 cm. long (8 in.); stems smooth or finely hairy, square, to 1 m. high (40 in.).

August–October

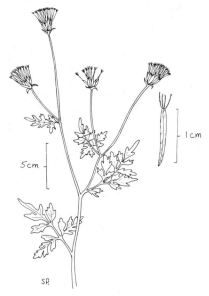

An annual with a well developed branched taproot. Each flower head produces several seed-like fruits with downward barbed awns. These readily become entangled in the fur of passing animals or the clothing of humans and are efficient adaptations for dispersal. It is easily distinguished from other members of the genus by its finely dissected leaves.

This species is most abundant in the southern states, where it sometimes becomes a troublesome weed in crop fields. Clean cultivation until late in the season and mowing before the seeds are formed have been used as control measures. The plants of this genus are of minor value as food for wildlife, but the seeds are eaten to a limited extent by the Ring-necked Pheasant, Bobwhite Quail, Purple Finch, Common Redpoll, and Swamp Sparrow.

Spanish Needles are found in cultivated fields and gardens, along roadsides, and in waste areas, often in sandy soil, from Massachusetts and Ontario to Illinois, south to Florida and Kansas.

✕ The natives of tropical west Africa are said to use the leaves of this plant as a potherb.

℞ In herbology the roots and seeds of *B. bipinnata*, *B. frondosa*, Beggar-ticks, and *B. coronata*, Swamp Beggar-ticks, not illustrated, are recommended for dispelling the mucus from the respiratory tract and promoting the menstrual discharge.

Bidens frondosa
Beggar-ticks*

160

*Also Sticktight

An annual with a shallow, branched taproot. The genus name is a Latin word meaning "two teeth," referring to the 2 barbed awns on the seed-like fruit. The plants of this genus are of minor importance as food for wildlife, but the seeds and foliage may constitute up to 2% of the diet of the Ring-necked Pheasant, Wood Duck, Purple Finch, and Cottontail Rabbit.

152

Flower heads orange, numerous, at ends of upper branches, subtended by 6–9 narrow leaf-like bracts; rays yellow-orange, often absent; leaves opposite, pinnately compound, with 3–5 lanceolate, toothed leaflets; stems smooth, somewhat 4-angled, branching in upper part, to 1.2 m. high (4 ft.).

August–October

Other common names for *B. frondosa* are Devil's Boot-jack and Pitchfork-weed, which refer to the very efficient mechanisms for seed dispersal. The seeds develop with the barbed awns directed outward, and seldom does an organism, including man, pass by without picking up some unwanted hitchhikers.

Beggar-ticks are found in moist pastures, roadside ditches, fields, and other damp open habitats throughout most of the United States and southern Canada.

℞ See *B. bipinnata*.

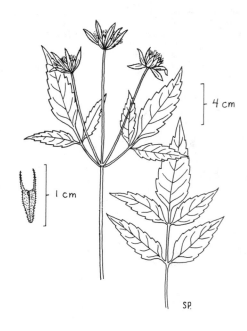

SP.

Centaurea jacea
Brown Knapweed

161

Asteraceae or Compositae
Aster Family

Flower heads rose-purple, to 4.3 cm. wide (1¾ in.), solitary, at ends of stems and branches; ray flowers absent, marginal disk flowers enlarged; bracts of flower heads tan or brown, toothed, not fringed; leaves alternate, toothed or shallowly lobed, basal leaves to 20 cm. long (8 in.), upper ones smaller; stems smooth or cobwebby, branched, to 1 m. high (40 in.).

June–September

A perennial introduced from Europe. Fertilization is the result mainly of cross-pollination by long-tongued bees. The dead stems and flower heads persist into the winter months and the seed-like fruits may be dispersed as wind ballistics. As the plant sways in the breeze the nutlets are thrown

C. maculosa

out of the receptacles. The seeds of some western species are important sources of food for several species of goldfinches.

This is a very large genus with its greatest area of concentration in the Mediterranean region. Of the approximately 12 species in North America only one, *C. americana*, Star Thistle, not illustrated, is a native. *C. jacea* is sometimes a troublesome weed in cultivated land, and the whole genus has been declared noxious weeds by the seed laws of 5 states.

C. maculosa, Spotted Knapweed, is similar to *C. jacea* but has the bracts of the flower head with black fringes (see illustration) and leaves that are deeply pinnately divided. It is a common biennial weed in

the same types of habitats as *C. jacea* throughout southern Canada, the northeastern states, and the Pacific coast states.

Brown Knapweed is found in open fields, along roadsides, and in waste areas from Quebec to Iowa, south to Virginia, Ohio, and Illinois.

℞ In herbology several species of this genus are described as having diuretic, astringent, and tonic properties. The roots and seeds have been used to make an ointment for wounds and sores. The crushed leaves have been applied externally to heal black and blue bruises of the skin. A solution made from the plant, taken internally, has been used to relieve coughs, asthma, and breathing difficulties.

Chrysanthemum leucanthemum
Ox-eye Daisy*

162

Asteraceae or Compositae
Aster Family

Also White Daisy, Whiteweed, Marguerite Flower heads with white rays, to 5 cm. wide (2 in.), solitary; ray flowers pistillate, disk flowers yellow, with both stamens and pistils; leaves alternate, dark green, pinnately lobed or coarsely toothed, basal leaves long stalked, to 15 cm. long (6 in.); stems smooth, erect, usually unbranched, to 75 cm. high (30 in.).

June–September

A perennial with a short thick rhizome, introduced from Europe. The 20–30 ray flowers and numerous disk flowers each produce a brown or black seed-like fruit with 10 white, lonitudinal ribs. The name daisy can be traced through ancient English to "day's-eye," referring to the bright yellow center of the flower.

It may become established as a weed in cultivated fields, especially those on which barnyard manure has been spread as fertilizer. In feeding experiments the seeds retained viability after passing through the intestinal tracts of horses, cows, swine and sheep. When it is eaten by cattle the plant gives an unpleasant flavor to milk. It has

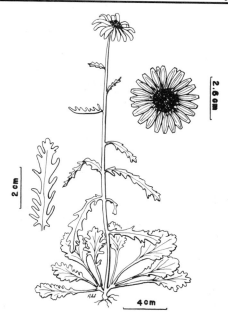

2.6 cm

2 cm

4 cm

been declared a noxious weed in the seed laws of 9 mostly northeastern states.

This is a genus of about 100 species native mostly to Europe and Asia. Several species

154

are very popular as ornamentals, including the florist's Chrysanthemum, C. morifolium, and the Shasta Daisy, C. maximum, not illustrated. Four species of the genus have been introduced into the Northeast, but C. leucanthemum is the most abundant.

Ox-eye Daisy is found in open fields and meadows, along roadsides, and in waste areas throughout most of the United States and southern Canada.

� The young leaves can be added to salads. They should be used with discretion because the flavor may not appeal to everyone. ℞ In herbology it is listed as a diuretic and a tonic. It has been used for whooping cough, asthma, and urinary disorders. A solution made from the fresh plant was added to ale as an early rural treatment for jaundice.

Cichorium intybus
Chicory*

163

Asteraceae or Compositae
Aster Family

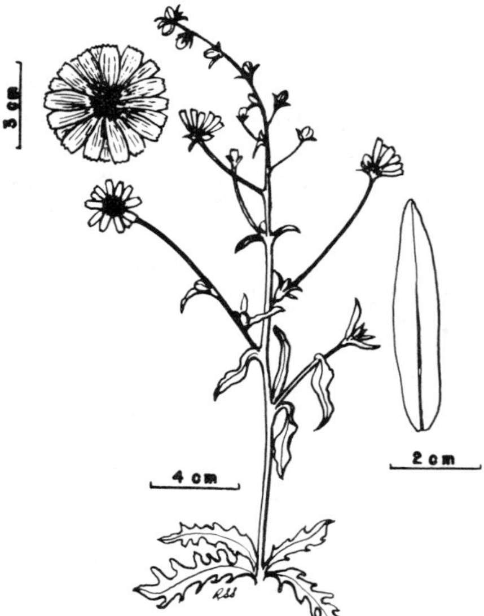

*Also Succory, Blue Sailors, Witloof
Flower heads blue, sessile or short stalked, in clusters of 2–3 on upper part of stem; ray flowers square tipped and fringed; stem leaves alternate, simple, sessile, basal leaves larger, lanceolate, pinnately lobed or toothed, in a rosette; stems branched, becoming woody with age, to 1.2 m. high (4 ft.).*

July–October

part. The common cultivated salad green, Endive, is C. indivia.

Chicory is common along roadsides and in open fields, pastures, lawns, and waste spaces, especially in limestone areas, from Nova Scotia to British Columbia, south to North Carolina and California.

✺ The young leaves can be cooked as greens, with several changes of water to reduce the bitterness, and the white underground parts of young basal leaves can be added to salads. This plant is cultivated in some areas for its root, which is ground up and used as a coffee substitute or additive. Many people prefer a mixture of Chicory and coffee to pure coffee.

℞ The dried root is described in herbology as a tonic, a laxative, and a diuretic. In describing a solution made from the dried root, *The Dispensatory of the United States of America* (1851) states, "It is said to be useful, if freely taken, in hepatic congestion, jaundice, and other visceral obstructions in the early stages, and is affirmed to have done good even in pulmonary consumption."

A perennial with a deep taproot, introduced from Europe. It is mostly a self-incompatible species with occasional self-pollination. The pollen grains are beautifully sculptured with long spines which are common in insect pollinated plants. Cross-pollination is effected mainly by bees. The flower heads last but one day and are usually closed in cloudy weather. Each flower head produces numerous longitudinally ribbed, finely wrinkled, seed-like fruits.

There are about 9 species of this genus, all natives of Europe, chiefly in the southern

Cirsium arvense
Canada Thistle*

164

Open Fields

Also Creeping Thistle
Flower heads rose-purple, numerous, fragrant, in terminal and axillary clusters; leaves alternate, lanceolate, spiny-margined, often white-woolly on underside; stems erect, ridged, branched in upper part, to 1.5 m. high (5 ft.).

July–October

A perennial with an extensively creeping rhizome, introduced from Europe. Staminate and pistillate flower heads are usually produced on different plants. Thus, an isolated plant commonly does not bear seeds; if it happens to be a staminate plant, it can never produce seeds; if it is a pistillate plant, no seeds will be present unless there is a male plant near enough for pollination to take place. When pollination does occur, the numerous gray to brown, finely ridged seeds are scattered by wind with the aid of a parachute of hairs. This thistle down is favored by the Common Goldfinch as a nest lining material.

This is a troublesome weed in many areas and very difficult to eradicate. In addition to reproducing by seeds, the spreading rhizome sprouts freely; thus the common name, Creeping Thistle. If it is cut into pieces by cultivation, each piece is capable of giving rise to a new plant. It has been declared a noxious weed in the seed laws of 37 states.

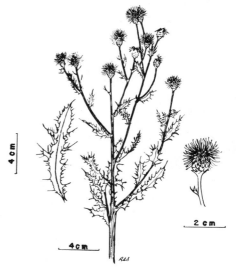

Canada Thistle is found in cultivated fields, pastures, meadows, and waste areas from Newfoundland to British Columbia, south to Virginia and California.

✂ The young leaves can be stripped of their spines and cooked as greens or used in salads. The young stems can be peeled and cooked as a potherb.

℞ According to some herbals, Canada Thistle has tonic, diuretic, and sweat-inducing properties. Boiled with milk, it has been used as a treatment for dysentery.

Cirsium vulgare
Bull Thistle*

165

Also Common Thistle
A biennial with a large fleshy taproot in the first year of growth, native to Europe and Asia. The sweet scented flower heads are rich in nectar, and pollination is effected mainly by bumblebees and butterflies. The numerous grayish seeds have dark longi-tudinal lines and a tuft of branched hairs that aids in wind dispersal.

This is an aggressive and troublesome weed, but it does not survive in cultivated fields. In areas that are badly infected, the plants should be removed by mowing or hoeing before the seeds mature.

Flower heads purple, numerous, at the tips of spiny branches; leaves alternate, pinnately lobed, white woolly or pale hairy on the underside, spiny margined; stems with prominent spiny ridges, to 1.8 m. high (6 ft.).

June–October

Bull Thistle is found in open fields and pastures, along roadsides, and in waste spaces from Newfoundland to British Columbia and throughout the United States. ✖ In autumn and through the winter, the first year fleshy root can be peeled and cooked. In the second year of growth, the young leaves, stripped of their spines and cooked like turnip greens, are said to be delicious. The young stems, when 30–35 cm. high (12–14 in.), can be peeled and eaten raw or cooked as a vegetable. The flowers are reported to have the property of causing milk to curdle.

Conyza canadensis
Horseweed*

166

Asteraceae or Compositae
Aster Family

Also Hog-weed, Butter-weed

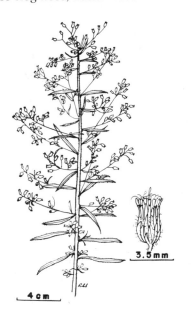

Flower heads white to pink, tiny, with vertical ray flowers, on numerous branches from upper axils; leaves alternate, narrowly lanceolate to linear, mostly entire and sessile, numerous, hairy; stems erect, unbranched at base, with stiff hairs, to 2 m. high (6½ ft.).

July–November

An annual of North America, introduced and widespread in Europe and Asia. Pollination is by bees and flies, but an equal number of seeds may be produced by self-pollination. Some of the pollen may become airborne, but it is not an important hayfever plant. Each seed has a tuft of white hairs that aids in wind dispersal.

Three intergrading varieties of this species have been recognized, based on hairiness of the stems and pigmentation of the bracts surrounding the flower heads. The species of this genus are very similar to the fleabanes (*Erigeron*), sometimes distinguished only by technical characters. *Co-*

nyza canadensis is also recognized as *Erigeron canadensis* by some botanists.

Horseweed is found in pastures, cultivated fields, and waste areas and along roadsides, usually on dry soils, throughout southern Canada and the United States.

℞ The whole plant, dried when in flower, has been recommended as a tonic, a diuretic, and a treatment for diarrhea and internal bleeding. Some tribes of American Indians are said to have used a tea made from the roots for menstrual irregularities and vaginal discharges.

The leaves are known to cause contact dermatitis in sensitive individuals.

Coreopsis lanceolata
Lance-leaved Coreopsis*   **167**   *Asteraceae or Compositae*
Aster Family

Also Tickseed

Flower heads with yellow rays to 6.3 cm. wide (2½ in.), solitary, on long terminal stalks, rays with 3–4 notches at ends; leaves opposite, lanceolate to narrowly spatula-shaped, often with 2 basal lobes, to 20 cm. long (8 in.); stems smooth, often with several flowering branches, to 60 cm. high (2 ft.).

May–July

A perennial with a short persistent stem base. Each flower head consists of about 8 ray flowers and many disk flowers. The seed-like fruits are black with thin marginal wings and 2 short teeth on the upper end. These may be dispersed by wind or by becoming attached to the coat of a passing animal.

The genus name is derived from Greek words that mean "bug-like" and refers to the shape of the seed. The genus includes more than 100 species, with about 11 native to eastern North America. The range of *C. lanceolata* has expanded as it has escaped from cultivation in areas beyond its natural region of occurrence. A related species, *C. major*, Greater Tickseed, not illustrated, has deeply cut, opposite leaves that appear as whorls of 6's. It occurs in dry woods from Pennsylvania and Ohio, south to Florida and Texas.

Lance-leaved Coreopsis is found in open fields and along roadsides from New England and Ontario to Wisconsin, south to Florida and Texas.

Erigeron annuus
Daisy Fleabane*

168

Asteraceae or Compositae
Aster Family

*Also Sweet Scabious, White-top

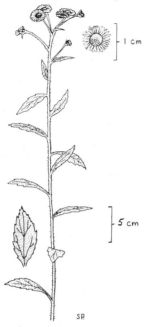

Flower heads with white, pale pink, or bluish rays, numerous, usually concentrated near the stem tip, rays 50–75; leaves alternate, lanceolate to oval, sharply toothed; stems hairy, hairs stand out at right angles to stem, to 1.5 m. high (5 ft.).

May–September

recommended control measure is close pasturing of sheep to eliminate the rosettes. Mowing or plowing infested areas, before the flowering stem can produce seeds, is also a recommended control measure. The mature plants do not make good hay and it is not valued by farmers, including sheep growers, as a forage plant. The common name is supposedly based on the belief that the dried flower heads, when placed in a living area, will drive away fleas. There is no evidence to support this contention.

The plants of this genus are very similar to some of those in both the genera *Aster* and *Conyza.* Consequently, botanists are not in agreement as to the genus designation for some species.

E. annuus has been introduced and is widespread in Europe and Bermuda. A closely related and similar species, *E. strigosus,* not illustrated, has entire leaves, has stem hairs parallel to stem, and reaches a height of 90 cm. (3 ft.). Although both species commonly produce seeds without fertilization, there are varieties of *E. strigosus* that are almost intermediate between the two.

Daisy Fleabane is found in open fields, meadows, pastures, and waste places from Nova Scotia to Manitoba, south to Georgia to Texas. On the Pacific Coast it ranges from British Columbia to lower California.

An annual or winter annual that frequently overwinters as a well developed rosette. The flower heads are visited by several species of bees and flies, but these visits only rarely result in pollination. Most of the seeds are produced without fertilization, and the plants that grow from them are genetic duplicates of the parent plants. The seeds soon lose their parachutes of hair but are so small they may be carried, dustlike, by the wind into new areas.

It sometimes becomes established as a weed in meadows and cultivated fields. It is favored by sheep as a forage plant, and a

Erigeron philadelphicus
Philadelphia Fleabane

169

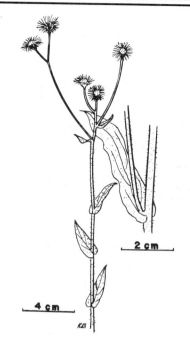

Flower heads with pink to white rays, to 2.5 cm. wide (1 in.), few to very numerous, rays 100–150; leaves alternate, clasping the stem; stems soft, hairy, to 1 m. high (40 in.).

April–July

A biennial or short-lived perennial with a short rootstock. Philadelphia Fleabane is found in moist meadows, fields, and open woods from Newfoundland to British Columbia, south throughout the United States.

For other information see *E. annuus.*

Open Fields

2 cm

4 cm

R.B

Eupatorium maculatum
Joe-Pye Weed

170

Flower heads light purple, in dense, terminal, flat-topped clusters; leaves in whorls of 4's or 5's, tapering at each end; stems purple or spotted with purple, usually not hollow, to 2 m. high (6½ ft.).

July–September

A perennial with a fibrous root system. Pollination is by insects which are attracted by the nectar at the bottom of the tube of each tiny floret that makes up the flower head. The seed-like fruits are black, 5-angled, and finely wrinkled, with a parachute of hairs that aids in wind dispersal.

It is usually not a problem as a weed because it does not survive where cultivation is frequent. It can be controlled in meadow or pastureland by mowing or improving the drainage.

This is a very large genus of at least 500 species throughout the world. The genus

3 cm

S.R.

was named for Mithridates Eupator, king of an area near the Black Sea, who reputedly used one species for medicine. Several species are widely recognized by herbalists today as medicinal plants.

There are about 23 species in the Northeast. These are closely related and very similar, with some showing the intergrading characteristics of hybridization.

Joe-Pye Weed is found in abandoned fields, moist thickets, wet meadows, and the edges of swamps and ponds, especially in limestone areas, from Newfoundland to British Columbia, south to Pennsylvania and Indiana, and in the mountains to North Carolina.

Eupatorium perfoliatum Boneset*	**171**	*Asteraceae or Compositae* Aster Family

*Also Thoroughwort, Agueweed, Feverweed

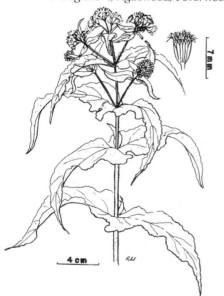

Flower heads white, in dense, terminal, more or less flat-topped clusters; leaves opposite, joined at the base around the stem, appearing wrinkled, hairy on underside, with long pointed tips; stems hairy, usually branched in upper part, to 1.5 m. high (5 ft.).

July–October

are known to be eaten occasionally by Wild Turkeys and Swamp Sparrows.

This is a highly variable species with several unusual varieties. A form with narrower leaves and less hair on leaves and stems has been identified along the margins of brackish tidal estuaries in Quebec and Maine. Some of the varietal forms may be hybrids between _E. perfoliatum_ and another species.

Boneset is found in damp fields and wet meadows, in swales, along the margins of marshes and ponds, and in roadside ditches from Nova Scotia and Quebec to Manitoba south to Florida and Texas.

For additional information see _E. maculatum_.

℞ This species has a long history of medicinal use by American Indians and is said to have been introduced into England as a medicinal herb in 1699. Its main uses were to induce sweating, as a treatment for fevers (thus the common names Agueweed and Feverwort), and, in stronger doses, as an emetic and a laxative.

The origin of the common name Boneset has two explanations in folklore. The belief that the plant is good for broken bones may

A perennial with a thick rhizome and fibrous roots. The tight clusters of tiny florets in each flower head produce abundant nectar which attracts honeybees, among others, as pollinators. Each flower head produces 10–25 black, 5-angled, finely dotted seeds, crowned with parachutes of hairs.

It sometimes invades cultivated areas or pastures but is usually not a problem. It can be controlled by mowing before seeds are formed, digging out individual plants, or by improving drainage. This is not a significant wildlife food genus, but its seeds and leaves

be an application of the "Doctrine of Signatures." The fusion of the leaves around the stem may have suggested that the plant might cause broken bones to reunite. Another explanation is that the plant was used in treating colds and influenza, which in early times were called "break-bone fevers."

Eupatorium purpureum
Sweet Joe-Pye Weed*

172

Asteraceae or Compositae
Aster Family

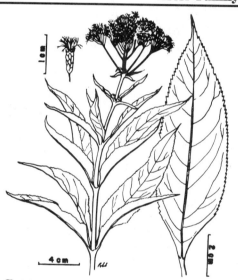

*Also Gravel Root, Queen of the Meadow
Flower heads pale pink or light purple, in dense, terminal, round-topped clusters; leaves in whorls of 3's or 4's, short petioled, sharply toothed; stems usually green, purplish or dark at nodes, not hollow, to 2 m. high (6½ ft.); plant smells like vanilla when bruised.*

July–September

A perennial with a densely fibrous root system. Since the flowers are self-incompatible, cross-pollination, often by honeybees, is necessary for the production of seeds. Each flower head, at maturity, bears 5–7 black, longitudinally ridged seeds with tufts of hairs that aid in distribution by wind.

It is usually not a problem as a weed in cultivated fields and pastures because it is easily controlled by mowing and improving drainage. The leaves and seeds are occasionally eaten by a few upland game birds and songbirds. The seeds may persist on the dead stems into the winter months.

Sweet Joe-Pye Weed is found in moist meadows, woodland borders, the edges of ponds and open swamps, and wet waste places from New Brunswick to Manitoba, south to Florida and Texas.

For further information see *E. maculatum* and *E. perfoliatum*.

℞ Many species of this large genus have been used medicinally. The common name is said to commemorate Joe Pye, an American Indian herb doctor in the Massachusetts Bay colony who used this plant to cure typhus fever. The root has diuretic and tonic properties and is reputed to be particularly useful in the treatment of kidney stones and other urinary problems. This plant is closely related to boneset, and they have been used interchangeably.

Galinsoga ciliata
Galinsoga*

173

Asteraceae or Compositae
Aster Family

*Also Quickweed, Peru Weed, French Weed

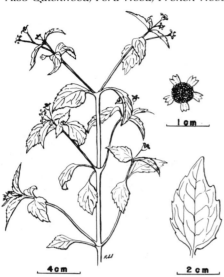

Flower heads with 5 tiny, 3-lobed, white ray flowers, disk flowers yellow, heads small, numerous on leafy branches; leaves opposite, with pointed tips, oval, coarsely toothed; stems hairy, freely branching, erect or sprawling, to 60 cm. high (2 ft.).

June–November

This species was present in New England in 1866 and has since spread throughout North America. It has also been introduced and is widespread in Europe and southern Asia. In England the name Galinsoga has been corrupted to Gallant-soldiers.

There are 3 species of this genus in the Northeast, all introduced from Central and South America. They have similar growth habits and are easily recognized. *G. parviflora*, Peru Weed, not illustrated, has smaller ray flowers, a less hairy stem and is not as common as *G. ciliata*. *G. caracasana*, not illustrated, is more abundant around large cities and has pink to purple ray flowers.

An annual introduced from South and Central America. Both ray flowers and disk flowers have stamens and pistils and both produce fertile seeds. Each seed is black with close lying stiff hairs and has a tuft of hairy or awn tipped scales. It often becomes established and forms dense clumps in gardens and cultivated fields. The most effective control measure is to plow or dig up the seedlings before seeds have developed.

Galinsoga is found in gardens, cultivated fields, roadsides, and waste areas from Quebec to Oregon, south to Florida, Mexico, and California.

✄ The plants of this genus are reported to be very palatable and nutritious as cooked greens. They do not create problems as weeds in southeastern Asia, where they are favored by the native population as potherbs.

Gnaphalium viscosum
Clammy Everlasting*

174

Asteraceae or Compositae
Aster Family

*Also Cudweed, Balsam Weed

An annual or sometimes a biennial, with a basal rosette of leaves that have their widest part above the middle. The flower heads contain numerous flowers, the outer ones pistillate, a few inner ones with both pistils and stamens. Each of the tiny, gray to brown, seed-like fruits has a tuft of hairs which is shed at maturity.

This is a very large genus of more than 100 species distributed over a wide geo-

graphic area from the equator to the arctic region. Several of the 8 species that are native to the Northeast have geographic ranges that extend beyond North America. The genus name is derived from a Greek word which means "soft down" and refers to the woolly hairs on the stems and leaves.

It sometimes becomes established as a weed in cultivated areas but does not persist where there is periodic plowing. In pastureland it can be controlled by mowing

Flower heads white, numerous, in branched terminal clusters; leaves alternate, narrow, entire, woolly white on under-side, clasping the stem and extending downward as a wing; stems hairy, upper part woolly, to 90 cm. high (3 ft.).

July–September

before the seeds are developed. Other names that have been used for this species are *G. decurrens* and *G. macounii.*

A closely related species, *G. obtusifolium,* Sweet Everlasting, not illustrated, is similar but has leaves that taper at the base and do not clasp the stem. It usually does not get as tall as *G. viscosum,* and it has a wider distribution in the southern United States.

Clammy Everlasting is found in abandoned fields, pastures, dry roadsides, and open waste areas from Quebec to British Columbia, south to Tennessee in the East, and to Arizona and Mexico in the West.

℞ Two species of the genus, *G. obtusifolium* and *G. uliginosum,* Low Cudweed, not illustrated, are listed as medicinal plants. They are characterized as astringents and are recommended for sores of the mouth and throat, intestinal irritations, diarrhea, and (applied externally) for bruises.

The Creek Indians used the leaves to make a poultice to relieve the swelling of glands associated with mumps.

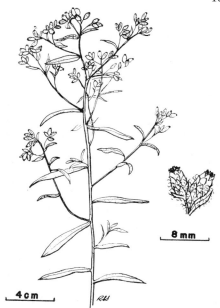

8 mm

4 cm

Helenium flexuosum **Purple-headed Sneezeweed***	**175**	*Asteraceae or Compositae* *Aster Family*

*Also False Sunflower

Flower heads with yellow, 3-lobed, drooping rays; flower heads numerous, solitary, at the tips of stems and branches, central disk dome-shaped, purplish brown; leaves alternate, usually angled upward, narrow, entire, base extending down stem as a wing; stems angled by wings from leaf bases, covered with fine hairs, to 90 cm. high (3 ft.).

June–October

A perennial with a fibrous root system. The flower heads are visited mainly by bees that may effect pollination. Each flower head produces numerous top-shaped, hairy, ridged, brown seeds which have tufts of

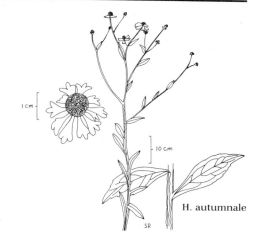

1 cm

10 cm

H. autumnale

164

pointed scales. These are windblown, and the plant sometimes becomes established as a weed in meadows and pastures. Improvement of drainage, cultivation, and early mowing are recommended methods of control.

This genus includes about 40 species of North and South America, 5 of which are native to the Northeast. The genus name was given to another plant in the ancient world in honor of Helen of Troy and was later assigned by Linnaeus to the present genus. The powdered flower heads are reputed to have been used medicinally to induce sneezing, thus the name sneezeweed. *H. flexuosum* is listed as *H. nudiflorum* in some systems of classification.

A closely related species, *H. autumnale*, Sneezeweed, is similar but has a yellow central disk, leaves oval to lanceolate and more numerous, and reaches a height of 1.5 m. (5 ft.). It occurs in similar habitats but has a more northerly distribution than *H. flexuosum*.

Purple-headed Sneezeweed is found in wet meadows, open fields, roadsides, and waste areas from New England to Michigan south to Florida and Texas. It is spreading northward.

☠ The plants of this genus are bitter and somewhat acrid. They have tonic properties and have been recommended for fevers. They are known to contain substances that cause contact dermatitis in sensitive individuals.

All of the species in this genus contain a toxic substance known as dugaldin, which is most concentrated in the mature flower heads. There are records of poisoning of cattle, sheep, horses, and mules from eating these plants. Sheep are particularly subject to poisoning because after once eating the plants they sometimes develop a taste for them. It was estimated in 1944 that in Colorado alone about 8000 sheep per year die from sneezeweed poisoning.

Helianthus tuberosus
Jerusalem Artichoke*

176

Asteraceae or Compositae
Aster Family

*Also Girasole

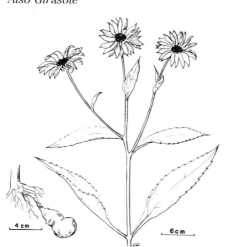

Flower heads with yellow rays, several at the tips of stems and branches, to 9 cm. wide (3½ in.), rays 10–20; leaves alternate or opposite, broadly lanceolate, toothed, to 25 cm. long (10 in.), stems coarse, hairy, branching in upper part, to 3 m. high (10 ft.).

August–October

A perennial with a long rhizome bearing tubers on short branches and at its tip. Self-pollination takes place only occasionally. Most flowers are self-incompatible, and cross-pollination is by insects. The ray flowers do not produce fruits, but each of the numerous disk flowers is fertile. The seed-like fruits are gray or brown, with longitudinal ridges, slightly downy and up to 8 mm. long (⅓ in.).

It may become a problem weed in some areas, but in others it is cultivated for its tuberous roots. A native of the north central states, it was cultivated for centuries by American Indians, who expanded its range to include all of eastern North America. It was early introduced into Europe and became popular in the Mediterranean region where Italians gave it the name "Girasole," which can be loosely translated as "it turns with the sun." The English transformed this to "Jerusalem," and the meaningless name Jerusalem Artichoke has persisted.

This species is found in moist soil of fields, gardens, roadsides, and waste areas from Nova Scotia to Manitoba, south to Georgia and Arkansas, and west to the Rocky Mountains.

✘ The tubers have a pleasant taste and can be prepared in the same ways as potatoes. They can be used fresh in salads and Oriental dishes and can be made into pickles. The tubers are produced up to 10 cm. (4 in.)

or more beneath the surface and can be harvested from fall to spring. According to one report, if they are harvested before they are to be used, they should be stored in sand or soil to prevent them from turning a dark color.

This plant was cultivated and used for food more extensively in Europe than it has ever been among European settlers in America. The tubers provided a very plentiful and inexpensive source of food for the common man in early to mid-seventeenth-century London. By some accounts they were eaten so often and in so many ways that the population eventually grew tired of them.

In improved strains of Jerusalem Artichoke, the tubers are up to 10 cm. long (4 in.) and weigh up to 113 grams (4 oz.). They often grow in areas where other crop plants cannot grow, and they constitute a potential but neglected source of food.

Hieracium aurantiacum
 Orange Hawkweed*

177

Asteraceae or Compositae
 Aster Family

Also Devil's Paint-brush, King Devil
Flower heads orange to reddish, few to several in a terminal cluster on a slender leafless stem; leaves numerous in a basal rosette, hairy; stems glandular-hairy, unbranched, to 60 cm. high (2 ft.).

June–September

A perennial with a long slender rhizome, introduced from Europe. The flower head produces two kinds of seeds, those resulting from fertilization and those that develop asexually. The relative number of each type varies from plant to plant. Each dull black, ribbed seed has a parachute of hairs that aids in dispersal by wind. The plant also spreads by coarse runners that produce roots at intervals.

It is capable of surviving on sterile soil that will support few other plants. Under these circumstances it may proliferate to the exclusion of other species. Although the

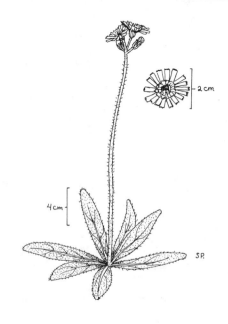

166

flowers are beautiful, the plant is sometimes a weed pest. Control measures include mowing before seeds are formed and adding fertilizer to the soil. With improved soil conditions native species will soon reclaim the area.

The genus name is derived from a Greek word for hawk and refers to an ancient belief that hawks ate this plant to sharpen their vision. This is a very large genus of temperate and boreal areas. European botanists have identified thousands of sub-species and varieties. There are about 17 species

in the Northeast, 8 of which are natives of Europe.

A related introduced species, *H. pratense*, King Devil, not illustrated, is very similar but has yellow flower heads and blackish glandular hairs on its stem. The leaves and seeds of the hawkweeds are eaten by the Ruffed Grouse, Wild Turkey, Cottontail Rabbit, and White-tailed Deer.

Orange Hawkweed is found in abandoned fields, meadows, and lawns and along roadsides from Newfoundland to Minnesota, south to Virginia and Illinois.

Hypochaeris radicata
Cat's ear*

178

Asteraceae or Compositae
Aster Family

*Also Gosmore

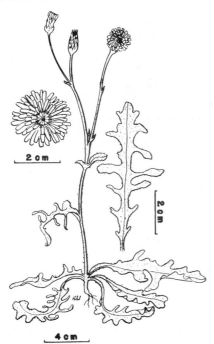

Flower heads yellow, usually several, each at the end of a branch, to 4 cm. wide (1½ in.); leaves mostly in a basal rosette, pinnately lobed, very hairy; stems usually branched in upper half, with scale-like leaves, to 45 cm. high (18 in.).

May–September

brown, longitudinally ridged, long beaked seeds, each with a parachute of hairs at the end of the beak. These are readily distributed by wind, and neglected fields may become badly infested. Frequent cultivation over a period of a year or more may be necessary before the field can be re-seeded. It has long-lived seeds which have shown a 71% rate of germination after being in dry storage for 5 years.

This genus is native to the Mediterranean region and the mountains of South America. *H. radicata* is the only species in the Northeast. The name Cat's-ear apparently refers to the hairy leaves that are supposed to resemble a cat's ears. The flowers are a fairly important source of summer food for upland game birds in some parts of its range.

A perennial with a persistent stem base and several large roots, introduced from Europe. The flowers are self-incompatible, and cross-pollination is effected by insects. The flower head produces numerous reddish-

Cat's-ear is found in fields, pastures, and lawns and along roadsides from Newfoundland to Ontario, south to North Carolina and Indiana. It is also very abundant on the Pacific Coast.

X The leaves may be cooked as greens or prepared as one would prepare a wilted lettuce salad. It is similar to dandelions and wild lettuce and can be prepared in the same ways as these plants. According to some reports, the leaves remain bitter-free all summer.

Inula helenium Elecampane*

179

Asteraceae or Compositae Aster Family

Also Horseheal, Yellow Starwort, Elf-Dock

Flower heads with numerous narrow, yellow rays, the string-like rays giving the flower head a straggly appearance; flower heads to 10 cm. wide (4 in.), at the tips of stems and axillary branches; leaves alternate, upper ones clasping the stem, woolly on underside, basal leaves with long petioles, oval, to 50 cm. long (20 in.); stems hairy, robust, to 1.8 m. high (6 ft.).

7 cm

July–September

A perennial with a thick, branched root, introduced from Europe. The flower heads are visited mostly by bumblebees, moths, and butterflies. Each head produces numerous brown, 4–5 sided seeds, with parachutes of hairs. The windborne seeds sometimes carry the plant as a weed to cultivated areas. However, it does not persist under cultivation, and where it has invaded meadowland and pastures it should be cut below the basal leaves before seeds mature.

There are about 100 species of this genus in Europe, Asia, and Africa. *I. helenium* is a native of Europe and northern Asia and was introduced as an ornamental for borders and flower gardens. Although other species of the genus have been collected in scattered local areas, this, apparently, is the only one that has become established in North America.

Elecampane is found in old fields, meadows, and pastures, along roadsides, and in waste areas from Nova Scotia to Minnesota, south to Georgia and Texas. It is also known in the Pacific Northwest.

X An aromatic candy can be made by boiling slices of the peeled root in sugar syrup.

R The dried second year root has a long history of medicinal use. According to Pliny, a writer in ancient Rome, it was valued as a remedy for indigestion. In the Middle Ages, it was used in external application to cure skin diseases in horses; thus the common name Horseheal. In some remedies it has magical properties which gave it the name Elf-dock.

In seventeenth-century England it was used chiefly for tuberculosis, coughs, bronchitis, and other pulmonary complaints. According to *The Dispensatory of the United States of America* (1851), "In this country it is chiefly used in chronic diseases of the lungs, and is sometimes beneficial when the affection of the chest is attended with weakness of the digestive organs, or with general debility."

Lactuca biennis
Tall Blue Lettuce

180

Asteraceae or Compositae
Aster Family

10 cm

SR.

Flower heads with blue rays, very numerous, in a branched terminal cluster; leaves alternate, pinnately lobed, with coarse sharp teeth; stems smooth, thick, with milky juice, to 4.5 m. high (15 ft.).

July–September

in the Northeast, 3 of which are native to Europe. The genus name is derived from a Latin word that means "milk," referring to the milky sap. The seeds of this plant are eaten by the Goldfinch and possibly other songbirds, and the vegetation is eaten by mammals such as the Cottontail Rabbit and White-tailed Deer.

This is the tallest of the wild lettuces in the Northeast. It is a highly variable species with several recognized varieties. It has cross-pollinated with *L. canadensis* to produce a fertile hybrid designated as *L. morssii*, not illustrated. In some systems of classification *L. biennis* is listed as *L. spicata*.

A biennial with a long sturdy taproot. Each flower head produces 20–35 longitudinally ribbed, mottled gray and brown seeds with parachutes of light brown hairs. It sometimes invades neglected fields and meadows as a weed. Control measures include close and frequent mowing or planting a row crop that requires several inter-row cultivations.

There are about 9 species of this genus

Tall Blue Lettuce is found in moist open areas, meadows, and damp thickets from Newfoundland to British Columbia, south to North Carolina, Tennessee, and Oregon. ✖ The wild lettuces are in the same genus as garden lettuce. They are all edible and the young tender leaves can be added fresh to salads or cooked as greens in one change of water. The older plants become too bitter for use.

Lactuca canadensis
Wild Lettuce

181

Asteraceae or Compositae
Aster Family

A biennial with a smooth stem and a thick, deep taproot. Pollination is by flying insects. Easily ruptured latex filled hairs on the bracts surrounding the flower head constitute a sticky trap that prevents ants from reaching the flowers. Ants consume pollen and nectar but have smooth hairless bodies and are poor pollinators.

The flower heads are about 6 mm. wide (¼ in.) and produce 20–25 seeds. Each of

these is flattened with a prominent ridge on each surface, black or mottled black and brown, and has a parachute of hairs at the tip of a slender beak. The airborne seeds are sometimes carried to cultivated fields, but the plant is usually not a serious weed pest. It can be controlled in the same manner as *L. biennis*.

This is the most variable species of the genus with 4 intergrading varieties based

*Flower heads small, with pale yellow rays,
numerous, in a branched terminal cluster;
leaves alternate, with a basal rosette, variable, unlobed to pinnately lobed; stems unbranched except at tip, with milky juice, to
3 m. high (10 ft.).*

July–September

mainly on leaf characteristics. The cultivated garden lettuce, a native of Europe and
Asia, belongs to this genus and is classified
as *L. sativa*, not illustrated. It sometimes escapes cultivation and occurs as a weed in
open spaces.

Wild lettuce is found along roadsides and
in fields, waste spaces, and open woods from
Quebec to Saskatchewan, south to Florida
and Texas.

⚒ See *L. biennis*.

℞ The milky juice was used by the Menominee Indian tribe as a remedy for poison
ivy, poison oak, and poison sumac.

Open Fields

Matricaria matricarioides
Pineapple-weed
182
Asteraceae or Compositae
Aster Family

*Flower heads yellow, cone-shaped, without
rays, several to numerous, at the ends of
stems and branches; leaves alternate, finely
1–3 pinnately divided, to 5 cm. long (2 in.);
stems smooth, branched, sometimes
sprawling, to 40 cm. high (16 in.); plant has
the odor of pineapple when bruised.*

June–October

An annual introduced and naturalized
from the Pacific Coast region. Each flower
head produces numerous smooth, 3–5
ribbed, olive or brown seeds which sometimes have 2 red lines. These are small
enough to be windblown even though they
do not possess the parachute of hairs that
characterizes many members of this family.
It sometimes infests cultivated fields as a
weed and can be controlled by early plowing, inter-row cultivation, and close mowing
before seeds are formed.

This genus consists of approximately 40
species which are natives mainly of the

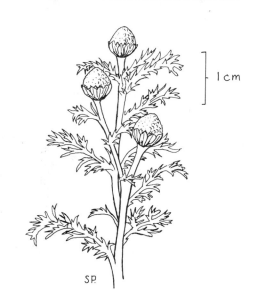

Mediterranean region, south Africa, and eastern Asia. Some species are grown as ornamentals, others are cultivated for matricaria oil, and many are strongly aromatic. Two of the 3 species that grow wild in northeastern United States were introduced from Europe. In older systems of classification, *M. matricarioides* may be listed as *M. suaveolens* or *Chamomilla suaveolens*.

Pineapple-weed is found along roadsides and in open fields, farmyards, and waste spaces in the west from Alaska to Baja, California, and in the east from Newfoundland to Manitoba, south to Pennsylvania and Missouri. It is spreading rapidly in the east and has been introduced into Europe.

The flower heads, fresh or dried, can be used to prepare an aromatic pineapple scented tea.

℞ The Blackfoot Indians of the Northwest made an insect repellent from the dried flower heads. A related species, *M. chamomilla*, German Chamomile, not illustrated, has been used as a breath sweetener, a hair conditioner for blond hair, and a remedy for stomach gas, earache, and other local aches and pains.

Picris hieracioides **Ox-tongue*** — 183 — ***Asteraceae or Compositae*** **Aster Family**

**Also Hawkweed Ox-tongue*

An annual or biennial with a thick root, introduced from Europe. It is self-incompatible, and most seeds are the result of cross-pollination by insects. Rarely seeds are developed without fertilization. Each flower head produces numerous horizontally ribbed, black or dark brown seeds. These have parachutes or hairs which drop off readily and may not function for dispersal.

Flower heads yellow, several, at the ends of stems and branches, on short stalks, composed of all ray flowers; basal leaves spatula-shaped with narrowly winged petioles, to 30 cm. long (1 ft.); stem leaves alternate, lanceolate, sessile and often clasping, irregularly toothed; all leaves covered with bristly hairs; stems bristly-hairy, branched, with milky juice, to 1 m. high (40 in.).

July–September

This is a genus of about 50 species native to Europe, Asia, and northern Africa. A related species, *P. echioides*, Bristly Ox-tongue, not illustrated, is very similar and is found over most of the range of *P. hieracioides*. These are the only two species of the genus in North America.

Ox-tongue is found in fields, along roadsides, and in waste places from New England to Michigan, south to New Jersey, Pennsylvania, and Illinois, and it may be spreading.

Rudbeckia hirta
Black-eyed Susan*

184

Asteraceae or Compositae
Aster Family

*Also Yellow Daisy, Coneflower

Ray flowers yellow, disk dome-shaped, purplish-brown; flower heads solitary, terminal, to 10 cm. wide (4 in.); leaves alternate, hairy, usually slightly toothed, lower ones long stalked and 3 ribbed; stems bristly-hairy, slender, with few or no branches, to 1 m. high (40 in.).

June–October

4 cm

S.P.

A biennial or short lived perennial with large, showy flower heads. It is self-incompatible and fertilization is the result of cross-pollination by bees and small butterflies. Only the disk flowers are fertile, and each flower head produces numerous black, oblong, finely nerved seeds without parachutes. The dead stems with seed bearing receptacles may persist into the winter months. It is the state flower of Maryland.

The most common variety of this species is a native of the western prairie that has spread eastward with the clearing of the land. It is sometimes grown in wild flower gardens. The seeds are eaten by the Common Goldfinch and other birds which probably contribute to its dispersal. It is usually not a serious weed in cropland because it yields readily to cultivation. Some botanists have recognized this variety as a distinct species, *R. serotina.*

Black-eyed Susan is found in meadows and fields, along roadsides, and in waste areas from Nova Scotia to Manitoba, south to Florida, Texas, and Colorado.

☠ *R. hirta* and *R. laciniata*, Green-headed Coneflower, are said to be responsible for poisoning hogs, sheep, cattle, and horses in the western states. In feeding experiments, symptoms consisting of lack of coordination, abdominal pain, and aimless wandering were produced but no deaths resulted.

Rudbeckia triloba
Thin-leaved Coneflower*

185

Asteraceae or Compositae
Aster Family

*Also Brown-eyed Susan

A biennial or short lived perennial with thin, bright green leaves. Although the flower heads may be visited by insects, most of the 4-angled seeds are produced without fertilization, and the new plants that grow from them are genetically identical to the parent plants. Unlike most members of this family, the seeds of the plants in this genus do not have parachutes of hairs.

There are at least 8 species of this genus in the Northeast and all are native to North

172

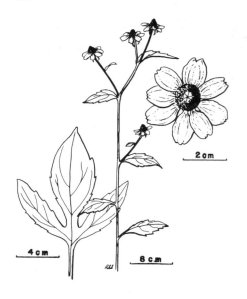

Ray flowers yellow, disk dome-shaped, purplish-brown; flower heads several to many, smaller than R. laciniata, ray flowers 6–12; basal leaves long petioled, broadly oval shaped, stem leaves alternate, usually some of the larger ones with 3 lobes; stems hairy, branched, to 1.5 m. high (5 ft.).

June–October

America. R. triloba is recognizable by the presence of some lower leaves that are palmately 3-lobed and ray flowers that are generally shorter, about 2.5 cm. long (1 in.) and wider than the other two species listed here. A variety with pinnately lobed leaves is sometimes observed from Kentucky southward.

Thin-leaved Coneflower is found in fields and waste areas, along roadsides, and in thickets and open woods from New York to Minnesota, south to Florida and Louisiana.

Solidago canadensis
Canada Goldenrod*

186

Asteraceae or Compositae
Aster Family

*Also Tall Goldenrod

Ray flowers yellow, flower heads numerous, in a dense branched terminal cluster; leaves alternate, crowded, narrowly lanceolate, sharply toothed, with 3 prominent veins, to 12.5 cm. long (5 in.); stems unbranched below flower cluster, upper part hairy, to 1.5 m. high (5 ft.).

July–October

duction. The longitudinally ribbed, slightly hairy seeds have tufts of hairs that facilitate dispersal by wind. The seeds and vegetation are eaten by several species of birds and small mammals.

It is host to the parasitic Goldenrod Gall Fly (Eurosta solidagins). The galls, which house the overwintering larval and pupal stages of the flies, are a characteristic feature of many stems, and when present they are a reliable guide to the identity of this species. The larvae are often a winter source of food for small woodpeckers and chickadees.

Most people who are allergic to ragweed pollen are also sensitive to goldenrod, but

A perennial with a creeping rhizome. Pollination is by insects, and since it is self-incompatible, pollen from another plant is necessary for fertilization and seed pro-

being insect-pollinated the latter produces relatively small amounts of airborne pollen. It is ordinarily not an important hayfever plant. However, since goldenrods are in flower at the time ragweed pollen is at a peak, they are often mistakenly credited as the causative agents in pollen allergies.

Canada Goldenrod is found in abandoned fields, meadows, fence rows, and waste places from Newfoundland to Manitoba, south to North Carolina and Tennessee. ℞ See *S. sempervirens*.

Sonchus oleraceus
Common Sow Thistle
187
Asteraceae or Compositae
Aster Family

Flowers heads yellow, to 2.5 cm. wide (1 in.), several in a branched terminal cluster; leaves spiny-toothed, lower ones pinnately lobed, crowded, upper ones smaller, unlobed, clasping the stem with sharp pointed, ear-like projections; stems smooth, branched, with milky juice, to 2 m. high (6½ ft.).

July–October

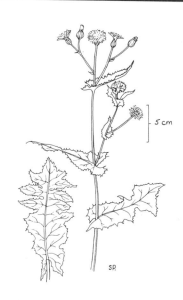

An annual with a short thick taproot, introduced from Europe. The pollen produced by this species is 90% abnormal and possibly sterile. This suggests that the seeds of *S. oleraceus*, like those of dandelion, may be produced without fertilization. The dark brown or black, longitudinally ridged seeds have parachutes of hairs that aid in dispersal by wind.

This is a genus of about 70 species native to northern Africa, western Asia, and Europe. Four species have been introduced into North America, all often troublesome weeds in cultivated fields. Since *S. oleraceus* is an annual, it is most easily controlled by mowing before seeds are formed and by clean cultivation.

A related species, *S. asper*, Spiny-leaved Sow Thistle, not illustrated, is similar but is more spiny, and the bases of the leaves clasping the stem have rounded ear-like projections. It has about the same geographic range as *S. oleraceus*, and it has been declared a noxious weed in the seed laws of 11 states.

Common Sow Thistle is found in open fields, gardens, and orchards, along roadsides, and in waste areas throughout the United States and southern Canada. It is most abundant along the Pacific Coast.

✖ The young leaves of all the species of this genus can be used in salads or cooked as greens after removing the prickly margins. They are much more widely used as food in Europe than they have ever been in North America.

℞ In herbology the milky juice is described as a bitter diuretic and a powerful cathartic.

174

Tanacetum vulgare
Tansy*

188

Asteraceae or Compositae
Aster Family

*Also Golden Buttons

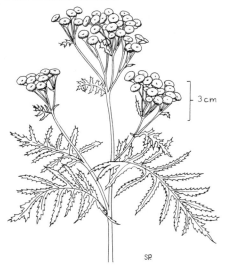

3 cm

SP

A perennial with a thick rhizome, introduced from Europe. Showy ray flowers in other members of this family usually function as attractants for insect pollinators. Their absence in this genus suggests a trend toward wind pollination. However, all members of the genus have spiny pollen grains and appear to be primarily insect pollinated. Each flower head produces numerous gray, glandular, 5-angled seed-like fruits.

This is a genus represented by only 2 species in the Northeast. *T. vulgare*, the most widespread, was introduced for medicinal uses and for seasoning. A single plant may produce up to 200 flower heads and the

Flower heads yellow, to 12 mm. wide (½ in.), button-like, ray flowers absent, in flat-topped terminal clusters; leaves alternate, numerous, 1 or 2 times pinnately divided, aromatic, to 20 cm. long (8 in.); stems erect, smooth, to 1 m. high (40 in.).

July–September

numerous seeds sometimes spread to cultivated fields. However, it yields readily to cultivation and is usually not a troublesome weed.

Tansy is found in open fields and neglected gardens, along roadsides, and in waste areas throughout most of the United States and southern Canada.

✖ The young leaves and flower heads can be used for seasoning as a substitute for sage. However, it has a powerful flavor and should be used according to taste. In some parts of the Northeast, tansy cheese is made by adding an extract of the plant to milk. *Caution:* This plant contains tanacetin, an oil that is toxic to humans if taken internally in large quantities.

℞ In herbology, it is described as a stimulating aromatic bitter. It has been used for intestinal worms and to promote the menstrual discharge. The powdered seeds and leaves have been used as an insecticide.

There are recorded instances of fatal poisoning in humans from overdoses of the medicinal extract. Abortion in cattle has been associated with the consumption of Tansy in Pennsylvania. As a result of its bitter taste, it is usually not eaten by animals.

Tragopogon pratensis
Yellow Goat's Beard*

189

Asteraceae or Compositae
Aster Family

*Also Meadow Salsify

A biennial with a thick taproot, introduced from Europe. On sunny days the flowers open in the early morning and are usually closed by noon. The pollen grains are very spiny, as is often the case with in-

sect-pollinated members of this family. Each flower head produces many longitudinally ribbed, long-beaked seed-like fruits, each with a parachute of white bristles.

This is a genus of 40–50 species native to central Asia and the Mediterranean region.

*Flower heads yellow, solitary, on long ter-
minal stalks, with ray flowers only, to 6.2
cm. wide (2½ in.), usually subtended by 8
green, pointed bracts; leaves alternate,
grasslike, clasping stem at base, to 30 cm.
long (12 in.); stems smooth, branched, with
milky juice, to 90 cm. high (3 ft.).*

June–October

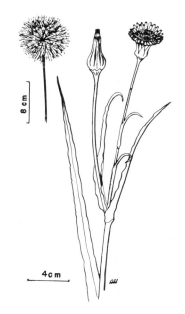

Three species have been introduced from
Europe into North America with *T. pratensis*
the most widespread. These species are very
similar and are known to hybridize occasionally when their ranges overlap. *T. porrifolius*, Salsify, not illustrated, differs from
the others in having purple flower heads.

Yellow Goat's Beard is found in neglected
meadows and old fields, along roadsides,
and in waste areas from New Brunswick and
Quebec to Manitoba, south to Georgia, Tennessee, and Missouri.

✂ The first-year root, collected before the
flowering stem starts to develop, can be prepared as a cooked vegetable and is said to
taste like oyster stew. The young stems and
basal leaves can be added to salads or cooked as greens. *T. porrifolius* is cultivated as a
root vegetable but has escaped and is often
found growing wild.

**Tussilago farfara
Coltsfoot***

190

**Asteraceae or Compositae
Aster Family**

*Also Coughwort, Ginger-root
*Ray flowers yellow, very numerous, appear-
ing before the basal leaves develop, flower
heads solitary; stem leaves alternate, scale-
like, basal leaves long-stalked, roundish,
heart-shaped at base, toothed and shallowly lobed, white-hairy on underside, to
20 cm. long (8 in.); flowering stem unbranched, to 45 cm. high (18 in.).*

March–June

A perennial with a horizontal creeping
rhizome, introduced from Europe. The ray
flowers are fertile but do not have stamens,
and the disk flowers have stamens but non-

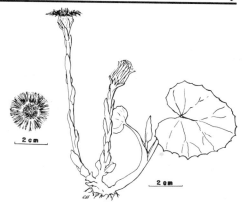

functional pistils. Pollen is shed in direct sunlight or in the absence of direct sunlight at a minimum air temperature of 10°C. (50°F.). Fertilization is usually the result of cross-pollination by early butterflies and bees. The seed-like fruits have tufts of hairs that aid in dispersal by wind. There are records of the slender, faintly ribbed seeds being carried up to 14 kilometers (about 9 miles) in storms. It may become over abundant and weedy on moist clay soil but can be controlled by improving drainage.

Coltsfoot is the only species of a genus that is native to Europe and Asia. It is well established in the Northeast and is found in moist heavy soil along roadsides and stream banks and in meadows and waste areas from Newfoundland and Quebec to Minnesota, south to West Virginia and Ohio. ✖ As emergency foods the young scaly stalks, before flowering, can be prepared like

asparagus, and the young basal leaves can be cooked as greens. These are reported to be passable but not necessarily good. The dried leaves can be used to make a tea that is said to be fragrant and pleasant tasting. A delicious candy can be made using sugar and the water from boiled fresh leaves.

℞ The genus name is derived from the Latin word for cough and refers to its supposed usefulness as a remedy for respiratory problems. In reference to this plant the 1851 edition of *The Dispensatory of the United States of America* states, "It is, however, demulcent and is sometimes used in chronic coughs, consumption, and other affections of the lungs." The leaves, flowers and rhizomes have all been used, but the dried leaves and flowers are the most popular. Hard candy made from an extract of the fresh leaves serves as delicious cough drops.

Vernonia noveboracensis
New York Ironweed

191

Asteraceae or Compositae
Aster Family

10 cm

S.P.

A perennial with a horizontal rhizome and dense fibrous roots. Fertilization is probably the result of both self-pollination and cross-pollination by insects. In a closely related species (see below), almost half of the seeds were shown to be the result of self-polli-

Flower heads purple, in an open terminal cluster, each head with 30–50 disk flowers only, bracts of flower heads with hair-like tips; leaves alternate, lanceolate, finely toothed, sessile, to 25 cm. long (10 in.); stems smooth or slightly hairy, usually unbranched below flower clusters, to 2 m. high (6½ ft.).

August–October

nation. The hairy-ribbed, seed-like fruits have parachutes of hairs that aid in dispersal by wind.

This is a very large genus that includes herbs, vines, and trees, mostly of tropical regions. The genus was named in honor of William Vernon, an English botanist who collected and studied the plants of colonial Maryland. The 7 species that are native to the Northeast are all herbaceous perennials that seem to hybridize freely when their ranges overlap.

A similar species, *V. altissima*, Tall Ironweed, not illustrated, occurs over much of the same area as *V. noveboracensis*. It is usually taller, to 3 m. (10 ft.), has 13–30 flowers per flower head, and the bracts subtending the flower heads do not have hair-like tips. Neither species is eaten by grazing animals, and they are thus good indicators of overgrazing.

New York Ironweed is found in meadows, moist bottomland, pastures, and waste areas from southern New England to New York, south to Georgia and Mississippi.

Xanthium strumarium
Cocklebur*

192

Asteraceae or Compositae
Aster Family

Also Clotbur, Sea Burweed

Flower heads unisexual, staminate heads in upper short terminal clusters, dropping after pollen has been released; pistillate heads in lower axillary clusters, each head with 2 flowers, developing into a single hard bur with hooked spines and 2 terminal, spine-like beaks; leaves alternate, long-stalked, usually heart-shaped at base, often with 3–5 shallow lobes, axils without spines, to 15 cm. long (6 in.); stems hairy, angled, often with purple spots, to 90 cm. high (3 ft.).

August–October

An annual with a thick, woody taproot, of North America and Europe. Pollen is spread by wind and it may be an important cause of hayfever in some areas. The burs readily become attached to passing animals for dispersal. Since they float, the burs may also spread in this manner from plants growing near water.

Each bur contains 2 seeds; one germinates the first year and the other remains dormant until the second year after it was produced. This plant is very sensitive to the length of the daylight period. It is a short-day plant, and, regardless of its size, flowering can be induced by exposing it to a single daylight period of less than 15.6 hours.

Gray's Manual of Botany recognizes 15 species of this genus, based mainly on variations in the mature bur, but the number is reduced to 2 in *The New Britton and Brown*

Illustrated Flora. X. strumarium, the most common species, is a very troublesome weed in, and probably a native of, the Mississippi Valley. It is also widespread in Europe. The other species, *X. spinosum*, Spiny Cocklebur, not illustrated, is similar but has only one or a few burs in the axils, and each axil has 3 yellow spines. It occurs throughout the Northeast but is much less abundant than *X. strumarium*; it also has a wide distribution in Europe.

Cocklebur is found along roadsides, in fields and flood plains, and on sandy shores and beaches throughout the United States and southern Canada.

℞ In herbology the plants of this genus are said to have tonic and alterative properties. They have been used to treat herpes and urinary disorders and to stop bleeding of cuts and scratches. *X. spinosum* has been used in the treatment of hydrophobia. *Note:*

The seeds and seedlings of these plants contain a poisonous substance, hydroquinone, which is toxic to all classes of livestock, especially swine. This poisonous property is believed to have evolved as a defense against insect feeding.

Campsis radicans **Trumpet Creeper***	**193**	***Bignoniaceae*** **Bignonia Family**

*Also Cow-itch

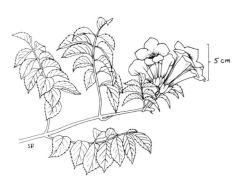

A perennial woody vine with vigorously spreading roots. The shape and color of the flower is typical of hummingbird pollinated plants. The elongate seed capsule splits along 2 seams to release numerous seeds,

Flowers orange-red, trumpet-shaped, 5-lobed, to 8 cm. long (3 in.), in terminal clusters; seedpods to 15 cm. long (6 in.), flattened, spindle-shaped; leaves opposite, pinnately compound, with 5–13, pointed, toothed leaflets; stems woody, climbing or trailing, to 12 m. long (40 ft.).

July–September

each of which has a broad membranous wing that aids in dispersal by wind.

There are only two species of this genus; the other is a native of Japan. Our species is cultivated as an ornamental but has escaped widely, often invading fields as a weed on the southern border of its range.

Trumpet Creeper is found along roadsides and in open fields and waste places on fences, stumps, posts, and old buildings from New England and New York south to Florida and Texas.

Echium vulgare **Viper's Bugloss***	**194**	***Boraginaceae*** **Borage Family**

*Also Blueweed, Blue Devil

A biennial with a fleshy taproot, introduced from Europe. Pollination is chiefly by honeybees and bumblebees. Although the flowers are odorless to most humans and to honeybees, they are detectable by scent to bumblebees. Honeybee visits to the flowers have been observed to be most concentrated at about 3:00 P.M. Each flower produces 4 small, gray-brown, wrinkled, 3-angled seeds.

This is a coarse and, when in flower, somewhat colorful weed. It is rare west of the Mississippi but sometimes invades cultivated fields in the East. Control methods include cutting the taproot below the leaf crown in autumn or spring, and mowing flowering stems before seeds are produced.

Viper's Bugloss is found in dry open fields, pastures, meadows, and waste areas, usually in limestone regions, from Nova Scotia,

Flowers bright blue, in dense coiled clusters that straighten with age, in upper axils; corolla funnel-shaped, with 5 lobes, stamens protruding, with red filaments; stem leaves alternate, basal leaves in a rosette, covered with bristly hairs; stems bristly hairy, to 90 cm. high (3 ft.).

June–September

Quebec, and Ontario south to Georgia and Alabama.

℞ The genus name is derived from a Greek word meaning "a viper." This refers either to the markings on the stems which were thought to look like snakes, or to the shapes of the seeds which supposedly resemble the heads of snakes. According to the "Doctrine of Signatures" this signified to the herbalists of yore that the plant should be used as a remedy for snakebites.

The dried leaves of the rosette have been used as a diuretic and a cure for headache and to soothe inflamed or irritated mucous membranes.

The plants of this species contain pyrrolizidine, a poisonous alkaloid that has caused death in livestock by its effect on the liver. Contact with the bristly hairs on the leaves and stems may cause itching and inflammation of the skin in sensitive individuals.

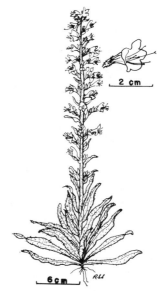

Lithospermum officinale **European Gromwell**   **195**   ***Boraginaceae*** **Borage Family**

A perennial with a semi-woody base and a large root, introduced from Europe. Each flower produces 1 to 4 teardrop-shaped, smooth, glossy white seeds. The genus name is derived from Greek words that mean "stone seed" and refers to the hard seed coats. However, in feeding experiments, the seeds of this species failed to germinate after passing through the digestive tracts of swine and sheep.

There are 5 native and 2 introduced species of this genus in the Northeast. A native of Europe and Asia, *L. arvense*, Corn Gromwell, not illustrated, is probably the most

180

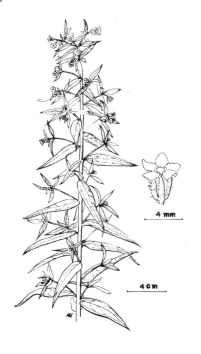

Flowers yellowish white, in axils of upper leaves or on axillary branches, solitary or in few-flowered clusters; corolla funnel-shaped, with 5 lobes; calyx deeply 5-parted, hairy, approximately same length as corolla tube; leaves alternate, lanceolate, numerous, prominently veined; stems erect, sturdy, freely branched, hairy, to 90 cm. high (3 ft.).

June–August

widespread member of the genus in North America. It is an annual with white flowers, coarsely pitted gray seeds, and leaves that are not conspicuously veined. Growing in the same types of habitat as *L. officinale*, it occurs throughout most of the United States and southern Canada.

European Gromwell is found in abandoned fields, gravelly pastures, and waste spaces, especially in limestone areas, from Quebec to Minnesota south to New Jersey and Illinois.

Symphytum officinale
Comfrey*

196

Boraginaceae
Borage Family

Also Healing Herb, Bruisewort

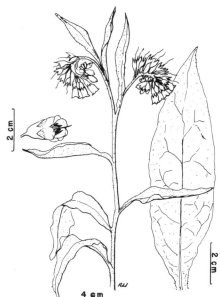

Flowers whitish, yellowish or pale blue, tubular, numerous, in curled axillary clusters; leaves alternate, hairy, extending down the stem from the point of attachment as 2 wings, the lower ones to 20 cm. long (8 in.); stems hairy, branched, to 90 cm. high (3 ft.).

June–September

A perennial with a thick mucilaginous root, introduced from Europe. Pollination is by insects, probably bees that must squeeze past the anthers, thus becoming dusted with pollen, to reach the nectar at the base of the flowering tube. Bumblebees sometimes by-pass this method by biting a hole in the base of the tube. Excessive nectar thievery of this sort in self-incompatible species could conceivably lead to their extinction, but *S. officinale* does not appear to be threatened. Each flower produces 4 nutlets which are glossy black and have a toothed fringe around the base.

This is a genus of about 20 species native to Europe and Asia. Three species have been introduced into eastern North America with *S. officinale* being the most widespread. A related species, *S. asperum*, Prickly Comfrey, not illustrated, is similar but the leaf stalks do not extend down the stem as wings, and the stem is prickly. It occurs in waste areas across southern Canada and northern United States.

Comfrey is found in moist meadows, roadside ditches, and waste places from Newfoundland to Ontario, south to Georgia and Louisiana.

✕ The young leaves can be prepared like spinach. Older leaves must be boiled in several changes of water to reduce the bitterness. Some reports praise this as a wild food but others are less enthusiastic. The dried leaves can be used to make a tea, and the roasted ground roots can be used as a coffee substitute.

℞ This plant was introduced into North America as a cultivated medicinal herb. It was early believed to be effective in the treatment of wounds but this was replaced by its use to soothe inflamed mucous membranes. The dried root is the part used medicinally. *The Dispensatory of the United States of America* (1851) states, "It is a very common ingredient in domestic cough mixtures, employed in chronic catarrh, consumption, and other pectoral affections." A poultice made by pulping the whole plant has been used for sprains, swellings, and bruises.

Alliaria officinalis
Garlic Mustard

197

Brassicaceae or Cruciferae
Mustard Family

Flowers white, to 8 mm. wide (⅓ in.), in dense terminal clusters, elongating in fruit; sepals 4, petals 4, stamens 6, 2 shorter than others; seedpod 4-angled in cross-section, to 6 cm. long (2½ in.), standing out from stem at maturity; leaves alternate, triangular, heart-shaped at base, toothed, with the odor of garlic, to 10 cm. long (4 in.); stems smooth, usually unbranched, to 1 m. high (40 in.).

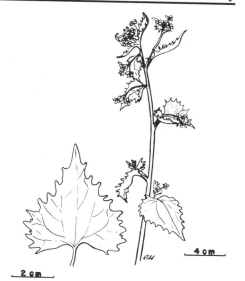

2 cm

4 cm

May–June

A biennial with a rosette of kidney-shaped basal leaves, introduced from Europe. The seedpod splits longitudinally into 2 sections, releasing several dark brown or black, ridged seeds. The genus name is derived from the Latin word for onion or garlic and refers to the odor of the leaves. *A. officinalis* is the only species of the genus in North America.

Garlic Mustard is found along roadsides and in neglected gardens, open fields, and waste areas, often near human habitations, from Quebec and Ontario south to Virginia and Kansas.

✕ The leaves may be used for seasoning in salads and to give flavor to milder cooked greens. An English publication of the early 1800's reported that the poor people of England ate the leaves with bread and salt fish. Since it was used as the solitary dressing for their meal it came to be known as Sauce

182

Alone. This plant sometimes occurs as an unwanted weed in meadows and pastures where, if eaten by cattle, it gives an undesirable flavor to milk and milk products.

Alyssum alyssoides
Madwort

198

2 cm

Flowers pale yellow to white, tiny, in terminal clusters that elongate in fruit; sepals 4, petals 4, slightly longer than sepals; stamens 6, 2 shorter than others; leaves alternate, covered with star-shaped hairs, lanceolate, with widest part near tip, to 16 mm. long (⅔ in.); stems erect, unbranched, covered with star-shaped hairs, to 30 cm. high (12 in.).

May–June

zontally into 2 sections, releasing 4 tiny, light brown seeds with winged margins that aid in dispersal by wind.

This is a genus of about 100 species, nearly all native to the Mediterranean region. The genus name is derived from Greek words that mean "without madness" and refers to a supposed use as a cure for hydrophobia, thus the common name. A related species *A. saxtile*, Golden Tuft, not illustrated, has bright yellow flowers and is grown in gardens as a border plant, but it is beginning to spread from cultivation.

An annual often with several stems that curve upward from a common base, introduced from Europe. The rows of lens-shaped fruits are more conspicuous than the small flowers. The circular seedpod splits hori-

Madwort is found in grassland, open fields, roadsides, and waste spaces from Quebec to British Columbia, south to West Virginia and Indiana, and on the Pacific coast to California.

Arabis lyrata
Lyre-leaved Rock Cress

199

A biennial or short-lived perennial with a basal rosette of leaves. The seedpod splits into 2 sections, releasing numerous light brown, finely pitted, egg-shaped seeds. The genus name is derived from the country Arabia, but the reason for its application to this genus is unclear. It includes more than

100 species widely distributed in the Northern Hemisphere.

There are about 14 species of this genus native to the Northeast, with a few, including *A. lyrata*, that occur also in Europe or Asia. A related species, *A. hirsuta*, Hairy Rock Cress, not illustrated, has hairy basal leaves

*Flowers white, about 6 mm. wide (¼ in.), in
terminal clusters; petals 4, stamens 6; fruit
a slender pod, curving upward, to 4 cm.
long (1½ in.); basal leaves usually pinnately
lobed, to 4 cm. long (1½ in.), stem leaves
alternate, very narrow, tapering to base;
stems hairy in lower part, usually branching from base, to 30 cm. high (12 in.).*

May–July

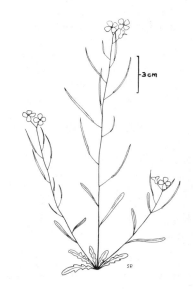

and seedpods that hug the stem. It occurs
in dry limestone woods from Quebec to
Alaska, south to Georgia, New Mexico, and
California.

Lyre-leaved Rock Cress is found in sandy,
gravelly, and rocky soils of fields and open
woods from New England and Ontario to
Minnesota, south to Virginia and Missouri
and in the mountains to Georgia.

✖ The rosette leaves, in early spring before
the stems appear, can be used on sandwiches and salads or cooked as greens. Later
in the summer they become bitter and must
be cooked in several changes of water.

Barbarea vulgaris
Common Winter Cress[*]
200
Brassicaceae or Cruciferae
Mustard Family

---

*Also Yellow Rocket
*Flowers yellow, to 8 mm. wide (⅓ in.), in
dense, cylindrical, terminal clusters; sepals
4, petals 4, stamens 6, 2 short; seedpods to
3 cm. long (1.2 in.); basal leaves pinnately
divided, with 1–4 pairs of lobes, long
stalked, to 12.5 cm. long (5 in.); stem leaves
alternate, upper ones smaller, clasping the
stem; stems branched, to 60 cm. high (2
ft.).*

April–August

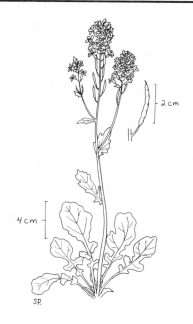

A biennial or short-lived perennial with a
sturdy taproot, introduced from Europe. It
has a basal rosette of leaves that remains
green into the winter months and grows
vigorously during warm periods. The flowers produce both pollen and nectar and are
regularly visited by bees and syrphus flies.
The seedpod splits from the base into 2 sections, releasing several glossy, brown or
black, pitted seeds.

184

The seeds may survive in the soil for several years. In germination studies they showed a viability rate of 33% after 5 years of dry storage. In feeding studies they remained germinable after passing through the intestinal tracts of horses, cows, swine, and sheep. They are eaten by several species of birds including the Mourning Dove and Pine Grosbeak.

A related species, *B. verna*, Early Winter Cress, not illustrated, is similar but has seedpods to 7 cm. long (3 in.) and basal leaves with 4–10 pairs of lobes. It is found in habitats similar to those of *B. vulgaris* and occurs throughout the Northeast.

Common Winter Cress is found in gardens, cultivated fields, wet meadows, and along roadsides from Newfoundland to Ontario, south to South Carolina and Arkansas. ✄ All the species of this genus are edible. The young leaves can be used in salads or cooked as greens during late fall, winter, and early spring. Depending on individual taste, they may be cooked in one or more changes of water to dispel some of the bitterness. During summer the leaves are too bitter for most tastes, but the flower buds can be prepared like broccoli by cooking in one or more changes of water.

Berteroa incana
Hoary Alyssum
201
Brassicaceae or Cruciferae
Mustard Family

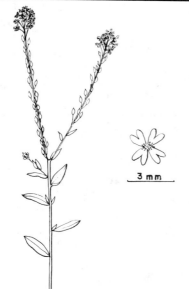

3 mm

4 cm

Flowers white, tiny, in dense clusters at the ends of stems and branches; petals 4, deeply cleft, stamens 6, 2 short, 4 long; leaves alternate, lanceolate, covered with fine grayish-white hairs, to 3.7 cm. long (1½ in.); stems erect, branched, covered with fine grayish-white hairs, to 70 cm. high (28 in.).

June–September

moderate breeze. The plant sometimes invades cultivated fields as a weed but it can be controlled by clean cultivation or mowing before seeds are formed.

This is a genus of 7 species native to Europe and Asia. It was named in honor of C. G. Bertero, an early nineteenth-century Italian botanist. *B. incana* is the only widespread species of the genus in North America. A related species, *B. mutabilis*, not illustrated, is very similar but has a seedpod that is wider and more flattened. It has been reported from Massachusetts and Kansas.

Hoary Alyssum is found on dry soils of roadsides, pastures, and waste areas from Nova Scotia to Minnesota, south to West Virginia and Missouri. It has also become established in the Pacific coast states.

An annual with a substantial taproot, introduced from Europe. The stalked, elliptical seedpod splits into 2 sections, releasing a few brown, flattened, circular seeds. These are small enough to become airborne in a

Open Fields

Brassica kaber
Charlock*

202

Brassicaceae or Cruciferae
Mustard Family

*Also Wild Mustard
Flowers yellow, to 12 mm. wide (½ in.), in
clusters at the ends of stems and
branches; petals 4, sepals 4, stamens 6;
fruit a slender, beaked pod, to 2.5 cm. long
(1 in.), constricted between seeds, beak
about ½ length of pod; leaves alternate,
lower ones often lobed, upper ones
smaller, toothed; stems hairy, at least in
lower part, branched toward top, to 75 cm.
high (30 in.).*

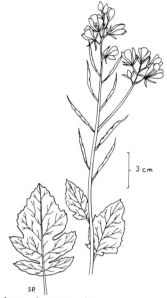

3 cm

SP.

May–July

An annual or winter annual with a well
developed taproot, introduced from Europe. The seed capsule splits from the base
into 2 sections, releasing several dark brown
or black, pitted seeds. The seeds of all the
plants in this genus are relished by several
species of birds including the Mourning
Dove, Ring-necked Pheasant, and Common
House Finch.

This plant may have been introduced accidentally as a contaminant in commercial
seeds. It is sometimes a serious weed pest
in grain fields, especially oats, where it is
an aggressive competitor. One plant of *B.
kaber* requires twice as much nitrogen and
phosphorus and four times as much potassium and water as one healthy oat plant.
It has been declared a noxious weed in the
seed laws of 19 states.

Charlock is found in open fields, cultivated areas, roadsides, and waste spaces
throughout the United States and southern
Canada.

The very young tender leaves of all the
species of this genus can be used in salads
or cooked as greens. As they get older they
become too bitter to use as food. The clusters of unopened flower buds can be prepared like broccoli. The seeds can be used
whole as pickling spice. Those of *B. nigra*
are ground and made into the familiar prepared mustard.

℞ See *B. nigra*.

Brassica nigra
Black Mustard

203

Brassicaceae or Cruciferae
Mustard Family

An annual or winter annual with a thick
taproot, introduced from Europe. The flowers are visited by small bees and flies that
probably effect pollination. The pistil exceeds the stamens in length, decreasing the
possibility of self-pollination. The seedpod
splits longitudinally into 2 sections, releasing several brown to black pitted seeds.

There are six species of this genus in eastern North America, all introduced from Europe or Asia. The genus includes such
common garden vegetables as cabbage, cau-
liflower, broccoli, radishes, and brussel
sprouts. The seeds of some species are capable of surviving long periods of storage.
B. nigra seeds were buried in bottles of sand
18 inches below the surface for over 50 years
and retained their germinability.

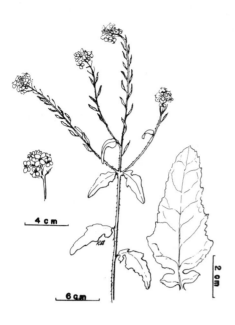

Flowers yellow, about 8 mm. wide (⅓ in.), in terminal clusters; petals 4, sepals 4, stamens 6; fruit a slender beaked pod, hugging the stem, to 2 cm. long (¾ in.); leaves alternate, lower ones pinnately lobed, upper ones lanceolate; stems green, hairy, usually branched, to 1.5 m. high (5 ft.).

June–October

Black Mustard is found in abandoned fields, pastures, roadside ditches, and waste areas throughout most of the United States and southern Canada.

✕ See *B. kaber*.

℞ The crushed seeds of species in this genus, applied as a chest plaster, are powerful skin irritants and have been widely used as a home remedy for pneumonia, bronchitis, and other respiratory ailments. For those who were reared in rural Appalachia and have experienced this remedy, the memory of the treatment has definitely outlived the memory of the ailment.

Capsella bursa-pastoris
Shepherd's Purse*

204

Brassicaceae or Cruciferae
Mustard Family

*Also Pick-pocket

Flowers white, tiny, numerous, in long clusters at the ends of stems and branches; petals 4, longer than sepals; fruits triangular with rounded margins; stem leaves alternate, arrow-shaped, clasping stem at base, basal leaves in a rosette, irregularly pinnately lobed; stems somewhat hairy, with a few branches toward tip, to 60 cm. high (2 ft.).

March–December

rope. Some forms of this species are self-pollinated, other forms are self-incompatible and are cross-pollinated by insects. Each cell of the 2-celled seedpod contains several tiny orange-brown seeds attached to the partition between the cells. At maturity the 2 halves of the seedpod are shed and the partition with the attached seeds becomes windborne, spreading the seeds as it is blown about.

Although flowering is mainly in the spring, flowers on this plant may appear almost any

An annual or winter annual with a slender branched taproot, introduced from Eu-

month of the year. The terminal rows of rounded triangular seeds, however, are more conspicuous than the flowers. The whole plant is eaten by chickens and small mammals, and the fruits are eaten by several species of birds. They may comprise up to 25% of the diet of the Common Goldfinch.

A related species, *C. rubella*, not illustrated, is very similar but has petals and seedpods with concave margins. A native of Europe, it occurs in the same types of habitats as *C. bursa-pastoris* but is more common in the southeastern states.

Shepherd's Purse is found in cultivated and abandoned fields and waste spaces and along roadsides throughout the United States and southern Canada.

✘ The young leaves have a mild cabbage flavor and may be cooked as greens or used fresh in salads. The dried seedpods can be used as a peppery seasoning.

℞ In herbology it is decribed as having diuretic and styptic properties. A tea made from the dried plant has been used in the treatment of both internal and external bleeding. During World War I the Germans are said to have used it for this purpose when other materials could not be imported.

Hesperis matronalis
Dame's Rocket*

205

Brassicaceae or Cruciferae
Mustard Family

*Also Dame's Violet

Flowers pink, purple or white, fragrant, in terminal clusters; petals 4, stamens and style enclosed in the corolla tube, usually not visible; seedpods to 10 cm. long (4 in.); leaves alternate, pointed, sessile or with short petioles, sharply toothed; stems erect, sometimes branched in upper part, to 90 cm. high (3 ft.).

May–August

A biennial or perennial introduced from Europe. The fragrance of the flowers are at a maximum in the evening hours. This is apparently an adaptation to attract evening or nocturnal insect pollinators. The long seedpod splits along each side to release numerous oblong, pitted, brown seeds.

This species is a native of the Mediterranean region and central Asia. It was introduced into North America as an ornamental garden plant and is commonly observed around old homesites. It has escaped widely and is very successful in competition with native species. The flower cluster resembles a phlox except that phlox has 5 rather than 4 petals.

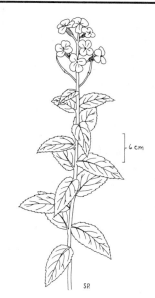

Dame's Rocket is found along roadsides and in open fields, vacant lots, fence rows, and waste areas from Newfoundland to Ontario, south to Georgia.

188

Lepidium densiflorum
Dense-flowered Peppergrass
206
Brassicaceae or Cruciferae
Mustard Family

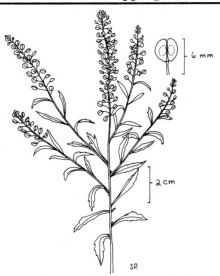

6 mm

2 cm

SP.

An annual or winter annual with a long slender taproot. A rosette of basal leaves may remain green throughout the winter months. The 2 stamens and the pistil are about the same length, facilitating self-pollination. At maturity the finely wrinkled light brown seeds have a narrow wing that may aid in dispersal by wind. They are also covered by a substance that becomes very sticky when wet. This may serve as a dispersal mechanism by causing them to adhere to dew or rain-dampened animal costs. By another interpretation, the sticky covering anchors the seed to the soil and aids in germination.

There are about 7 species of this genus

Petals white or none, if present, shorter than sepals; flowers and fruits numerous, in elongate clusters at the ends of stems and branches; fruits flattened, roundish, with a notch at the top, 2-seeded; leaves alternate, narrow, toothed, tapering at both ends; stems branched, to 50 cm. high (20 in.).

May–August

in the Northeast, 5 of which have been introduced from Europe and Asia. *L. densiflorum* and several other species sometimes invade cultivated areas and become trublesome weeds. Control measures include clean cultivation from early in the season until late autumn.

A related species, *L. virginicum*, Poorman's Pepper, not illustrated, is similar but has petals equal to or longer than the sepals. Sometimes petals are absent in both *L. densiflorum* and *L. virginicum* and the only way they can be distinguished is by microscopic characteristics of their seeds. *L. virginicum* has about the same geographic range as *L. densiflorum*.

Dense-flowered Peppergrass is found along roadsides and in neglected fields, vacant lots, and meadows from Quebec to Montana and Oregon, south to Florida, Texas, and California.

✖ The young leaves can be cooked as greens or added raw to salads. The seedpods are peppery to the taste and can be used for seasoning and as a dressing for meats.

Raphanus raphanistrum
Wild Radish*
207
Brassicaceae or Cruciferae
Mustard Family

*Also Jointed Charlock

An annual with a thick taproot, introduced from Europe. It is mostly self-incompatible and cross-pollinated by insects, but occasionally self-pollination occurs. The slender seed capsule contains up to 10 seeds and at maturity is strongly constricted between the seeds. It eventually breaks at the

constrictions, with each segment including one or two seeds.

This is a genus of 8–10 species native to Europe and eastern Asia, including the widely cultivated garden radish, *R. sativa*, not illustrated. Some botanists consider the garden radish to be a variety of *R. raphanistrum*. The latter is a common weed, often

Flowers yellow, becoming pale or white with age, in terminal clusters that lengthen as the season progresses; petals 4, with purple veins; fruit a slender beaked pod with constrictions between the seeds; leaves alternate, pinnately lobed, lower ones to 20 cm. long (8 in.), upper ones smaller; stems hairy, thick, to 80 cm. high (32 in.).

June–August

infesting cultivated fields where clean cultivation and frequent disking and harrowing are required for its eradication.

Wild Radish is found in fields and gardens, along roadsides, and in waste places from Newfoundland to British Columbia, south to Virginia, Indiana, Texas, and California.

℞ The fresh juice of the root of the cultivated radish is said to be an old European home remedy for coughs, bronchitis, intestinal gas, diarrhea, and insomnia.

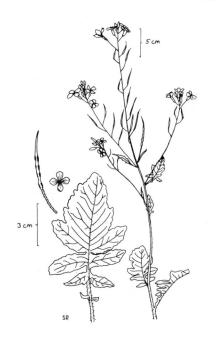

Open Fields

Sisymbrium altissimum
Tumble Mustard

208

Brassicaceae or Cruciferae
Mustard Family

Flowers pale yellow, to 1.5 cm. wide (¾ in.), in numerous long clusters at the ends of stems and branches; petals 4, longer than sepals; fruit a many-seeded pod, to 10 cm. long (4 in.); leaves alternate, petioled, deeply pinnately divided, with very slender segments; stems erect, usually bushy branched, to 1 m. high (40 in.).

May–August

An annual or winter annual with a deep taproot, introduced from Europe. If the plant is left to mature it breaks loose at the base and rolls as a tumbleweed. The seedpods split from the base and the seeds are dispersed as the plant is blown about by the wind.

This is a genus represented by 4 species in the Northeast, all natives of Europe and Asia. The most common are *S. altissimum* and related *S. officinale*, Hedge Mustard, not illustrated. The latter has shorter seedpods, less than 2.5 cm. long (1 in.), and leaves much

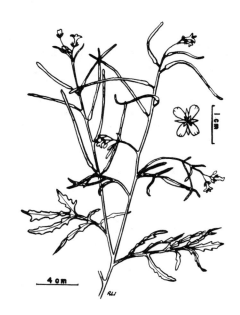

less dissected. Both species are common weeds in cultivated fields and require clean cultivation for eradication.

Tumble Mustard is found in grain fields, gardens, and waste areas from Quebec to British Columbia, south to Florida, Texas, and California.

✴ The young spring leaves of all the plants in this genus can be cooked as greens, and the seeds are said to be fairly good seasoning. ℞ Hedge Mustard is a popular European springtime home remedy for coughs, chest congestion, and laryngitis.

Cassia fasciculata
Partridge Pea*

209

Caesalpiniaceae
Caesalpinia Family

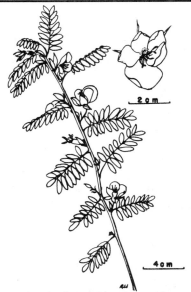

An annual of variable size with several recognized varieties. Although some seeds are produced by self-pollination, most are the products of cross-pollination by bees. The purple anthers apparently contain food pollen, and in the process of acquiring it the bees become dusted with pollination pollen from the yellow anthers. The dark brown, flattish, pitted seeds provide up to 10% of the diet of the Bobwhite Quail in some areas.

A related species *C. nictitans*, Wild Sensitive Plant, not illustrated, is similar but has only 5 stamens and very small flowers. It occurs in some of the same habitats of *C. fasciculata* with about the same geographic range.

*Also Prairie Senna, Golden Cassia, Locust-weed

Flowers yellow, showy, to 3.8 cm. wide (1½ in.), in axils of leaves, 5 unequal petals; stamens 10, 4 with yellow anthers, 6 with purple anthers; seedpod pea-like, to 7 cm. long (3 in.); leaves alternate, pinnately compound, sometimes folding when touched or disturbed; stems erect or tending to droop, hairy, to 90 cm. high (3 ft.).

July–September

Partridge Pea is most commonly found in dry sandy soil of prairies, open fields, and roadsides from southern New England to Minnesota, south to Florida and Texas.

✴ Some members of the genus have records of use for food. The seeds of *C. occidentalis*, Coffee Senna, not illustrated, an introduced species found mostly south of Virginia and Indiana, have been roasted and used as a substitute for coffee.

C. tora, Sicklepod, not illustrated, a native species occurring from Virginia and Kentucky to Florida and Texas, is said to be edible if the young shoots and leaves are cooked in two or three changes of water. A disagreeable odor in the fresh plants is somewhat reduced when they are cooked. ☠ Referring to *C. marilandica*, not illustrated, a native species occurring from Pennsylvania and Iowa to Florida and Texas, *The Dispensatory of the United States of America* (1851) states, "American senna is an efficient and safe cathartic, closely resembling the imported senna in its action, and capable of being substituted for it in all cases in which the latter is employed. . . . It is habitually used by many practitioners in the country."

The seeds and leaves of *C. occidentalis* and *C. tora* have been used to treat ringworm, eczema, and other skin diseases.

The plants of this genus are toxic to livestock but no deaths have been reported in the United States. Fatalities to sheep, cattle, and hogs have been reported in other countries, and human deaths have resulted from overdoses of home remedies using *Cassia*. Most species contain compounds that are toxic in overdose, and some are known to contain hydrocyanic acid.

Campanula rapunculoides
Creeping Bellflower
210
Campanulaceae
Harebell Family

Flowers blue, to 3.7 cm. long (1½ in.), in a long, often one-sided, terminal cluster; corolla bell-shaped, nodding, with 5 sharp lobes, style white and protruding; leaves alternate, lower ones heart-shaped, long stalked, upper ones lanceolate, short stalked or sessile, to 10 cm. long (4 in.); stems erect, unbranched, to 90 cm. high (3 ft.).

July–September

A perennial with a slender rhizome, often growing in clumps, introduced from Europe. Bumblebees and honeybees are frequent visitors to the flowers. The seed capsule opens by 3 basal pores from which the numerous finely ridged, slightly winged, light brown seeds may be thrown as the plant sways in the wind.

This species is the most common of the introduced bellflowers. When it grows in a habitat where one side receives more light, as in a hedgerow or near a building, there is a pronounced tendency for the flowers to be clustered on the side of the stem toward the light. However, when it grows in an open field the one-sided growth habit is scarcely noticeable.

Creeping Bellflower was introduced as an

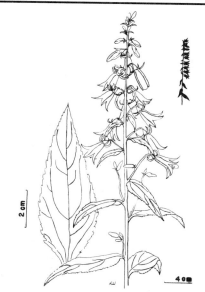

2 cm

4 cm

ornamental but is more often found along roadsides and in open fields and waste areas from Nova Scotia to Minnesota, south to West Virginia, Ohio, and Illinois.

✕ As the season advances, after flowering, the rhizome produces fleshy branches that are tender and may be added to salads or cooked as a parsnip-like vegetable.

192

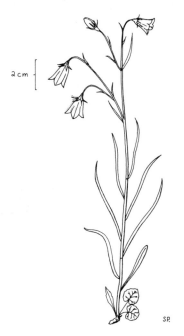

2 cm

S.P.

A perennial of North America, Europe, and Asia. Fertilization is the result of cross-pollination mainly by bumblebees, but the flowers are visited also by other smaller bees and bee-like flies. The top-shaped seed cap-

*Also Bluebell
Flowers violet-blue, bell-shaped, nodding, to 2.5 cm. long (1 in.), occurring singly on slender stalks at ends of stems and branches; basal leaves roundish, appearing early and usually withering by the time the flowers appear; stem leaves alternate, very narrow, to 10 cm. long (4 in.), upper ones smaller; stems slender, smooth, to 50 cm. high (20 in.).

June–September

sule opens from the base, releasing numerous brown, finely grooved, slightly winged seeds. These are small enough to become windborne in a slight to moderate breeze.

This is a highly variable species with regard to branching, number and size and shape of flowers, and leaf characteristics, depending on environmental conditions. A semi-double form occurs in which the flower has many fine divisions that extend almost to its base. Above the timberline in the mountains of New England a dwarf form occurs that bears only one flower and usually does not exceed 15 cm. (6 in.) in height.

Harebell is found in meadows and dry open woods and on rocky banks and shores throughout Canada and southward to New Jersey, West Virginia, Illinois, and Texas.

An annual with milky sap reproducing by seeds. It has two kinds of flowers: those in the lower axils which never open and are self-pollinated, and those in the upper axils that open wide and are cross-pollinated by bees and bee-like flies. The oval seed cap-

sule splits into 3 sections to about the middle, releasing numerous lens-shaped, reddish-brown, shiny seeds that are small enough to become windborne in a moderate breeze.

The genus name is derived from a Latin

Flowers blue, single, in axils of leaves, to 2 cm. wide (¾ in.), deeply 5-lobed, stamens 5, stigma 3-lobed; leaves alternate, almost round, clasping the stem, numerous, to 2.5 cm. long (1 in.); stems erect, usually un-branched, angular, hairy, to 75 cm. high (30 in.).

June–August

word that means "a mirror" and may refer to the shiny mirror-like seeds. A related species, *S. biflora*, not illustrated, is very similar but has flowers only at the tip of the stem and leaves that are not clasping. It occurs in the same type of habitats as *S. perfoliata* from Virginia and Kentucky south to Florida and Texas. These are the most widespread members of the genus in the Northeast.

Venus' Looking-glass is found in old fields, pastures, waste areas, and dry open woods from Maine and Ontario to British Columbia, south to Florida, Texas, and Oregon.

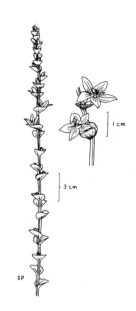

Open Fields

Dianthus armeria
Deptford Pink

213

Caryophyllaceae
Pink Family

Flowers pink, in clusters at the tips of the stems and axillary branches; petals 5, toothed; leaves opposite, stiff, narrow and pointed, tending to hug the stem; stems hairy, to 60 cm. high (2 ft.).

June–September

A biennial weed introduced from Europe. The seed capsule opens by 4 teeth to release numerous curved, pointed, roughened, dull black seeds.

Three other species of this genus occur in the Northeast. All were introduced from Europe and are well known as ornamentals but have escaped cultivation to become established locally. The only species native to North America is *D. repens*, not illustrated, normally found in Alaska.

The genus name is Greek and means "God's flower". This name was given by Theophrastus some 2300 years ago to the Carnation, *D. caryophyllus*, not illustrated, the best known species. The carnation has numerous varieties and is one of the oldest

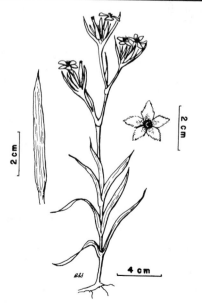

cultivated ornamentals. It is the state flower of Ohio.

Deptford Pink is found along roadsides and in open fields, meadows, and waste spaces from Quebec to British Columbia, south to Georgia and Missouri.

Lychnis alba
White Campion*

214

Caryophyllaceae
Pink Family

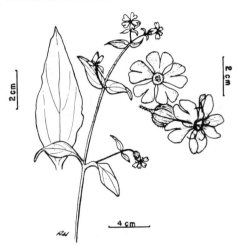

A biennial or short-lived perennial with a substantial taproot, introduced from Europe. Since the male and female flowers are on different plants, cross-pollination is maximized. The flowers open in the evening, and it has been described as a typical, sweet smelling, moth-pollinated species. However, the flowers often remain open during the day and may also be visited by other insects. The seed capsule opens from the top by 10 teeth to release numerous

*Also White Cockle, Evening Lychnis
Flowers white, unisexual, petals 5, prominently notched, styles 5; calyx inflated at maturity; leaves opposite, entire, elliptical in outline; stems sticky-hairy, branched, erect or sprawling, to 1 m. high (40 in.).

June–September

kidney-shaped gray seeds covered by concentric rows of small knobs.

It sometimes invades fields of legumes and grains and becomes a troublesome weed. One control measure is to plant a dense cover crop such as clover or alfalfa. Fields that are badly infested can be plowed and planted with a crop that requires several cultivations. It is more of a problem as a weed in cultivated fields from Iowa to Pennsylvania northward.

A related species, *L. dioica*, Red Campion, not illustrated, is similar but has red, odorless flowers that typically are closed at night and open when the sun rises. *L. alba* is also similar to *Silene noctiflora*, Night-flowering Catchfly, not illustrated, but the latter has three styles rather than five.

White Campion is found in well drained soil along roadsides and field borders and in gardens, meadows, and waste areas from Nova Scotia and Quebec to British Columbia, south to Georgia and California.

Lychnis flos-cuculi
Ragged Robin*

215

Caryophyllaceae
Pink Family

*Also Cuckoo-flower

A perennial with a short rootstock, introduced from Europe. Cross-pollination is effected mainly by bumblebees and butterflies. Since the anthers and stigmas mature at different times, self-pollination is nearly im-

possible. The seed capsule opens from the top by 5 teeth to release numerous, kidney-shaped, purplish-gray seeds covered by concentric rows of small wart-like projections.

There are about 40 species of this genus,

Flowers pink, numerous, in terminal branched clusters; each petal dissected into 2 large and 2 small lobes, giving the flower a ragged appearance; stem leaves opposite, 4–5 pairs, narrowly lance-shaped, basal leaves spatula-shaped; stems slender, erect, upper part slightly sticky, to 60 cm. high (2 ft.).

May–July

mostly in the mountainous and northern regions of the Northern Hemisphere. Of the 8 species that occur in North America, 6 are natives of Europe and Asia. The genus name is derived from a Greek word meaning "lamp" and refers to the brilliant flowers of some of the European species.

    L. flos-cuculi seems to be spreading in the Northeast. It may become established in cultivated areas but does not appear to be a serious weed pest. It is unable to compete with close growing cover crops, and it yields readily to cultivation. If undisturbed this species often grows in dense colonies, and when in flower the masses of pink provide a spectacular display. The dissected petals give the flower a tattered appearance that makes it an easy plant to identify.

4 cm   2 cm

Ragged Robin is found in moist meadows, swales, pastures, and waste areas from Nova Scotia to Quebec, south to Pennsylvania and Iowa.

Saponaria officinalis
Bouncing Bet*
216
Caryophyllaceae
Pink Family

---

*Also Soapwort

Flowers pink to white, in crowded terminal and axillary clusters, fragrant, sometimes double; petals 5, notched, calyx tubular; leaves opposite, with 3–5 prominent longitudinal veins, oval-lanceolate, to 10 cm. long (4 in.); stems smooth, slightly swollen at nodes, usually branched, to 60 cm. high (2 ft.).

July–September

    A perennial with a short horizontal rhizome, introduced from Europe. The flowers stay open at night and are visited by hawkmoths, which may be the chief pollinating insects. The flowers are also visited by bumblebees, that sometimes steal the nectar by biting a hole in the base of the calyx tube. The seed capsule opens from the top by 4 teeth, releasing numerous, black, kidney-

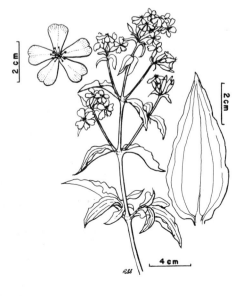

2 cm   2 cm   4 cm

shaped seeds with concentric rings of short wart-like projections.

This is a genus native to Europe and Asia. *S. officinalis*, the most abundant representative in North America, was probably introduced by early settlers as a garden ornamental. A related species, *S. vaccaria*, Cow Cockle, not illustrated, is similar but has pale red flowers. It is reported occasionally in our range but is not abundant.

Bouncing Bet is often observed growing in thick masses along roadsides and in open fields and waste areas throughout the eastern United States and southern Canada and along the Pacific coast.

℞ In herbology it is reported to be useful for treating respiratory congestion and con-

stipation. The roots have been used to make a poultice to reduce the discoloration of bruised or black eyes. *Note:* In feeding experiments with sheep, the animals died within four hours when fed a dry quantity equal to 3% of their body weights.

℞ This plant contains the glucoside saponin, which when mixed with water has cleansing properties. Soap-like suds are formed when the wet flowers are rubbed between the hands. It is said to have been used at one time as a substitute for soap in England. The leaves can be used as an ingredient in a natural shampoo. The name Bouncing Bet was applied to washerwomen of yore.

Silene cucubalis		
Bladder Campion*	**217**	*Caryophyllaceae* **Pink Family**

2 cm

4 cm

Also Maiden's Tears

Flowers white, 5 deeply lobed petals, styles 3, stamens 10, protruding; calyx swollen and bladder-like, prominently veined; leaves opposite, often clasping the stem, tapering to pointed ends, to 8 cm. long (3¼ in.); stems usually smooth, sometimes sprawling at base, often branched, to 50 cm. high (20 in.).

April–September

lination by bees and butterflies. The 3-celled seed capsule opens by several teeth at the top, releasing numerous gray-black, kidney-shaped seeds covered with concentric rows of finger-like projections.

The common name refers to the balloon-like calyx. A closely related species, *S. cserei*, not illustrated, is very similar but has slightly wider leaves and a less inflated calyx. It is a native of Europe and is often mistaken for *S. cucubalis*. Both species occur in the same habitats, and while *S. cserei* is less abundant it is spreading rapidly.

Bladder Campion is found in open fields and waste areas and along roadsides from Newfoundland to British Columbia, south to Virginia, Tennessee, and Missouri.

A hardy perennial with a creeping rhizome, introduced from Europe. Some plants have bisexual flowers, but very often individual plants will have either staminate or pistillate flowers but not both. Fertilization is thus commonly the result of cross-pol-

✂ The young shoots when only 2–3 inches high are said to be very tasty, with a flavor similar to green peas. They have a slightly bitter taste caused by small amounts of a chemical known as saponin. It is present in harmless quantities in the young shoots of Bladder Campion but is poisonous in large quantities.

Stellaria graminea Stitchwort — 218 — *Caryophyllaceae* Pink Family

Flowers white, numerous, at the ends of stems and branches; petals 5, so deeply cleft there appear to be 10, sepals 5, shorter than the petals; leaves opposite, narrowly lanceolate, to 5 cm. long (2 in.) but usually less; stems weak, 4-angled, smooth, often reclining, to 50 cm. long (20 in.).

May–September

A perennial with weak roots and a creeping stem, introduced from Europe. The star-shaped flowers are visited by small bees and flies that may effect pollination. The one-celled seed capsule opens from the top by about 6 teeth to release numerous dark brown, oval, coarsely ridged seeds.

It often forms tangled, spreading mats. The seeds and vegetation are eaten by several species of birds and small mammals and some of these no doubt contribute to seed dispersal. This species may invade cultivated fields, but it yields readily to cultivation and is usually not a troublesome weed.

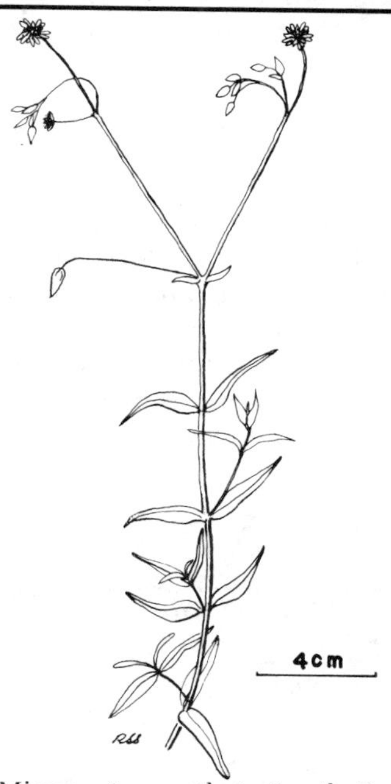

Stitchwort is commonly found in grassy areas and open fields, along roadsides, and in sandy soil from Newfoundland and Quebeç to Minnesota, south to South Carolina and Missouri.

✂ The young tender stems and leaves can be added to salads or cooked as greens, but are said to be rather tasteless.

Stellaria media Common Chickweed — 219 — *Caryophyllaceae* Pink Family

An annual or winter annual with a fibrous root system, introduced from Europe. Fertilization is mainly the result of self-pollination. The one-celled seed capsule splits into 5 sections, releasing numerous reddish-brown, slightly kidney-shaped, knobby seeds. It grows most vigorously in cool moist weather and may produce seeds all winter in protected locations or during mild seasons.

This is an important wildlife food species. The seeds are a highly favored food of the

198

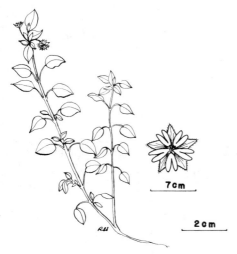

Mourning Dove, Bobwhite Quail, Common Goldfinch, Chipping Sparrow, White Crowned Sparrow, and many other species of birds. The whole plant is eaten by several species of small mammals. In feeding studies the seeds retained viability after passing

Flowers white, solitary or in few-flowered clusters at ends of branches and stems; petals 5, so deeply cleft there appear to be 10, sepals 5, longer than petals; leaves opposite, oval, with pointed tips, lower ones long stalked; stems weak, round in cross-section, branched, sprawling, rooting at nodes, forming mats, to 60 cm. long (2 ft.).

February–December

through the digestive tracts of horses, cattle, sheep, and swine. Thus the birds and mammals that eat the plant may serve as agents of dispersal.

Common Chickweed is found in open fields and lawns, along roadsides, and in waste areas throughout the United States and southern Europe.

✖ See *S. graminea.*

℞ A poultice made from the bruised fresh leaves or an ointment made by crushing the leaves in vaseline or lard has been used as a cooling application for skin irritations, sores, eye infections, and hemorrhoids.

Chenopodium album Lamb's Quarters*	220	*Chenopodiaceae* Goosefoot Family

*Also Pigweed, Goosefoot

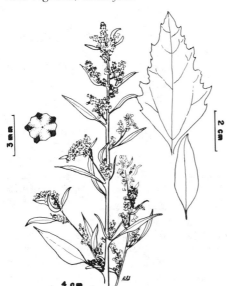

Flowers greenish, small, in dense terminal and axillary clusters, turning reddish late in season; seeds shiny black, lens-shaped; leaves alternate, whitish and mealy on the underside, upper ones lanceolate, entire, lower ones larger, unevenly toothed, diamond shaped; stems smooth, freely branched, to 2 m. high (6½ ft.).

June–October

An annual with a deep taproot, introduced from Europe and Asia. Pollination is by wind, and although this plant is of little importance as a hayfever plant in the East, it is believed to be a significant cause of hayfever in the western states. The seeds are known to pass unharmed through the intestinal tracts of grazing animals, an effective means of seed dispersal.

Of the nearly 75,000 seeds that may be produced by a single plant, all do not germinate at the same time, and some may lie dormant for years. Under experimental conditions, a 51% level of germinability was achieved after 5 years of dry storage.

This genus is an important source of food for wildlife. The seeds are eaten by several species of upland game birds, songbirds, and small mammals, constituting up to 25% of the diet of the Common Redpoll and the White-crowned Sparrow. The plants are browsed by White-tailed Deer in some parts of the East.

Lamb's Quarters is found along roadsides and in abandoned and cultivated fields, gardens, and waste areas throughout southern Canada and the United States.

The young leaves and stem tips can be cooked as greens. They cook down to one half, or less, of their original bulk and some wild plant food enthusiasts claim they are superior to spinach. The seeds can be boiled and mashed into a porridge-like dish, or dried and ground into a flour that tastes somewhat like buckwheat flour. The black seed coats give this color to the flour, and reportedly in hard times Napoleon lived on the black bread made from it.

Several species of *Chenopodium* are edible, but 2 species should be avoided. *C. ambrosioides*, Mexican Tea, not illustrated, is a native of tropical America and is more common in the southern part of the United States. *C. botrys*, Jerusalem Oak, not illustrated, is introduced from Europe and is more common in the central and northern part of our range. Both of these species are strongly aromatic and have caused poisoning in livestock.

Commelina communis **Asiatic Dayflower** — 221 — ***Commelinaceae*** **Spiderwort Family**

Flowers blue, in terminal clusters, subtended by heart-shaped spathes; petals 3, lower one smaller, white; fertile stamens 3, long, sterile stamens 3, shorter, with cross-shaped anthers; leaves alternate, oval to lanceolate, base of leaf forming a sheath around the stem; stems reclining, succulent, rooting at lower nodes, to 90 cm. long (3 ft.).

June–October

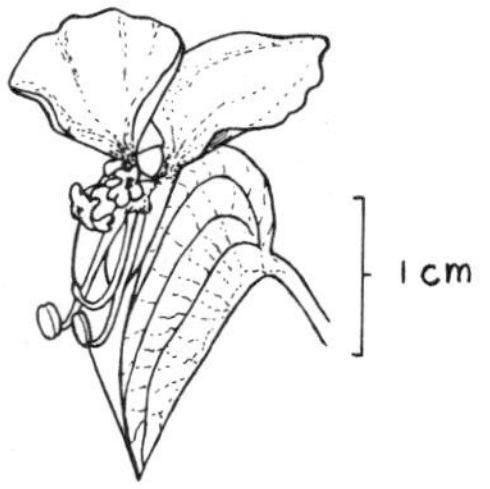

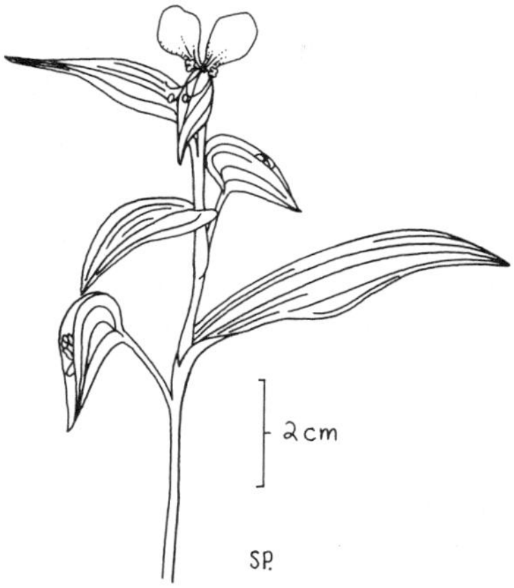

An annual introduced to North America from Eastern Asia. Each flower usually produces 4 grayish brown, pitted or coarsely wrinkled seeds with one side flat, the other convex. These are eaten by the Mourning Dove, Bobwhite Quail, Redwing Blackbird, and Cardinal. The plant is sometimes grazed by the White-tailed Deer.

The two large and one small petals of the flower inspired Linnaeus to name this genus after the three Commelijn brothers, who were Dutch botanists of the seventeenth century. Two of the brothers wrote extensively on botany, but the third died before he achieved a botanical accomplishment. The common name refers to the ephemeral nature of the flowers, which usually do not last more than one day.

A related native species, *C. virginica*, Virginia Dayflower, not illustrated, is very common from New Jersey to Illinois, south to Florida and Texas. It is a perennial with the lower petal blue, leaves to 20 cm. long (8 in.), and an erect stem to 1.2 m. high (4 ft.).

Asiatic Dayflower is found in moist shaded gardens, roadside ditches, neglected fields, waste areas, and moist woods from Massachusetts to Wisconsin, south to Florida and Texas.

✕ The young stems and leaves of Asiatic Dayflower and Virginia Dayflower can be used in salads or cooked as greens. The natives of the East Indies are reputed to use plants of this genus by steaming them and eating them with rice.

℞ Older men and women of the Navajo Indian tribe in the Southwest are said to have used a solution made from dayflower to improve their sexual performance.

Convolvulus arvensis Field Bindweed*	**222**	*Convolvulaceae* Morning-glory Family

*Also Small Bindweed

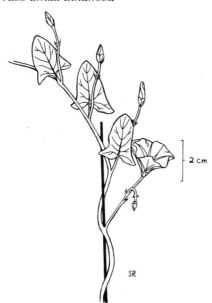

2 cm

SP.

A perennial with a deep root system, introduced from Europe as a contaminant in commercial seeds. Each flower has five white lines which by odor, and in pink forms by color, serve as nectar-guides for bees which effect pollination. The 2-celled seedpod contains 4 dark brown, finely wrinkled, granular seeds. Germination experiments have demonstrated that these may remain viable after 10 years of burial in the soil.

Flowers white to pink, funnel-shaped, less than 2.5 cm. long (1 in.), 1 or 2 on long axillary stalks; 2 small bracts 1–2 cm. (¾ in.) below calyx; leaves alternate, usually blunt, arrow-shaped, shorter than flower stalks; stems long, slender, spreading on the ground or climbing, intertwining to form dense mats, to 3 m. long (10 ft.).

June–September

This is a very serious weed in cultivated areas and is difficult to control, particularly in the western states, because of its extensive root system which may penetrate the soil to a depth of 5 meters (16 ½ft.). It is one of several noxious weeds specified in the Federal Seed Act and, as such, the degree to which its seeds are allowed as contaminants in imported commercial seeds is federally regulated.

There are about 200 species of this genus widespread in temperate regions, but most of them occur in southern Europe and southwestern Asia. Only 2 species are native to northeastern North America.

Field Bindweed is found in cultivated and abandoned fields, in waste areas, and along roadsides throughout southern Canada and the United States.

℞ The root is said to be a purgative and a diuretic. There are reported cases of mild poisoning in hogs from eating the roots. Toxic concentrations of nitrates have been measured in species of this genus.

Convolvulus sepium
Hedge Bindweed*

223

Convolvulaceae
Morning-glory Family

*Also Wild Morning-glory
Flowers pink or white, funnel-shaped, solitary, on long axillary stalks, to 7 cm. long (3 in.); 2 heart-shaped bracts at base of calyx; leaves alternate, arrow-shaped with blunt lobes at base; stems trailing on the ground or twining on other plants, to 3 m. long (10 ft.).

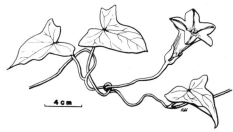

4 cm

May–September

A perennial with an extensive creeping root system, of North America and Europe. The European variety is pollinated by a hawkmoth and its distribution there is reputed to coincide with that of the Bindweed Hawkmoth. However, this variety also occurs in North America and, along with native varieties, is cross-pollinated by honeybees and bumblebees or is self-pollinated. Each seed capsule contains 4 black, coarsely granular, finely wrinkled seeds.

This is a highly variable species with 4 recognized native varieties in addition to one introduced European variety. The twining growth habit and the flowers are similar to the morning-glories (*Ipomoea sp.*) and, like them, the flower commonly closes in the afternoon.

This species is often a serious weed in cultivated fields, especially in the northeastern and north central states. Control methods include cultivation at 3–6 day intervals with cutting blades from 3–6 inches below the surface, followed by planting a dense cover crop such as millet or alfalfa. Chemical herbicides are sometimes used but these should be applied with caution since they may contaminate surface and ground water.

Ipomoea purpurea
Common Morning Glory

224

Convolvulaceae
Morning-glory Family

Flowers blue, purple, pink or white, to 7.5 cm. wide (3 in.), in clusters of 1–5 on axillary stalks; corolla funnel-shaped, stigma 1, with 3 lobes; leaves alternate, broadly heart-shaped, entire, to 12.5 cm. long (5 in.); stems twining, hairy, to 3 m. long (10 ft.) or more.

3 cm

July–October

An annual introduced from tropical America as an ornamental. The funnel-shaped flowers with long tubular bases are typical of hummingbird-pollinated flowers.

The globular seed capsule splits from the top into 3 sections, releasing usually 6, dark

202

brown or black, hairy, 4-angled seeds. The plants are of little value as food for wildlife but the seeds are occasionally eaten by the Ring-necked Pheasant and Bobwhite Quail.

This is a genus of 400 species or more, mainly of tropical regions. Four species occur in the Northeast, 2 introduced and 2 native. *I. purpurea* has been a popular trellis plant with its large colorful flowers throughout the summer and fall. However, it has escaped widely from cultivation and is often a troublesome weed.

A related species, *I. pandurata*, Wild Sweet Potato, not illustrated, has white flowers with purple centers and a large, deep, vertical root that has been recorded up to 2 feet

long and weighing as much as 20 pounds. It is a native species that occurs in dry woods from Connecticut and Ontario south to Florida and Texas.

Common Morning Glory is found in open fields and gardens, along roadsides, and in waste areas throughout the eastern United States and southeastern Canada and along the Pacific coast.

✖ The root of *I. pandurata* is said to have been used by American Indians for food. It is reputed to taste like a slightly bitter sweet potato and may require cooking in several changes of water. It may be a purgative if eaten raw. A related species is the familiar cultivated sweet potato (*I. batatas*).

Sedum acre
Mossy Stonecrop*   **225**   **Crassulaceae**
Orpine Family

Also Wallpepper, Love-entangle

| cm

A perennial with prostrate rooting stems introduced from Europe. Nectar scales at the base of each of the several pistils may function to attract insect pollinators. The numerous longitudinally ridged, slightly glossy, brown seeds are small enough to become windborne in a moderate breeze. The seed capsules open during rain showers, assuring sufficient moisture for germination.

Flowers yellow, to 12 mm. wide (½ in.), petals usually 5, pointed; leaves tiny, overlapping, very thick; stems creeping, forming low mats, flowering branches to 10 cm. high (4 in.).

June–July

This species is often cultivated as an ornamental in rock gardens and as an indoor potted plant. It is available from commercial greenhouses and is easy to establish and transplant. The creeping, mat-forming growth habit makes it a good soil stabilizer. It has escaped widely and has become naturalized in many areas of the Northeast.

The genus name is derived from a Latin word meaning "to sit" and apparently refers to the spreading growth habit of some species. The specific name has the same root as "acrid" and describes the pungent taste of the leaves. Although the bright yellow flowers are attractive, this plant sometimes becomes a troublesome garden weed.

Mossy Stonecrop is found in dry, open, rocky areas and sandy fields and beaches from Quebec to Minnesota, south to North Carolina and Illinois.

Sedum telephium
Live-forever*

226

Crassulaceae
Orpine Family

*Also Garden Orpine, Frogplant
Flowers red-purple, to 8 mm. wide (⅓ in.)
in dense branched terminal clusters; petals
5, longer than sepals, stamens 10; leaves
alternate, succulent, bluish-green, often
coarsely toothed, lower ones larger, to 6
cm. long (2½ in.); stems thick and fleshy,
smooth, erect, to 45 cm. high (18 in.).

July–September

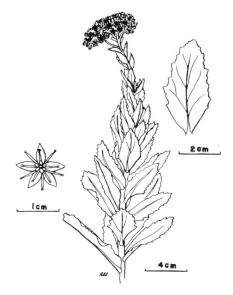

A perennial with a thick rhizome and a persistent stem base, native to western Asia. This plant was introduced as an ornamental for flower gardens but has escaped widely. In gardens it is easy to establish for a quick cover, but the spreading rhizome makes it difficult to eradicate.

The upper and lower epidermal cells of the leaf can be separated by exerting pressure on the sides. By blowing into the opening at the stalk end the sides can be distended somewhat like the throat of a croaking frog, thus the name Frogplant. This is a favorite game of children where the species is abundant.

This is the largest genus of the family, with several hundred species mainly of the North Temperate Zone and mountainous tropical areas. Many species are cultivated as potted plants or rock garden ornamentals. It is a very complex genus and species are sometimes difficult to distinguish. Some botanists recognize one variety of *S. tele-*

phium as a distinct species, *S. purpureum*.

Live-forever is a highly variable species that has widely escaped cultivation to the margins of fields, rocky exposures, and roadsides from Newfoundland to Minnesota, south to Virginia and Indiana.

✗ The very young stems and leaves can be used in salads or cooked as greens. The spring and late autumn tuberous roots can be cooked as a vegetable or made into pickles.

Echinocystis lobata
Wild or Prickly Cucumber*

227

Cucurbitaceae
Gourd Family

*Also Balsam Apple

A vigorously growing, annual, climbing vine. Cross-pollination is mainly by bees and wasps. The inflated 2-celled fruit produces

4 roughened black seeds that are expelled through terminal pores.

The star-shaped leaves and the luxuriant growth of this plant make it attractive to

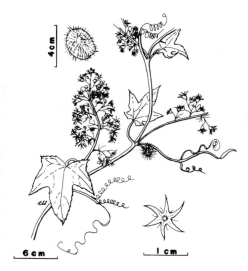

4 cm

6 cm   1 cm

Flowers greenish-white, unisexual, corolla lobes 6; staminate flowers numerous, on axillary stalks; pistillate flowers 1 to few, on shorter stalks in same axils, fruit inflated, bladder-like, covered with weak spines, about 5 cm. long (2 in.); leaves alternate, long petioled, with 5 pointed lobes; stems climbing, with branched tendrils, to 7.5 m. long (25 ft.).

June–September

This is the only species of the genus in the Northeast, but there are about 12 others in the western states. The genus name is a combination of Greek words that mean "hedgehog" (porcupine) and "bladder," referring to the inflated spiny fruit. With some imagination, one can see a resemblance of the fruit to a small spiny cucumber. The reason for the common name Balsam Apple is unknown.

Prickly cucumber is found in open thickets and fence rows, along stream banks and roadside ditches, and in waste areas from New Brunswick to Saskatchewan, south to Florida and Texas.

some as an arbor vine. When it is cultivated for this use, it escapes readily and is often observed growing profusely on vegetation around human habitations. It does not appear to seriously damage the plants that support its growth. When it invades cultivated fields it can be controlled by clean cultivation and hand hoeing.

Dipsacus sylvestris
Teasel

228

Dipsacaceae
Teasel Family

A biennial introduced from Europe. Pollination is by bumblebees, which are attracted to the flowers by nectar that is 26% sugar. The prickly flower head may aid in the mechanical dispersion of seeds. It has been suggested that the head may grip a passing animal, bending the stem, which snaps back as the animal passes, flinging the seeds in the opposite direction.

It develops a hardy rosette and a deep root system during the first year of growth. The flowering stem is beautiful and it remains so even in the winter as the dead stem sways above the snow. The dead stem tips with their flower heads are sometimes collected and dyed for use as winter ornamentals.

The genus consists of 2 additional intro-

Flowers lavender, tiny, 4-lobed, crowded, in a terminal bristly head, subtended by several slender pointed bracts; leaves opposite, entire, lanceolate, underside of midvein with prickles, stems stiff, prickly, to 1.8 m. high (6 ft.).

July–September

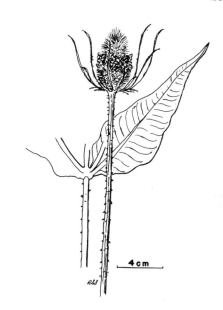

duced species, neither of which is as widespread as *D. sylvestris. D. fullonum,* Fuller's Teasel, not illustrated, was used for fulling wool in early wool industries. *D. lacinatus,* not illustrated, has prominent cups at each node formed by the fused bases of two leaves. These normally fill with rainwater, and some botanists have theorized that this is a mechanism that has evolved to prevent crawling insects from reaching the flower heads.

Teasel is found in pastures, old fields, roadsides, and waste spaces from Quebec to Michigan, south to North Carolina and Tennessee.

Knautia arvensis
Field Scabious* **229** *Dipsacacea*
Teasel Family

*Also Bluebuttons

Flowers blue or lilac, numerous, in a dense hemispherical cluster, to 5 cm. wide (2 in.), on a long terminal stalk; leaves opposite, pinnately dissected, upper pairs smaller; stems erect, hairy, often unbranched, to 90 cm. high (3 ft.).

June–August

A perennial with a substantial taproot, introduced from Europe. Although the flowers produce nectar with an average 19.5% sugar content that attracts honeybee and bumblebee pollinators, some seeds may be produced without fertilization. Each of the numerous tiny flowers bears a single, relatively large, greenish-yellow, 4-ribbed, hairy nutlet.

This species is the only North American member of a mainly Old World genus. The genus was named in honor of Christian Knaut, a seventeenth-century German physician and botanist. *K. arvensis* was introduced as an ornamental but has escaped

206

widely in the Northeast, where it is sometimes a weed in pastures and meadows. Control methods include early mowing before seeds are formed, followed by plowing the stubble. In an older system of classification it was listed as *Scabiosa arvensis*.

Field Scabious is found in old fields, dry pastures, and meadows, along roadsides, and in waste areas from Newfoundland and Quebec south to Pennsylvania.

| *Euphorbia corollata* Flowering Spurge* | **230** | *Euphorbiaceae* Spurge Family |

*Also Tramp's Spurge, Wild Hippo

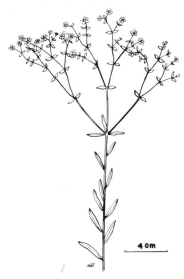

4 cm

Flowers white, about 6 mm. wide (¼ in.), numerous, in an open terminal cluster; petals none, 5 petal-like bracts surrounding minute flowers; leaves alternate below branches, in a single whorl subtending branches, narrowly elliptical, to 6 cm. long (2½ in.); stems smooth, unbranched below flower cluster, juice milky, to 1 m. high (40 in.).

June–September

is highly variable, exhibiting different structural characteristics in different environments. It sometimes infests pastures and croplands as a weed but can be controlled by mowing or clean cultivation. A related species, *E. pulcherrima*, not illustrated, is the popular Christmas Poinsettia, with large red leaf-like bracts surrounding the tiny flowers.

A perennial with a short rhizome and a deep stout root. Insects are attracted to this plant by nectar that is equal parts fructose and glucose. Each flower produces 3 oval, brown, finely pitted seeds. These may be dispersed by ants that are attracted by a small edible appendage at the base of each seed. As a source of food, the seeds are popular with several species of upland game birds and songbirds.

This is a very large genus mostly of tropical and desert regions with about 36 species in eastern North America. *E. corollata*

Flowering Spurge is found in old fields, pastures, and waste areas, along roadsides, and in open woods from New England and Ontario to Wisconsin, south to Florida and Texas.

☠ In herbology several species of this genus are recommended for the initiation of vomiting and as purgatives. However, the milky juice of these plants contains toxic compounds, and they should not be used in home remedies. The sap may cause blistering and inflammation of the skin in sensitive individuals. Death in cattle has been reported from eating hay that contained spurges.

Euphorbia cyparissias
Cypress Spurge

231

Euphorbiaceae
Spurge Family

Flowers greenish yellow, in a terminal cluster subtended by a whorl of leaves; the "petals" are 2 large bracts with a tiny flower in the center; leaves alternate, very narrow, numerous, to 2.5 cm. long (1 in.); stems smooth, green, branched or unbranched, juice milky, to 40 cm. high (16 in.).

April–July

A perennial with a branched, creeping, horizontal rhizome, introduced from Europe. The flowers are self-incompatible, and cross-pollination is effected by bees and butterflies. Each flower produces 3 dull gray or brown, finely pitted seeds. These have a relatively large appendage on one end that may attract ants as agents of dispersal. This plant was introduced as a flower garden ornamental but has escaped widely.

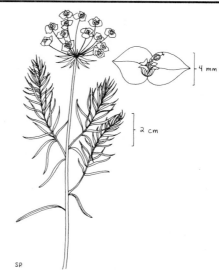

This species usually grows in dense colonies often from the same branched and spreading rhizome. It is widely cultivated as an ornamental, but less toxic plants are available that are equally attractive. The name spurge is probably derived from the Latin word *expurgare*, which means "to purge," and refers to the purgative properties of plants in this genus. A related species, *E. splendens*, not illustrated, is the popular house plant Crown-of-Thorns.

Cypress Spurge is found in open fields and cemeteries, along roadsides, and in waste areas from Maine to Minnesota, south to Virginia and Missouri.

℞ See *E. corollata*.

Euphorbia maculata
Spotted Spurge*

232

Euphorbiaceae
Spurge Family

Also Eyebane, Wartweed, Milk Purslane

An annual with a shallow branching taproot. Pistillate and staminate flowers occur on the same plant and pollination is accomplished by small insects. Some flowers have both stamens and pistils and may be self-pollinated. Each flower produces 3 black, pitted seeds with small soft appendages that may attract ants for dispersal. The seeds are a favored food for several species of upland game birds and songbirds.

The seeds of this species may lie in the soil for more than a year before germination; thus it is difficult to eradicate as a weed. In heavy infestations, repeated cultivations before it produces seeds are the best control

Flowers greenish, inconspicuous, at the ends of branches or in upper leaf axils; leaves opposite, oval, with a purplish spot in center, to 3.7 cm. long (1½ in.); stems hairy, usually sprawling, freely branched, forming flat mats, to 40 cm. long (16 in.).

June–October

measures. It has been introduced as a weed and is widespread in other parts of the world.

Spotted Spurge does not compete successfully with other plants where the soil is rich, so it is commonly found on sterile soil. It is often abundant as a weed in lawns, gardens, and dry fields, along roadsides, and in waste areas from Quebec to North Dakota, south to Florida and Texas.

℞ See *E. corollata*.

| *Coronilla varia*
Crown Vetch* | **233** | *Fabaceae* or *Leguminosae*
Bean Family |

*Also Axseed

Flowers pink and white, in umbels of 10–15, on long axillary stalks; leaves alternate, pinnately compound, leaflets 10–25; stems spreading over the ground, forming mats, flowering stalks erect, to 25 cm. high (10 in.).

June–September

len only. The 4-angled, linear seedpod is cross partitioned at intervals into 3–7 joints, which at maturity separate into one-seeded segments.

This is a genus of about 25 species native to the Mediterranean region, Europe, northern Asia, and Africa. *C. varia* is the most widely distributed species in North America. Several yellow-flowered species are grown as ornamentals on sloping areas in gardens and parks. These rarely escape cultivation and are seldom seen growing wild.

A perennial with a horizontal rhizome and deep roots, introduced from Europe. Since the flower does not produce nectar, pollinators, mainly bumblebees and honeybees, are apparently attracted to the plant for pol-

This is an excellent plant for soil stabilization and enrichment, and it is used extensively in the Northeast and Midwest to reduce erosion in roadcuts and on strip-mined areas. It is not widely grown for forage in North America because it does not regenerate after mowing or plowing. However, this species is cultivated for hay in the Soviet Union. Its roots have nitrogen-fixing nodules that are globular when young but become branched or Y-shaped in older plants.

Crown Vetch is found along roadsides, around old home sites, and in waste areas from Maine to South Dakota, south to North Carolina and Kansas.

☠ This plant is listed in early books of poisonous plants. The seeds contain a substance called coronillin and are thought to be poisonous by some European writers. Little has been written about the toxic properties of the species.

Lathyrus latifolius
 Everlasting Pea

234

Fabaceae or Leguminosae
 Bean Family

Flowers purple, pink or white, to 2.5 cm. long (1 in.), in a cluster of 4–10, on a long axillary stalk, to 20 cm. long (8 in.); leaves alternate, pinnately compound, with a single pair of leaflets, to 9 cm. long (3½ in.), terminated by a branched tendril, petioles winged; stipules lanceolate, with pointed basal lobes; stems prostrate or climbing, smooth, broadly winged, to 2 m. long (6½ ft.).

June–September

A perennial often forming tangled masses, introduced from Europe. Fertilization in many species of this genus is the result of self-pollination. The seedpod splits longitudinally into 2 sections, releasing several globular, dark brown, coarsely wrinkled seeds. These are eaten to a limited extent by the Ruffed Grouse and other game birds.

Two related species are very similar to and are sometimes confused with *L. latifolius*. *L. sylvestris*, Flat Pea, has narrow leaflets, to 15 mm. long (¾ in.), and a less prominently winged stem. *L. odoratus*, Sweet Pea, usually has only 2 very fragrant flowers per cluster; neither of the two is illustrated. The latter two are both introduced ornamentals that occasionally escape from cultivation to roadsides and waste areas.

Everlasting Pea is found in vacant lots, along roadsides, and in waste areas from New England to Indiana, south to Virginia and Missouri.

☠ Everlasting Pea is one of several species

4 cm

of this genus, including *L. sylvestris* and *L. odoratus*, that have been shown to have toxic substances in their seeds. When consumed in excess they cause a type of poisoning in man and livestock known as lathyrism. Among livestock species, horses seem to be the most susceptible, but fatal poisoning has occurred also in pigs and domestic birds.

In humans, young adult males are more often poisoned than children, females, or older people. Symptoms of poisoning are pains in the back and paralysis of the muscles below the knees. Throughout history, outbreaks of lathyrism have accompanied periods of famine.

210

Lotus corniculatus
Birdsfoot Trefoil

235

Fabaceae or Leguminosae
Bean Family

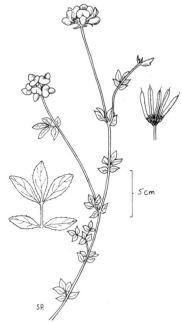

Flowers bright yellow, pea-like, in umbels at the tips of branches; seedpods slender, to 4 cm. long (1½ in.); leaves alternate, pinnately compound, with 3 terminal and 2 basal leaflets; stems freely branched, sprawling or erect, to 60 cm. long (2 ft.).

June–September

of a bird's leg and foot; thus the common name. It is a valuable browse and pasture plant for both domestic livestock and wildlife species. The leafy foliage is nutritious and palatable and makes good hay and silage.

The extensive root system has numerous small nitrogen-fixing nodules. It is a good cover crop for road cuts and slopes because it stabilizes and enriches the soil at the same time. Chemical analysis of the nodules has indicated a nitrogen content of 8–9 per cent. If the seeds are to be planted in an area where the plant has not previously grown, they may require inocculation with *Lotus* nitrogen-fixing bacteria for maximum nodule formation.

A deep rooted perennial native of Europe and Asia. The flowers are self-incompatible and cross-pollination is effected by honeybees and bumblebees. Bees are attracted to the flowers by nectar that may have a sugar content of 50 per cent or higher. The seedpod splits along two seams to release numerous nearly round, smooth, brown or black mottled seeds.

The clusters of 3–10 flowers are replaced by slender seedpods which with the supporting section of stem give the appearance

Germination studies at the Kew Gardens in England reported that seeds of this species remained viable after storage for 80 years or more. It is a drought-resistant, winter-hardy plant that makes an attractive ornamental in rock gardens and borders.

Birdsfoot Trefoil is found along roadsides and in meadows, fields and waste areas from Newfoundland to Minnesota, south to Virginia, West Virginia, and Ohio, and it seems to be spreading.

Medicago lupulina
Black Medick*

236

Fabaceae or Leguminosae
Bean Family

*Also Nonesuch, Hop Clover

An annual or winter annual with a shallow taproot, introduced from Europe. Fertilization is the result mainly of self-pollination. Each flower head produces several black one-seeded capsules; thus the

name Black Medick. The seeds are oval, smooth, and green to brown. In some members of the genus, the seeds are very long-lived, retaining viability after 80 years or more of storage.

This species is sometimes planted solely

Flowers yellow, in dense round or short cy-lindrical heads, on axillary stalks; flower heads soon replaced by clusters of black, veined, 1-seeded pods; leaves alternate, compound with 3 leaflets, leaflets spatula-shaped, often bristle-tipped; stems pros-trate, branched, to 60 cm. long (2 ft.).

May–October

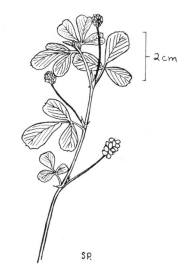

2 cm

SP.

as a forage crop or as green manure. The presence of nitrogen-fixing nodules on its roots enriches and improves the structure of the soil. It is sometimes considered an objectional weed when it occurs as an im-purity in Alsike Clover fields or invades lawn and gardens.

This genus includes about 50 species na-tive to Europe, Asia, and Africa. The genus name is Greek and refers to Media, an an-cient country (present day northwestern Iran) in which Alfalfa supposedly origi-nated. Of the seven species that have been introduced into the Northeast, only *M. lu-pulina* and *M. sativa*, alfalfa, are widespread and well known.

Black Medick is found in open fields, meadows, pastures, lawns, and waste areas and along roadsides from Nova Scotia to British Columbia, south to Florida, Mexico, and California.

✻ The seeds can be roasted and eaten di-rectly or ground into a flour.

Medicago sativa
Alfalfa*

237

Fabaceae or Leguminosae
Bean Family

*Also Lucerne

Flowers blue, in short clusters, to 3 cm. long (1.2 in.), clusters at ends of axillary stalks about the same lengths as the sub-tending leaves; leaves alternate, pinnately compound with 3 leaflets, leaflets toothed at tips; stems sprawling or erect, branched, to 90 cm. high (3 ft.).

May–September

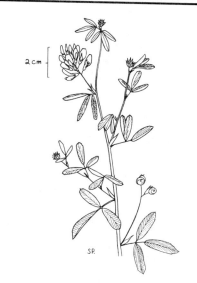

2 cm

SP.

A perennial with a deep taproot, and a native of central and western Asia. Fertil-ization is the result mainly of cross-polli-nation by bees. The weight of the insect on the flower triggers the release of a cloud of pollen and brings the stigma into contact with the back of the insect's thorax. The mature seedpod is coiled into 2–3 complete revolutions somewhat like an abbreviated corkscrew. This has been interpreted by some botanists as an adaptation for dispersal by rolling with the wind.

This species prefers well drained, sandy or loamy soils that are approximately neutral with regard to acidity. It is drought-resistant and in dry areas is noted for its very deeply penetrating root system. It is thus a particularly important crop in the arid and semi-arid western states.

Its very early cultivation in the Lucerne District in Switzerland resulted in its wide acceptance in Europe and the common name by which it is known in many parts of the world. It is said to be the only forage crop plant that was cultivated before recorded history.

For the livestock farmer this plant is a very flexible crop plant. In addition to pasture forage, it can be mowed for hay two or more times per year for several years, used for silage, and plowed under as green manure. It is a rich source of protein, calcium, magnesium, phosphorus, and Vitamins A and D for all classes of livestock. Nitrogen-fixing nodules are formed by the same species of bacteria that forms the nodules in the sweet clovers. When planted in new fields, inocculation with the bacteria may be necessary for good nitrogen-fixation.

Many species of birds and mammals favor the seedpods and leaves of Alfalfa throughout its range. Other species of the genus, called burclovers, are also very important wildlife food plants, especially in the western states. Alfalfa has escaped widely from cultivation and is found commonly in old fields, along roadsides, and in waste areas throughout the United States and southern Canada.

✂ The dried young leaves and flower clusters can be used to make a bland tea. The young leaves can be used fresh as an additive for salads. Fresh Alfalfa sprouts can be purchased in health food stores and many supermarkets. This plant has the same food value for humans as for farm animals.

Melilotus alba — **238** — ***Fabaceae or Leguminosae***
White Sweet Clover* — **Bean Family**

**Also White Melilot*

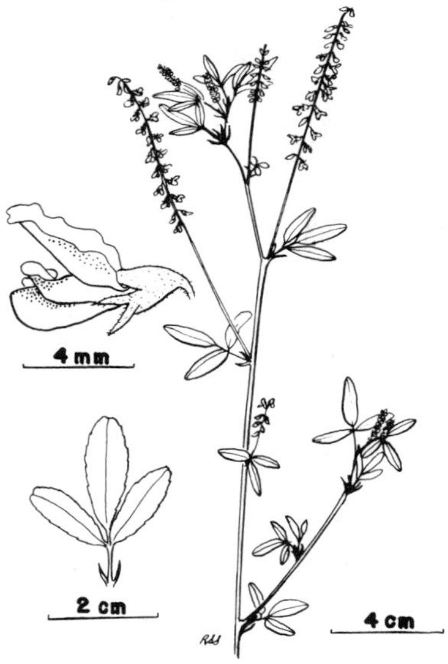

Flowers white, numerous, small, in long axillary clusters, to 20 cm. long (8 in.); leaves alternate, compound with 3 leaflets, fragrant when dried or crushed; stems smooth, freely branching, to 2.4 m. high (8 ft.).

May–October

A biennial with a deep taproot, introduced from Europe. Fertilization is the result of both self-pollination and cross-pollination by insects. Honeybees are important agents of pollination, and the plant is highly valued for the quality of the honey made from its nectar. The seed capsule contains 1–2 smooth, green to brown seeds and may persist unopened or open only late in the season.

It has nitrogen-fixing nodules on its roots and is often planted to improve soil fertility. Under the name of Bokhara Clover, it is sown in some areas for pasture forage or green manure and as a cover crop for erosion control. It is moderately winter-hardy and drought-resistant with a preference for the soils of limestone areas.

This is a genus of about 20 species native to southern Europe, western Asia, and northern Africa. Four species have been introduced into eastern North America, but only *M. alba* and *M. officinalis* are widespread and well known. One variety of *M. alba* is an annual and is known as Hubam Clover.

Yellow Sweet Clover, *M. officinalis*, not illustrated, is closely related and very similar to *M. alba* but has yellow flowers, is mostly insect-pollinated, and usually is not as tall. It is not as abundant as *M. alba* and is more concentrated in the Midwest than in the Northeast.

White Sweet Clover is found in open fields, along roadsides, and in waste areas throughout the United States and southern Canada.

✖ The spring leaves, before flowering, can be added to salads or cooked as greens. The young dried leaves can be used as a flavoring for sweets.

℞ It has been used, mainly without modern medical justification, for a variety of internal and external problems. These include the use of the dried flowering plant in salves for boils, in extracts for stomach gas and respiratory congestion, and as a diuretic. These remedies may have been suggested by the plant's somewhat bitter taste and the presence of coumarin, found also in the vanilla bean, which gives the fragrant odor to its leaves.

As the result of the action of a fungus, coumarin is converted into dicoumarin, which is an anti-clotting agent. Death to numerous cattle from internal and external hemorrhaging has resulted from eating moldy sweet clover hay.

Trifolium agrarium
Hop Clover*

239

Fabaceae or Leguminosae
Bean Family

*Also Yellow Clover

Flowers yellow, in dense, cylindrical clusters, on stalks in upper leaf axils, after maturing the flowers turn brown and fold down; leaves alternate, palmately compound, with 3 long and narrow leaflets, leaf petioles to 12 mm. long (½ in.); stems smooth, freely branched, usually erect or nearly so, to 40 cm. high (16 in.).

June–September

An annual often observed on poor soils, introduced from Europe. Each flower produces 1 or a few smooth, brown or straw-colored seeds. The seeds and the vegetation of the plants in this genus are eaten by numerous species of birds and mammals that no doubt contribute to seed dispersal. It is a good cover plant for poor soils since it has nitrogen-fixing nodules on its roots that enrich the soil where it grows.

Of the approximately 300 species of this genus in the North Temperate Zone, only 4

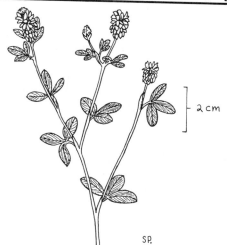

2 cm

SP.

are native to eastern North America. The western mountain states and the Pacific coast have over 100 mostly native species. The clovers are important throughout the world as honey plants, forage, hay, and soil conditioners. Although *T. agrarium* is not as abundant as *T. pratense*, Red Clover, and *T. repens*, White Clover, it provides a considerable amount of forage in some areas.

Hop Clover is found along roadsides, in old fields, and on dry, acid soils from Newfoundland to British Columbia, south to South Carolina and Arkansas.

✂ See *T. pratense*.

Trifolium pratense — Red Clover — 240 — *Fabaceae or Leguminosae* — Bean Family

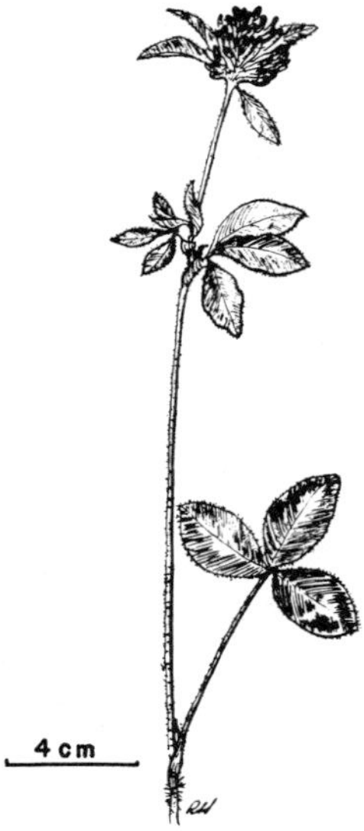

Flowers rose-purple, in dense globular clusters, usually subtended by a pair of leaves; leaves alternate, long stalked, palmately compound, with 3 leaflets, a whitish inverted V in center of each; stems soft-hairy, erect or with erect terminal part, to 60 cm. high (2 ft.).

May–September

A short-lived perennial with a deep taproot, introduced from Europe. The flowers produce an abundance of nectar at the bottom of the long, narrow, basal tube. The proboscis of the honeybee is too short to reach this, and butterflies are ordinarily not heavy enough to depress the flower to the degree necessary to expose the anthers. Long-tongued bumblebees are thus the most effective pollinators. This plant is self-incompatible, and a full crop of seeds is usually not achieved because of a lack of pollinators.

Each flower usually produces 1–2 seeds with very hard coats. This is a favored food plant not only of domestic livestock but also of many species of birds and mammals that aid in spreading the seeds. It is the second most important forage legume (the first is alfalfa) in the United States, and it is the state flower of Vermont.

T. pratense originated in the countries that border the Mediterranean Sea, and when introduced into European agriculture is said to have had a greater affect on civilization than did the potato or any other forage plant. It may have been introduced into America as early as 1625 on a ship that brought livestock and seeds from the Netherlands. The first definite reference to clover in the Colonies was about 1663, and the first newspaper advertisement of clover seed for sale appeared April 10, 1729.

Red Clover has escaped widely from cultivation and is found along roadsides and in fields and waste areas throughout the United States and southern Canada.

✂ The young leaves and flower heads can be added to salads or cooked as greens. They are rich in protein if not overly tasty. The roots can also be used as an emergency food. The flower heads and the seeds can be ground and used as a substitute for flour, and the dried heads are a suitable additive to other teas.

℞ The use of clover as a medicinal plant may date to a time when certain plants were thought to have magical powers. A 4-leaf clover is still looked upon by some as a good luck charm, and one species of *Trifolium*, the Shamrock, is especially prized in Ireland. In herbology, the dried flower heads are recommended for whooping cough and for external application to sores and ulcers of the skin.

Trifolium repens White Clover · 241 · *Fabaceae or Leguminosae* Bean Family

Flowers white, sometimes tinged with pink, in globular terminal clusters, on leafless stalks; leaves long stalked, palmately compound, with 3 leaflets, each faintly marked with inverted V, terminal one often notched; stems prostrate, creeping, to 90 cm. long (3 ft.), erect flowering stalks to 15 cm. high (6 in.).

May–October

A perennial rooting at the nodes of a sprawling stem, native to Europe and possibly also North America. The flowers produce an abundance of nectar, and it is an important honey plant. It is usually self-incompatible and requires cross-pollination for the production of seeds. This is a very important food plant for numerous upland game birds, songbirds, small mammals, and browsers.

Like other members of this genus, White Clover has nitrogen-fixing nodules on its roots that enrich the soil where it grows. It is a valuable hay and pasture plant, and it is often mixed with grass seed for lawn plantings. When planted together it improves the quality of the grass. It is very similar to the Shamrock of Ireland.

White Clover is found in fields and lawns and along roadsides throughout the United States and southern Canada.

✖ See *T. pratense*.

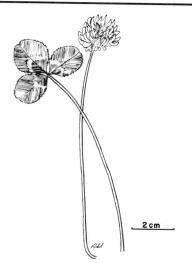

2 cm

Vicia cracca Tufted Vetch* · 242 · *Fabaceae or Leguminosae* Bean Family

*Also Canada Pea, Cow Vetch

A perennial with a horizontal rhizome, of North America and Europe. The plants in the northern part of the range are native to the Northeast, but southern plants appear to be introduced. The pea-like seedpod splits longitudinally, releasing several smooth, round, olive green seeds. The seeds and leaves of this species, and the approximately 16 other vetches in the Northeast, are eaten to a limited extent by the Mourning Dove, Ruffed Grouse, Bobwhite Quail, Wild Turkey, Song Sparrow, and other species of birds and small mammals.

An interesting adaptation of plants in this genus, apparently as protection against being overeaten, is the presence of a toxic non-protein amino acid called B-cyanoalanine. This substance has been shown to be fatal to rats in doses of 200 milligrams per kilogram of body weight. Its major role,

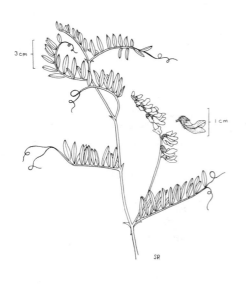

3 cm

1 cm

SP

Flowers blue, in long, one-sided, crowded, axillary clusters; flower clusters equal to or longer than subtending leaf; leaves alternate, pinnately compound, terminated by a tendril, leaflets 5–12 pairs, with pointed tips; stems slender, trailing or climbing, with hairs that lie against the stem, to 1 m. long (40 in.).

June–August

however, is probably protection against consumption by insects.

Tufted Vetch is a good stabilizer for the loose gravelly or sandy soils on which it often occurs, and since it has nitrogen-fixing nodules on its roots it also enriches the soil. It is found in open fields and meadows and along shores and roadsides from Greenland to British Columbia and Washington, south to Virginia and Indiana.

Vicia dasycarpa
Vetch

243

Fabaceae or Leguminosae
Bean Family

1 cm

4 cm

eu

Flowers blue or white, in a long, one-sided, crowded, axillary cluster of 10–30; leaves alternate, pinnately compound, terminated by single or branched tendrils, leaflets 5–10 pairs; stems trailing, smooth or with few hairs that lie against stem, to 1 m. long (40 in.).

May–September

climbing vines, abundant on all continents except Australia. They have been extensively used as cover crops to control erosion, for soil improvement and livestock forage, and as wildlife food and shelter. There are about 25 species native to North America and at least 10 others that have been introduced.

A related species, *V. villosa*, Hairy or Winter Vetch, not illustrated, is very similar to *V. dasycarpa* but has stems covered with spreading hairs. It was introduced from Europe as a forage crop but has escaped to roadsides and abandoned fields from Nova Scotia to British Columbia, south to Georgia, Texas, and California.

Similar to *V. cracca* but introduced from Europe and with no native forms. This is a genus of at least 150 species, mainly tendril

Vetch (*V. dasycarpa*) is found in abandoned fields, along roadsides, and in waste areas from Maine to Montana, south to Georgia and California.

Centaurium umbellatum — 244 — *Gentianaceae*
Centaury — *Gentian Family*

Flowers pink, to 12 mm. long (½ in.), in somewhat flat-topped terminal clusters; corolla tubular, with 5 lobes, calyx lobes 5; stem leaves opposite, oval or oblong, without stalks, to 4 cm. long (1½ in.); basal leaves in a rosette, stalked; stems smooth, erect, branched at tip, to 40 cm. high (16 in.).

July–September

An annual or biennial introduced from Europe. The waxy-pink symmetrical flowers are delicate and attractive. They usually open only during daylight on sunny days, and fertilization is the result mainly of self-pollination. The elliptical, single chambered seed capsule splits into 2 sections, releasing numerous tiny seeds. This is a small plant and usually not a troublesome weed.

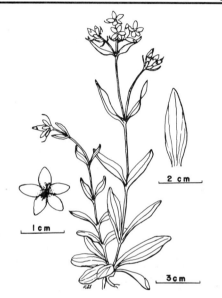

2 cm

1 cm

3 cm

Gentiana andrewsii — 245 — *Gentianaceae*
Closed Gentian — *Gentian Family*

A perennial with a thick persistent stem base and long stout roots. The closed flower is mostly self-pollinated but bumblebees often force their way into the flower and may effect cross-pollination. The seed capsule splits into 2 sections, releasing numerous shiny, white, winged seeds that are dispersed by wind or water.

This plant is remarkable for its beautiful deep blue, unopened, bottle-shaped and odorless flowers. It is the most common of the gentians in the Northeast. It was named in honor of Henry Andrews, a botanical artist of the nineteenth century. The genus includes about 400 species, most of which occur in the North Temperate and Arctic Zones.

A closely related species and one that *G. andrewsii* is often confused with is *G. clausa*, Bottle Gentian, not illustrated. In *G. clausa* the tips of the Corolla lobes extend beyond the point of attachment to the intervening membrane.

Closed Gentian is found in moist meadows, woodland borders, and damp woods from Quebec to Saskatchewan, south to Kentucky and Missouri, and in the mountains of Georgia.

℞ The genus is named for Gentius, a king of Illyria, who is credited with discovering the medicinal value of Gentian roots. The roots of several species are sources of bitters, that in times past were very popular

218

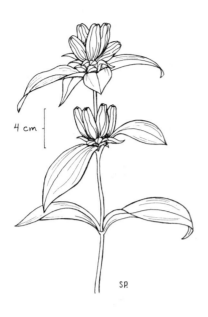

4 cm

S.P.

Flowers blue, usually in dense terminal and upper axillary clusters; flowers closed, whitish membrane between lobes fringed and exceeding them, to 4 cm. long; leaves opposite, lanceolate, sessile, to 15 cm. long (6 in.); stems unbranched, smooth, to 80 cm. high (32 in.).

August–October

for improving the appetite and promoting digestion.

A European species, *Gentian lutea*, Yellow Gentian, not illustrated, was used in the treatment of gastrointestinal disorders such as stomachache, indigestion, and diarrhea. American species have similar properties.

The roots of *G. andrewsii* were used by some Indians of eastern North America in an antidote for snakebite. Other tribes used the roots of *G. catesbaei*, Blue Gentian, not illustrated, in a remedy for backache.

The native wild gentians are rare and should not be picked. *G. andrewsii* is protected by law in New York State.

Gentiana crinita
Fringed Gentian

246

Gentianaceae
Gentian Family

5 cm

S.P.

Flowers violet-blue, solitary, on long terminal or axillary stalks; corolla 4-lobed, lobes conspicuously fringed; leaves opposite, sessile, oval to lanceolate; stems branched in upper half, to 90 cm. high (3 ft.).

September–October

An annual or biennial native of North America. The flowers open only in bright sunlight and are closed tightly at night or in dense shade. The spindle-shaped seed capsule splits into 2 sections, releasing numerous tiny, brown, hairy seeds that are spread by wind.

A related species, *G. procera*, Smaller-Fringed Gentian, not illustrated, is very similar but has corolla lobes that are toothed rather than fringed, and leaves that are very narrow. Its habitat is somewhat similar to that of *G. crinita* and it ranges from Ontario to Alaska, south to western New York, Indiana, and Wisconsin.

G. crinita is the most famous species of the genus in North America mainly because of William Cullen Bryant's poem, "To a Fringed Gentian." However, there are about 18 species in the Northeast, most of which are native to North America. It is a very large genus and is most abundantly represented in European and Asian mountainous regions.

Like other native gentians, *G. crinita* is usually not abundant. Since it reproduces by seeds only, extensive collecting of the plant may result in extinction from local areas. It can be most appreciated when it is undisturbed in its natural habitat. In some states of the Northeast it is protected by law.

Fringed Gentian is found in moist meadows, around natural springs, and in wet woods from Quebec to Manitoba, south to Pennsylvania and Indiana, and along the mountains to Georgia.

℞ See *G. andrewsii*.

Sabatia angularis
Rose Pink*

247

Gentianaceae
Gentian Family

*Also Bitter-bloom

Flowers pink with a greenish-yellow star in the center, to 4 cm. wide (1½ in.), several in a branched terminal cluster; leaves opposite, rounded at the base, with a pointed tip, usually with 5 veins, to 4 cm. long (1½ in.); stems sharply 4 angled, freely branched, to 80 cm. high (32 in.).

July–September

A biennial with numerous opposite branches, each bearing one or more flowers. These are delicately fragrant with 5 long rounded petals and 5 prominent stamens. The flowering stem arises from a basal rosette of leaves that are broadest toward the tip. The one-celled oval seed capsule splits into 2 sections, releasing numerous seeds. A white-flowered form is occasionally observed.

This is a genus of about 20 species native to North America, Mexico, and the West Indies. Most of the 10 species native to the Northeast occur near salt or fresh water along the Atlantic coastal plain. *S. angularis* and *S. campestris*, not illustrated, are usually found in upland habitats. The genus was named in honor of Liberato Sabbati, an

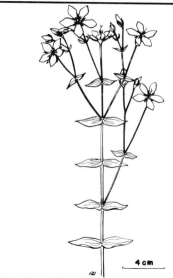

4 cm

Italian botanist of the sixteenth century.

Rose Pink is found in fields, disturbed areas, thickets, and open woods from Connecticut to Ontario and Wisconsin, south to Florida and Louisiana.

℞ A bitter tonic made from the yellow fibrous root has been used to improve the appetite and to dispel indigestion.

Hypericum perforatum
Common St. John's-wort*

248

Hypericaceae
St. John's-wort Family

*Also Klamathweed, Goatweed

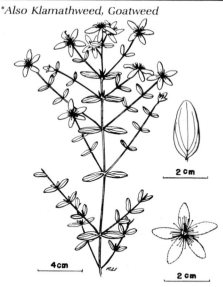

A perennial with a deep, branching root system and short runners, introduced from Europe. Although it produces pollen and is visited by bees and other insects, the flower is nectarless and seeds develop without fertilization. The seedpod splits into 3 sections, exposing many small, shiny, black, pitted seeds that are scattered as the plant sways in the wind.

Of the approximately 24 species of this genus in the Northeast, *H. perforatum* is the most common. In germination studies its seeds were found to be viable after 10 years of burial in the soil. As with many introduced species, it is much more of a weed pest in North America than in its native land. It has been introduced into Australia and is a troublesome weed there also. Control measures include clean cultivation and repeated mowings in pastures to prevent seed formation.

Common St. John's-wort is found in ne-

Flowers yellow, numerous, in a branched terminal cluster; petals 5, with black dots along margins, stamens numerous; leaves opposite, with translucent dots, sessile, oblong, to 4 cm. long (1½ in.); stems freely branched, ridged below the base of each leaf, to 80 cm. high (32 in.).

June–September

glected meadows, fields, and pastures and along roadsides from Newfoundland to Manitoba, south to Florida and Texas, and in the far west from British Columbia to central California.

☠ The common name of this plant apparently refers to its supposed use by St. John the Baptist. Perhaps because of this, it has historically been given many magical and medicinal uses. It is said to have been used in exorcisms in England, and to ward off witches in Norway and Sweden. The dried plant has been used in remedies for a variety of ailments including intestinal problems, congestion of the lungs, anemia, insomnia, and menstrual difficulties. In external applications it has been recommended for sores, bruises, and skin problems.

There are numerous cases on record of poisoning of livestock from eating this plant. It contains a substance, hypericin, which enters the bloodstream and sensitizes the nerve endings in the skin to sunlight. Subsequently, when the animal is exposed to bright sunlight, especially if it has a white or unpigmented skin, the result is dermatitis, blistering and falling hair. Among humans, photo-sensitivity in susceptible individuals may result from physical contact with the plant.

It is likely that other species of *Hypericum* may cause the same types of reactions in white-skinned livestock and in humans.

Sisyrinchium montanum
Blue-eyed Grass*

249

Iridaceae
Iris Family

Also Satin Flower

Flowers blue with yellow centers, in few-flowered, terminal clusters, overtopped by a pointed bract; flowers with 6 bristle-tipped petal-like divisions, to 19 mm. wide (¾ in.), stamens 3; leaves basal, grasslike; stems unbranched, flattened, to 50 cm. high (20 in.).

May–July

A grasslike perennial with a fibrous root system. Fertilization is most often the result of cross-pollination by bees and bee-like flies. The globular 3-celled seed capsule splits into 3 parts to release the dull black, finely wrinkled seeds. The plant is eaten by the Wild Turkey and the Ruffed Grouse.

There are about 10 species of this genus in the Northeast and all are native to the region. They all have attractive 6-pointed flowers and the most common color is blue. Among tropical species, the most common color is yellow. The different species are often difficult to distinguish from one another, and botanists are not in agreement as to the most important characteristics.

Blue-eyed Grass is found in sunny open fields and meadows and on sandy shores from Quebec to British Columbia, south to New Jersey and Pennsylvania, and in the mountains to North Carolina.

Glecoma hederacea
Ground Ivy*

250

Lamiaceae or Labiatae
Mint Family

Also Gill-over-the-ground, Creeping Charlie, Alehoof

A perennial with shallow roots, introduced from Europe. Two types of flowers are produced; some plants have only pistillate flowers, others are more robust and have larger flowers with both pistils and stamens. Cross-pollination is by bumblebees, which are regular visitors to the bisexual flowers. The ovary is deeply 4-lobed with each lobe becoming a brown, glandular-dotted seed.

This is a genus of only 5 species, all native to Europe and Asia. *G. hederacea* is the only representative of the genus in eastern North America. A form with variegated leaves is sometimes cultivated as a flower garden ornamental. In older systems of classification it is listed with the catnips as *Nepeta hederacea*.

Ground Ivy is a common weed in lawns, orchards, shady open spaces, and moist woods from Newfoundland to Minnesota, south to Georgia, Alabama, and Missouri.

✖ The dried leaves can be used to make an aromatic herbal tea reputed to be rich in vitamin C.

℞ In herbology, a tea made from the leaves is recommended for coughs and pulmonary complaints, as a tonic for the kidneys, and for intestinal problems resulting from lead poisoning. The fresh juice has been

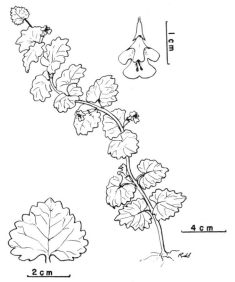

used for bruises and black eyes and has been snuffed up the nose to relieve nasal congestion and headache.

Flowers blue, to 2 cm. long (¾ in.), in axillary clusters of 3's; corolla 2-lipped, upper lip 2-lobed, lower lip with 3 larger lobes; leaves opposite, round or kidney-shaped, with broad shallow teeth, to 4 cm. wide (1½ in.); stems square, trailing over the ground, rooting at nodes, flowering branches rising to 20 cm. high (8 in.) or more.

April–August

Fatal poisonings of horses in Europe and one instance in North America have been reported as the result of ingestion of large quantities of this plant. It is toxic if eaten fresh or dried in hay. Poisoning in other livestock has not been reported.

Before the introduction of hops, this plant was used in Europe to clarify beer; thus the names Alehoof and Gill-over-the-ground. Gill is derived from a French word that means "to ferment."

Lamium maculatum
Spotted Dead Nettle

251

Lamiaceae or Labiatae
Mint Family

Flowers rose-purple, to 2.5 cm. long (1 in.), in axillary clusters; corolla tubular, 2-lipped, upper lip entire, lower lip 3-lobed, stamens 4; leaves opposite, stalked, triangular, heart-shaped at base, toothed, often with a pale blotch on upper surface, to 5 cm. long (2 in.); stems often prostrate at base with erect branches, smooth, square, to 50 cm. high (20 in.).

April–September

A perennial with a creeping stem, often rooting at nodes, introduced from Europe. The shape of the flowers is typical of those that are bee-pollinated. Each flower produces 4 seeds with soft appendages that may attract ants as agents of dispersal. It may become a weed in cultivated fields but yields to clean cultivation or in competition with a smother crop.

There are 4 species of this genus in the

223

Northeast, all introduced weeds. The name Dead Nettle apparently refers to the similarity of the leaves to those of Stinging Nettle but without the stinging hairs. A white flowered form of *L. maculatum* is occasionally observed. A related species, *L. amplexicaule*, Henbit, not illustrated, has smaller flowers, to 15 mm. long (⅔ in.), and smaller leaves, to 18 mm. long (¾ in.), that are palmately veined and clasp the stem. It occurs in open spaces throughout our range.

Spotted Dead Nettle is found in abandoned gardens and along roadsides and in waste areas from New England and Ontario, south to North Carolina and Tennessee.

✄ The young leaves of Henbit (*L. amplexicaule*) can be cooked as greens or added to salads.

Leonurus cardiaca
Motherwort

252

Lamiaceae or Labiatae
Mint Family

Flowers pink to light purple, in dense clusters in upper axils, 2-lipped, upper lip bearded; calyx lobes persistent, spine-like, lower 2 bent backward; leaves opposite, palmately 3–5 lobed, with sharp teeth, long petioled; stems thick, square, often branched, to 1.5 m. high (5 ft.).

June–August

A perennial with an evergreen basal rosette of leaves, introduced from Europe. Pollination is by bees, and each flower produces 4 three-sided, finely pitted, slightly glossy, black seeds. It sometimes spreads as a weed to cultivated fields, but it cannot survive clean cultivation.

This plant is an alternate host for cucumber mosaic disease, which may justify its elimination around vegetable gardens.

This is a genus of 10 species native to central and western Asia. One of three species that have been introduced into North America, *L. cardiaca* is the most widely distributed. It was introduced early into herb gardens as a medicinal plant but soon escaped. It has a very strong odor, and the spiny calyx teeth make it prickly to touch.

A related species, *L. siberica*, not illustrated, is similar but has leaves that are very deeply palmately 3-lobed. It occurs in waste areas locally throughout the Northeast and

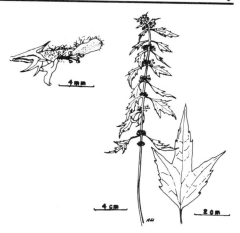

in tropical America but is much less abundant than *L. cardiaca*.

Motherwort is found along roadsides and in neglected fields, gardens, and waste areas from Nova Scotia and Quebec to Montana, south to North Carolina and Kansas.

℞ The dried tops and leaves have been recommended for a variety of ailments. Chief among these are disorders of the womb, hence the common name. A tonic made from the leaves is recommended, in old writings, for heart ailments and hysteria. Contact with the plant may cause dermatitis in susceptible individuals.

Mentha piperita
Peppermint

253

Lamiaceae or Labiatae
Mint Family

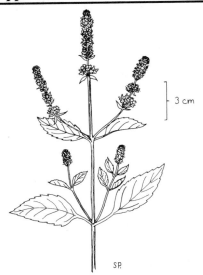

3 cm

S.P.

A perennial spreading by runners which root at intervals, introduced from Europe. Fertilization is the result of self-pollination, and each flower produces 4 tiny brown seeds. It may occasionally invade wet fields and pastures as a weed and may be controlled by improving drainage and close mowing.

This species has been cultivated for oil of peppermint in North America for over 100 years. It grows best on deep loam and mucklands. In the United States the leading peppermint-producing state is Oregon, which supplies one-third of the peppermint oil produced in this country. Eastern states

Flowers purplish-pink, in dense, rounded, elongate clusters, to 8 cm. long (3 in.), at the ends of stems and branches; leaves opposite, with a distinct petiole, to 8 cm. long (3 in.), lanceolate, hot tasting; stems square, smooth, purplish, branched, to 1 m. high (40 in.); the whole plant smells strongly of peppermint.

July–September

that produce peppermint oil are Indiana and Michigan. Menthol, the ever popular cough drop ingredient, is made from peppermint oil.

This is a genus of about 15 species native to Europe, Asia, and Australia. About 10 species occur in the Northeast but only two, *M. piperita* and *M. spicata*, spearmint, are well known and widespread. Both are native of Europe, and *M. piperita* is thought to have originated as a hybrid of *M. spicata* and *M. aquatica*, not illustrated, a native of Europe and Asia.

Peppermint is found in damp fields, wet meadows, and stream banks, along roadsides, and in waste areas throughout the United States and southern Canada.

℞ Oil of peppermint has been used medicinally from the time of the ancient Egyptians. It has been taken internally for nausea, stomach disorders, and abdominal pains and to relieve intestinal gas. Applied externally it has been used as a treatment for rheumatism, toothache, headache, and other local pains.

Mentha spicata
Spearmint

254

Lamiaceae or Labiatae
Mint Family

A perennial with a creeping rhizome, introduced from Europe. Each flower produces 4 tiny, dull brown seeds that are small enough to be windblown. It can be distinguished from *M. piperita* by its sessile or short-stalked leaves and slender interrupted flower clusters.

It has been cultivated for its oil from ancient times and is the mint of the kitchen garden in England. One of its most important uses is as a flavoring agent for candy and chewing gum. It is grown commercially in the United States in Indiana, Michigan, and Washington State.

Flowers pinkish to pale lavender, in dense, often interrupted, elongate clusters to 12 cm. long (4.8 in.), at the ends of stems and branches; leaves opposite, sessile or with a very short stalk, to 6 cm. long (2½ in.); stems square, green, smooth, branched, to 50 cm. high (20 in.); the whole plant has the odor of spearmint.

July–September

This is a very complex genus in which hybridization and chromosomal aberrations have produced a wide variety of forms. According to Greek legend, the genus name is derived from Minthe, a nymph who was changed into an aromatic herb because she displeased the gods. The name spearmint, as reported in a sixteenth-century herbal, refers to the sharp pointed leaves of the plant.

Spearmint is found in moist pastures, field margins, abandoned gardens, old home sites, roadsides, and waste areas throughout the United States and southern Canada. ✕ ℞ See M. piperita.

3 cm

SP.

Monarda fistulosa
Wild Bergamot*

255

Lamiaceae or Labiatae
Mint Family

---

*Also Horsemint

Flowers pale lavender or lilac, 2-lipped, in a dense, crown-like, terminal cluster; leaves opposite, grayish-green, triangular or lanceolate, aromatic, small leaves subtending flower head often lilac-tinged; stems square, hairy, often branched above middle, to 1 m. high (40 in.).

July–September

A perennial with a deep, branched root system. Pollination and seed characteristics are similar to M. didyma. In the Northeast, this is the most common representative of the genus. It is somewhat variable, with a western variety that is shorter and less branched than the one in the East. A form with white flowers is occasionally observed.

This is a North American genus of 17 species. It was named in honor of Nicolas Monardes, a sixteenth-century Spanish physician and botanist who wrote a book on the me-

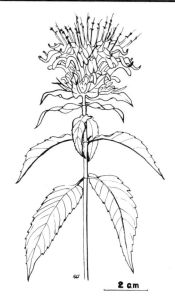

2 cm

dicinal and other useful plants of America. The 8 species that occur in the northeastern states are tall, aromatic, herbaceous plants. *M. fistulosa* often occurs in dense showy masses on dry, sloping roadsides and hillside pastures.

Wild Bergamot is found in well drained pastures, prairies, thickets, and woodland borders from Quebec to British Columbia, south to Georgia, Texas, and Arizona.

� ℞ As with other species in this genus, the leaves can be used to make a fragrant tea. The American Indians and early settlers used these plants for a variety of ailments including upset stomach, intestinal gas, headache, colds, sore throat, and bronchial problems, and externally for pimples and other skin problems.

Nepeta cataria Catnip*	**256**	*Lamiaceae* or *Labiatae* Mint Family

*Also Catmint

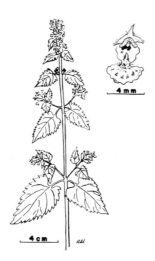

4 mm

4 cm   *RLL*

A perennial with a strong root system, introduced from Europe. Fertilization is most commonly the result of cross-pollination by insects, mainly bees. Each flower produces 4 tiny, smooth, dull brown or black seeds, with two eye-like spots at one end. This plant sometimes invades cultivated fields but usually does not constitute a weed problem because it is easily controlled by clean cultivation.

The specific and common names both refer to the fascination for this plant shown by cats. The leaves have special scent glands which contain volatile oils including a sub-

Flowers white to pale violet, the lower lobe with pink or purple spots, in dense cylindrical or interrupted clusters, at the ends of stems and branches; leaves opposite, aromatic, whitish underneath, heart-shaped at base, long stalked, with coarse teeth; stems square, hairy, usually branched, to 1 m. high (40 in.).

June–September

stance known as nepetalactone. In laboratory studies this substance was found to have repellent action on most of the insects with which it was tested. This suggests that nepetalactone evolved as a protection against insect predation and that its ability to attract cats is a coincidence.

This is a genus of about 150 species native to Europe and Asia. A few species were formerly cultivated for medicinal uses and some are grown for ground cover and as ornamentals. *N. cataria* is one of the best known species and probably the most abundant one in North America. It was introduced as a medicinal plant for herb gardens but has escaped widely.

Catnip is found along roadsides and fence rows and in neglected yards, open fields, and waste areas from New Brunswick to Oregon, south to Georgia and Texas.

✕ An aromatic tea can be made from the dried or fresh leaves. A mint flavored nibble can be made by dipping the leaves in egg white, sprinkling with sugar and allowing to dry.

℞ In former times a tea made from the dried

flower clusters and leaves was used to relieve colic in children. In southern Appalachia, catnip tea has been used since the eighteenth century as a remedy for colds. It has also been used for upset stomach and intestinal gas.

Prunella vulgaris
Heal-all*

257

Also Self-heal, Carpenter-weed, Woundwort
Flowers blue, in dense clusters at the tips of stems and branches; corolla 2-lipped, upper lip forming a hood, lower lip fringed; leaves opposite, stalked, oval to lanceolate; stems square, sprawling at base with erect tips, sometimes rooting at nodes, usually branched, to 60 cm. long (2 ft.).

May–October

A perennial with a short rootstock of North America, Europe and Asia. Fertilization is the result of cross-pollination usually by bumblebees and butterflies. Each flower produces 4 tiny, brown, angular, finely ridged seeds which are enclosed in the persistent calyx. An interesting type of seed dispersal that may function here has been refered to as a rain-ballistic method. A raindrop striking the ridged calyx tube bends it slightly; as it springs back the seeds are hurled out the open throat of the tube.

When it is persistently mowed, trampled or grazed, it forms spreading mats, the individual plants low with smaller leaves and flower clusters. As a weed in lawns, patches may have to be dug out and the lawn reseeded. In fields and pastures, clean cultivation or a close growing crop to crowd out the weed is recommended as a control measure.

The genus consists of about 6 species with only 1 native to North America. The genus name may be derived from the German word *bräune*, the name for a throat disease for which the plant was used as a remedy. An earlier spelling of the genus was *Brunella*.

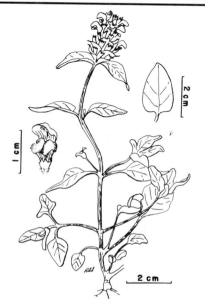

P. vulgaris is a highly variable species with several varieties and forms in the Northeast. A white-flowered variety is occasionally observed.

Heal-all is found along roadsides and in pastures, lawns, meadows, and waste spaces throughout the United States and southern Canada.

℞ The plant is bitter with slight astringent properties. It has been used for external wounds, ulcers of the mouth and throat, and diarrhea. In the seventeeth century it was highly regarded as a wound herb, and when mixed with other herbs was thought to help the body overcome all sorts of internal and external afflictions; thus the names Heal-all, Self-heal, and Woundwort.

228

Satureja acinos
Mother-of-thyme*

258

Lamiaceae or Labiatae
Mint Family

*Also Basil Thyme

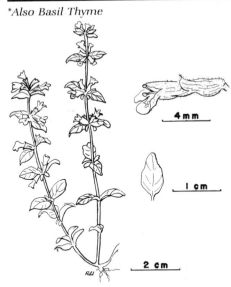

4 mm

1 cm

2 cm

R.L.L

An annual or short-lived perennial introduced from Europe. Each of the 4 seeds produced by the numerous flowers is pointed and has a light spot on one end and a pitted brown surface. These are small enough to be windborne in a moderate

Flowers light purple, in clusters of 1,s–3,s, in upper axils, corolla 2-lipped, upper lip notched or entire, lower lip 3-lobed; calyx 2-lipped, about half as long as corolla tube, base of lower side enlarged; leaves opposite, oval, usually entire, to 12 mm. long (½ in.); stems square, hairy, usually branched from base, without runners, to 40 cm. high (16 in.).

June–September

breeze. Although it is often observed in orchards and fields, it is usually not a troublesome weed.

This is a large genus mostly of tropical and subtropical areas with a large concentration in the Mediterranean region. It includes many aromatic herbs and semishrubs, some of which are cultivated as border plants and pot herbs. There are 5 species in the Northeast, 3 introduced from Europe, and several others in the southeastern United States.

Mother-of-thyme is found in open fields and waste places and along roadsides from Quebec and Ontario south to New Jersey, West Virginia, and Michigan.

Satureja vulgaris
Wild Basil*

259

Lamiaceae or Labiatae
Mint Family

*Also Dogmint, Field Basil

A perennial with short, creeping runners of North America and Europe. Nectar with a sugar content of 85% in some plants of this genus suggests pollination by bees. Each flower produces 4 egg-shaped, smooth, brown seeds. These are small enough to be windborne in a moderate breeze as they are

shaken from the persistent calyx.

The plants in the southern part of the range of this species were probably introduced from Europe, where it is abundant, but those in the northern part are native to North America. The native variety is less hairy and sometimes goes by the name Dogmint.

Flowers pink-purple, crowded in terminal and upper axillary, dome-shaped clusters; corolla 2-lipped, upper lip notched, lower lip 3-lobed; calyx tubular, hairy, about as long as corolla tube; leaves opposite, oval, entire or slightly toothed, to 4 cm. long (1½ in.); stems square, hairy, often branched, to 60 cm. high (2 ft.).

June–September

Wild Basil is found in old fields, pastures, and meadows, along roadsides, and in open woods from Newfoundland and Quebec to Manitoba, south to North Carolina, Tennessee, and Kansas. It also occurs in Colorado, New Mexico, and Arizona.

✖ The fresh or dried leaves can be used to make a tea, and the dried leaves can be used as a mild substitute for commercial basil. A related species, *S. hortensis*, Savory, not illustrated, a native of Europe, is cultivated in several states and occasionally escapes. It is used as a flavoring for dressing, sauces, and gravies.

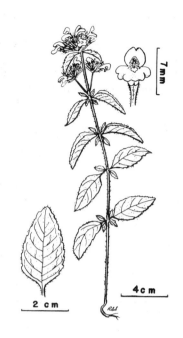

2 cm   4 cm   7 mm

Hemerocallis fulva
Orange Day Lily

260

Liliaceae
Lily Family

Flowers orange, funnel-shaped, with 6 backward curving segments, to 10 cm. wide (4 in.), in a terminal cluster on a leafless stalk; leaves basal, grasslike, to 2.5 cm. wide (1 in.), tapering to a sharp point, to 60 cm. long (2 ft.); stem leafless, to 1.8 m. high (6 ft.).

June–August

A perennial with a branching rhizome and tuberous roots, introduced from Europe. Each showy flower opens for a single day, then wilts. This is noted in the genus name which is derived from Greek words that mean "beautiful for a day." The flowers are odorless and do not produce seeds.

This is a genus of about 15 species native to eastern Asia. It was introduced into Europe at an early date and is now distributed from France to Japan. Of the two species in the Northeast, both were introduced as cultivated ornamentals but have escaped and can now be observed in the wild.

4 cm

The related species, *H. flava*, Yellow Day Lily, not illustrated, is fragrant, is cross-pollinated by insects, and has yellow flowers.

Orange Day Lily has more widely escaped from cultivation than Yellow Day Lily. It is often observed in very dense colonies in which there is a complete exclusion of other species. These clusters very often mark the location of an old homesite. This plant is an attractive border plant in gardens but requires attention to prevent it from getting out of control.

Orange Day Lily is found along roadsides, borders of fields, untended meadows, and waste areas from New Brunswick south to Virginia and Tennessee.

✖ The young shoots can be used in salads or prepared as asparagus. The flower buds and the newly expanded flowers can be cooked as vegetables, dipped in batter to make fritters, added to soups, cooked with meat dishes and noodles, or prepared as a side dish, like mushrooms. They can also be dried and preserved for later use as an additive and a flavoring herb. The clusters of white tubers can be cooked as a vegetable and have a taste somewhat like corn.

Lilium superbum
Turk's-cap Lily

261

Liliaceae
Lily Family

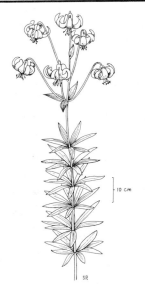

Flowers red-orange with purple spots and green star-shaped centers, one-several, nodding; perianth parts 6, curled back to base of flower, stamens 6, orange-red, protruding; principal stem leaves in whorls of 3's–8's, tapering at both ends; stems sturdy, unbranched below flower cluster, to 2.1 m. high (7 ft.).

July–August

This is a genus of about 80 species of the Northern Hemisphere. Several species, such as Easter Lily and Madonna Lily, are widely cultivated as ornamentals. Of the 8 species that occur in the Northeast, 6 are native to the region. Turks-cap Lily, as the specific name indicates, is a superb plant and a treat to observe in its natural setting. A variety of this species, in which the green star-shaped center is smaller and less defined, is classified by some botanists as *L. michiganense*, Michigan Lily.

Turk's-cap Lily is found on alluvial flats and in wet meadows, swales, and other low ground from New Brunswick to Manitoba, south to Florida and Missouri.

A perennial with a globular white bulb that develops runners in spring. The flowers are self-incompatible and cross-pollination is effected mainly by bees and butterflies. Many closely stacked, flat seeds are produced in the oblong, 3-lobed seedpod. A thin papery wing surrounding each seed aids in dispersal by wind.

✖ The bulbs of all native species of lilies are edible, and several were used as food by American Indians. Today, however, unless they are cultivated for that purpose,

231

there is no reason, except in emergencies, to collect these plants for food. To do so may result in their extinction from local areas.

Ornithogalum umbellatum / Star-of-Bethlehem — **262** — *Liliaceae* / Lily Family

Flowers white on inside, in a branched terminal cluster of 5–6, on a leafless stalk; petals 6, with green stripes on the underside, stamens 6; leaves basal, grass-like, with pale midrib; plant to 30 cm. high (1 ft.).

May–June

A perennial with a fleshy, onion-like bulb, introduced from Europe. The lower flowers have longer stalks than the upper, forming a somewhat flat-topped cluster of blossoms that open only in bright sunlight. The 3-lobed seed capsule splits into three sections to release a few black, coarsely granular seeds. In some species of this genus the seed coat contains a food substance that attracts ants, which serve as agents of dispersal.

This is a genus of about 100 species native mainly to the tropical areas of the Old World. The nonsensical genus name is derived from Greek words meaning "bird's milk." Several species have been introduced into North America as flower garden ornamentals, but only O. umbellatum has escaped widely.

A related species, O. nutans, not illustrated, is similar but has nodding flowers in a cylindrical cluster and attains a height of 60 cm. (2 ft.). It has escaped cultivation and occurs in open areas from New York to Maryland and the District of Columbia.

Star-of-Bethlehem is found along roadsides and in grassy areas and wastelands

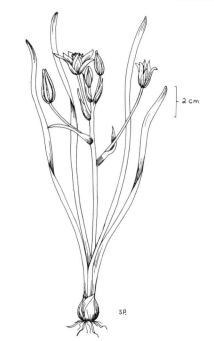

from Newfoundland to Ontario and Nebraska, south to North Carolina, Mississippi, and Missouri.

All parts of the plant contain toxic compounds called cardiac glycosides, that can cause nausea and irritation of the intestinal tract. In one case on record from Maryland, 1000 sheep died in one year after eating the bulbs.

Open Fields

232

*Also Harebell, Squill

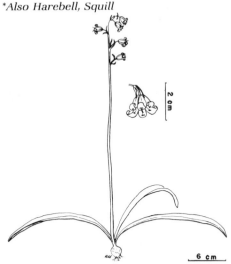

2 cm

6 cm

A perennial with an underground bulb, introduced from Europe. It is self-pollinated, and the 3-celled seed capsule splits into 3 sections, releasing a few to several seeds. In some species of the genus a soft appendage on the seeds attracts ants and results in their being planted in ant hills.

Flowers blue, sometimes white, bell-shaped, nodding, in a loose terminal cluster of 6–12 or more, on a leafless stalk, each flower subtended by 2 long bracts; leaves basal, long and narrow, grass-like; flowering stem smooth, to 30 cm. high (1 ft.).

May–June

The species includes lilac and pink horticultural varieties.

A related species, *S. sibirica*, Siberian Squill, not illustrated, has a shorter flowering stem, to 15 cm. (6 in.), and usually 3 deep blue flowers. It is a native of Russia and southwestern Asia and also exists in several colored horticultural varieties. Its flowering period lasts for only a few days.

English Bluebell and Siberian Squill were both introduced as early flowering ornamentals. They are most attractive when growing in dense masses and are particularly effective as border plantings. Both have escaped from cultivation to open fields, vacant lots, and roadsides and are spreading in the Northeast.

An annual or winter annual that may be dwarfed and unbranched on poor soils. The enlarged, 2-celled seedpod opens at the top, exposing numerous tiny, spindle-shaped, light brown seeds that become airborne as the stem sways in the wind. It may invade cultivated areas as a weed but yields to plowing and mowing.

This genus was named in honor of M. Lobel, a sixteenth-century Belgian botanist and physician to James I of England. There are about 15 species native to the Northeast and several others in the southern and western states. The common name for *L.*

inflata originated when early American settlers observed that the leaves of this plant were dried and smoked by the Indians.

L. inflata is the most common of the lobelias in the Northeast. Other interesting common names by which it has been known are Asthma-weed, Gag-root, and Puke-weed. These are expressive reminders of its medicinal uses in the past.

Indian Tobacco is found in fields, meadows, roadsides, open woods, and waste areas from Prince Edward Island to Minnesota, south to Georgia and Mississippi.

Flowers light blue, 2-lipped, lower lip 3-lobed, upper lip 2-lobed; flowers single in the axils of small leaves or bracts at the ends of stems and axillary branches; the ovary becoming greatly enlarged (inflated) in fruit; leaves alternate, oval in outline, hairy on underside, toothed; stems usually branched, hairy, to 90 cm. high (3 ft.).

July–October

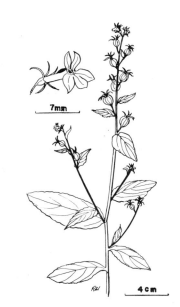

☠ It has been used for a variety of convulsive disorders: spasmodic asthma, epilepsy, diptheria. Externally it has been used as an application for sprains, bruises and skin disorders. However, several toxic substances have been identified in the plant that, in overdoses, cause death in humans. This plant should not be experimented with in home remedies.

Abutilon theophrasti Velvet Leaf*	**265**	*Malvaceae* Mallow Family

*Also Butter-print, Pie-marker

Flowers yellow, usually borne singly or a few on stalks in axils of leaves, to 2.5 cm. wide (1 in.), petals 5; fruit a cup-shaped capsule with a ring of prickles and 12–15 pointed lobes; leaves alternate, heart-shaped at base, pointed, soft-hairy, to 25 cm. long (10 in.); stems thick, often un-branched, to 1.5 m. high (5 ft.).

July–September

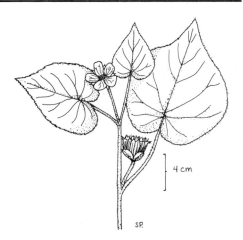

An annual with a deep taproot, introduced from India. The lifespan of each flower is about 3½ hours, and, although they are visited by several species of potential insect pollinators, fertilization is mainly the result of self-pollination. The seed capsule opens from the top, releasing numerous dull black, finely pitted seeds. These are very long-lived and have been shown to retain their germinability after dry storage or burial in soil for over 50 years. The names Butter-print and Pie-marker refer to the peculiar shape of the seed capsule.

This is a genus of over 100 species widely distributed in the tropical regions of the world. The genus name is of Arabic origin and was given by Avicenna, an Arabic scholar of the tenth century. *A. theophrasti* is the only species in the Northeast. As a result of the longevity of its seeds, it is difficult to eradicate when it becomes established as a weed. The best control measure is to eliminate the first plants before they produce seeds.

Velvet Leaf is found in fields, gardens, vacant lots, and waste areas throughout most of the United States and southern Canada.

Malva moshata
Musk Mallow

266

Malvaceae
Mallow Family

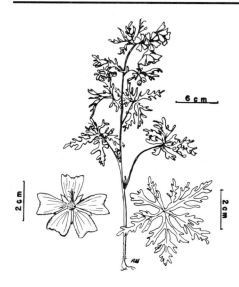

6 cm

2 cm

2 cm

Flowers pink or white, to 5 cm. wide (2 in.), with 5 slightly notched petals, on axillary stalks and in terminal clusters; filaments united into a tube around the style; fruits referred to as cheeses, circular with numerous segments; stem leaves alternate, lower and basal ones 5-lobed, upper ones deeply palmately lobed; stems hairy, to 60 cm. high (2 ft.).

June–October

which has given the plant its common name. The occurrence of white flowered forms is not uncommon, and they are often observed mixed with the pink forms.

A related species, *M. alcea*, Vervain Mallow, not illustrated, is very similar but has star-shaped hairs on its stems and leaves that are not as deeply lobed. It has escaped cultivation as an ornamental but is not as common or as widely distributed as *M. moschata*.

A perennial with a deep taproot, introduced from Europe. Self-pollination is avoided by the maturation of the stigma before pollen is released. The pollen grains are very spiny and tend to cling together in masses, as is typical for many insect pollinated plants. The round flat fruits consist of 10–20 segments, each with a single seed.

It flowers from early summer to mid-fall and is often an ornamental in flower gardens and borders. It has escaped widely and is an attractive roadside weed. Since it is capable of surviving in tall grass, it sometimes invades hay fields where heavy infestations reduce the quality of the hay. Control measures include digging out scattered plants and mowing badly infested areas before seeds are developed.

The large showy flowers are made even more attractive by the faint odor of musk,

Musk Mallow is found along roadsides and in grassy fields, meadows, and waste areas from Nova Scotia to British Columbia, south to Virginia, Tennessee, and Oregon.

✗ The plants of this genus have a high mucilage content; thus the young leaves can be used as a thickener for soup-like dishes. They can also be cooked as bland potherbs, but the mucilage content gives them a slippery quality. The young flat fruits, or cheeses, can be eaten as roadside nibbles or added to salads.

Another European species of this family, *Althaea officinalis*, Marsh Mallow, not illustrated, less common North America than *M. moschata*, was the original source of the mucilaginous base for the commercial marshmallow.

℞ See *M. neglecta*.

Malva neglecta
Common Mallow*

267

*Also Cheeses

Flowers pale pink to white, in axillary clusters; petals 5, notched, filaments united into a tube around style; fruits referred to as cheeses, round and flattened, with about 15 one-seeded segments; leaves alternate, long stalked, roundish, with 5–7 shallow lobes; stems prostrate, creeping, erect at tips, to 60 cm. long (2 ft.).

May–October

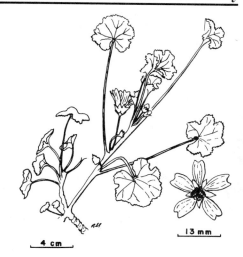

4 cm

13 mm

An annual or biennial with an unbranched taproot, introduced from Europe. Pollination and fruit characteristics are similar to those of *M. moschata*.

This is a genus of about 30 species native to Europe, northern Africa, and Asia. The genus name is derived from an ancient Greek word meaning "softening" and apparently refers to the emollient quality of the leaves of some species. Six species have been introduced, accidentally or as ornamentals, into eastern North America.

M. neglecta is probably the most common and widespread member of the genus in North America. It may invade cultivated fields and lawns where its deep taproot gives it an advantage during droughts. This advantage is particularly demonstrated by its lush greenery in unwatered lawns during dry periods. Control measures include digging out individual plants and clean cultivation before seeds have developed.

A closely related species, *M. rotundifolia*, Round-leaved Mallow, not illustrated, is very similar but usually has only about 10 sections in the fruit and petals that are only slightly longer than the sepals. It is more common in the western and southern states than in the Northeast.

Common Mallow is found in cultivated and neglected fields, gardens, roadsides, and waste areas throughout the United States and southern Canada.

✖ See *M. moschata*.

℞ The mucilage content of members of this genus has contributed to their medicinal uses. They have been used in remedies for coughs and colds, and applied externally for skin irritations.

One species in Australia, *M. parviflora*, not illustrated, has been shown to be responsible for a sometimes fatal disease in livestock called "shivers" or "staggers" when the green plant is eaten in large quantities.

Cannabis sativa
Hemp*

268

*Also Marijuana

An annual with a much branched taproot, introduced from Asia. Pistillate and staminate flowers are usually on different plants, and although nectaries are present, pollen is produced in great quantities and it is wind-pollinated. The nectaries may be relicts of an earlier evolutionary form that was insect-pollinated. Each of the numerous pistillate flowers produces a single yellow to olive-brown seed with a network of veins on its surface.

Open Fields

236

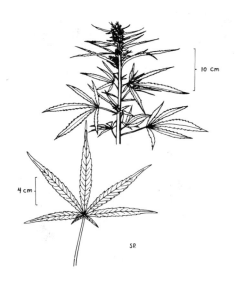

This is a genus of one species native to central Asia. It is the source of a fiber for ropes and other cordage, and the hallucinogenic drug tetrahydrocannabinol (THC). It has been cultivated from antiquity, mostly for the fibers in the stems. Brought to North America by the earliest settlers, this plant was at one time an important fiber crop in Kentucky. In the 1930's and 40's Wisconsin was the leading producer in the United States. The best quality of fibers is obtained

Flowers greenish, inconspicuous, in clusters in axils of upper leaves; lower leaves opposite, upper ones often alternate, palmately compound, with 3–7 narrow, pointed, coarsely toothed leaflets, to 15 cm. long (6 in.); stems rough to the touch, hollow, unbranched below flower clusters, to 3 m. high (10 ft.).

June–October

from staminate plants harvested when they are in full flower.

The drug THC is concentrated in the pistillate flowers as a resin-like substance called "hashish." It is present in lower concentrations in the leaves of both pistillate and staminate plants that are smoked or eaten for their drug effect. There are several geographic strains with varying concentrations of THC, from practically none in Turkish strains to high concentrations in Mexican strains. THC is toxic to many insects and is believed to have evolved as a defence against insect grazing.

It is illegal in the United States today for a private citizen to grow Marijuana, or to have parts of the plant in his or her possession.

Hemp or Marijuana is found in neglected fields, on moist flood plains, along roadsides and in waste areas throughout most of the United States and southern Canada.

Epilobium angustifolium
Fireweed*

269

Onagraceae
Evening Primrose Family

*Also Great Willow Herb, Wickup

A perennial with a creeping rootstock, of North America, Europe, and Asia. Cross-pollination is favored by the maturation of the stamens before the pistil. However, as the flower ages, the stigmas curve backward and may come into contact with the anthers, effecting self-pollination. The seedpod splits from the tip by 4 valves, releasing numerous nearly smooth, brown seeds, each of which has a parachute of white silky hairs at one end.

The genus name is a Greek word meaning "upon the pod" and refers to the location

of the flowers at the tip of the long seed capsule. This is a very large genus of about 100 species, most common in the cooler portions of the North Temperate Zone. Botanists are in disagreement as to the taxonomic status of several of the species.

E. angustifolium is an early successional species. It is commonly observed in burn scars of the northern United States and boreal Canada, giving beauty and color to an otherwise desolate landscape; thus the common name Fireweed. The tufted windborne seeds enable it to spread rapidly under these circumstances. The name Willow

Flowers pink to purple, numerous, in an elongate terminal cluster, 4 unequal petals, upper flower buds drooping; leaves alternate, narrow lanceolate, with conspicuous venation; stems erect, smooth, to 2 m. high (6½ ft.).

June–September

Herb is evidently a reference to the similarity of the shape of the leaves of this genus to those of the willow tree.

Fireweed is found in open fields, recently cleared land, and burned forest areas from Newfoundland to Alaska, south to New England and California, and in the eastern mountains to West Virginia and North Carolina.

✗ The young shoots can be used as a substitute for asparagus, and the leaves, before flowering, can be cooked as greens. American Indians of the Northwest used the fresh pith from large stems to make a thick soup or to eat directly. The dried mature leaves have been used in England and Russia as a substitute for tea.

℞ The roots and leaves have been used to prepare a tonic with astringent properties. This is reputed to be useful in the treatment of whooping cough, hiccoughs, asthma, and irritated mucous membranes.

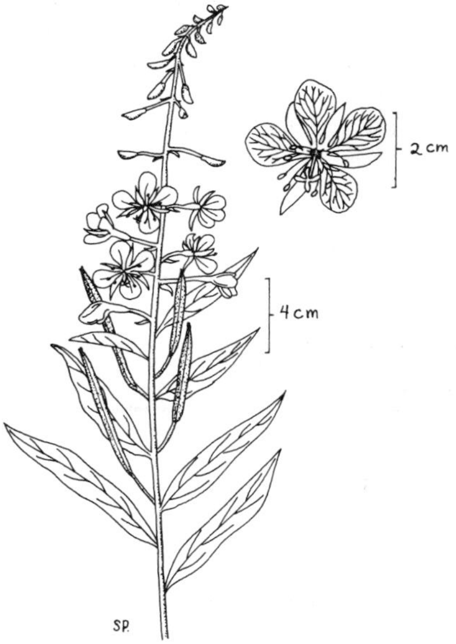

Epilobium hirsutum 270 *Onagraceae*
Hairy Willow Herb — Evening Primrose Family

Flowers pink to purple, in clusters, in upper leaf axils, petals 4, notched; leaves opposite, hairy, toothed; stems freely branched, hairy, to 1.5 m. high (5 ft.).

July–September

A perennial with a thick rhizome and underground runners, native of Europe, Asia, and northern Africa. The seed capsule is up to 8 cm. long (3 in.) and splits into 4 sections, releasing a large number of brown, longitudinally ridged seeds, each with a tuft of nearly white hairs.

Although it is recognized as an introduced weed, the rather large showy flowers of this species give color to some otherwise uncolorful areas without seeming to be a threat to native species. In cultivated areas where it has become established, it can be controlled by improving the drainage.

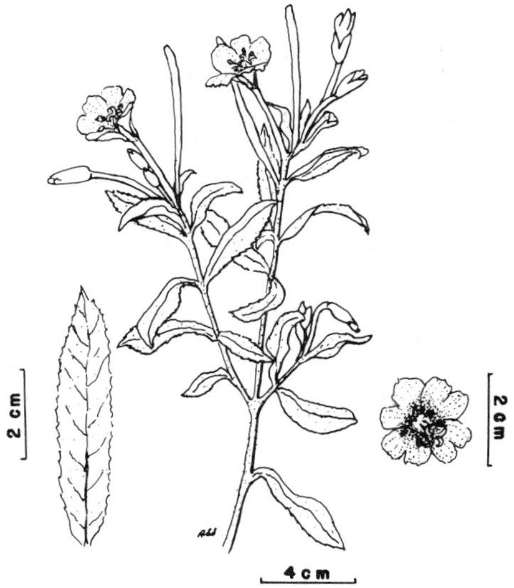

238

Hairy Willow Herb is found in roadside ditches, wet meadows, and low waste places from Quebec to Illinois, south to New Jersey and Ohio.

For other ecological information see *E. angustifolium*.

✖ ℞ It is unclear whether the food and medicinal information for *E. angustifolium* is valid for this species. There seems to be no record of poisoning ascribed to this genus.

Oenothera biennis
Evening Primrose

271

Onagraceae
Evening Primrose Family

2cm

s.p.

Flowers yellow, numerous, often crowded at the ends of stems and branches; stigma cross-shaped, petals 4, sepals 4, turned back; leaves alternate, lanceolate, often wavy-margined; stems often tinged with red, to 1.8 m. high (6 ft.).

June–October

Several species of this genus in the Northeast exist as numerous local races. These persist normally by self-pollination but are very similar and interfertile. Sometimes cross-pollination between two races occurs and a new race is formed. Thus it is not uncommon to find intermediate forms that do not fit the descriptions given in plant manuals.

This is a very large genus native to North and South America, mostly in the Temperate Zone. The number of species in western North America is greater than in the Northeast. *O. biennis* has been introduced into Europe and is widespread there as a weed. The Mutation Theory of evolution, developed by Hugo De Vries in 1900, was based on experiments with this and closely related species.

In North America, *O. biennis* is found along roadsides and in abandoned fields and waste areas from Labrador to Minnesota, south to Florida and Texas.

A biennial which normally develops a pinkish rosette of leaves and a fleshy taproot in the first year of growth. It is normally self-pollinated, but the fragrant flowers, which usually open near sunset and fade in the heat of the following day, are visited by moths, bumblebees, and honeybees.

The 4-angled elongate seed capsule opens from the top to release many tiny, irregular, brown seeds. These have been demonstrated to have great longevity, showing germinability after storage for 70 years in the soil. They are eaten and perhaps dispersed by the Common Goldfinch. The plants sometimes spread to cultivated fields as weeds. Control measures include plowing or cutting the first-year rosette and close mowing the second-year stems before seeds are formed.

✖ The first-year roots, collected in late fall or early spring, can be cooked in two or three changes of water as a root vegetable. The young leaves of the rosette can be cooked in two or three changes of water or used in salads, but are too bitter for the taste of some.

℞ As with many bitter plants, it is considered in herbology to have medicinal properties. The dried plant is recommended as a remedy for asthmatic coughs and whooping cough, and as an ointment it has been used for skin irritations.

Spiranthes cernua
Nodding Ladies' Tresses*

272

Orchidaceae
Orchid Family

*Also Autumn Ladies' Tresses, Screw-auger

Flowers white, with 3 petals and 3 petal-like sepals, with a slight tendency to bend downward (thus nodding); flowers in 3 spiral rows like the threads of a screw, with a light vanilla odor; leaves mostly basal, narrow and grasslike, to 30 cm. long (12 in.); flowering stem with leaves reduced to sheathing bracts, to 60 cm. high (2 ft.).

August–September

A perennial with tuberous fleshy roots. Some forms of the species produce seeds without fertilization. In some cases this may occur when the flower has not been pollinated or has been pollinated with pollen from another species. Other forms are known to reproduce sexually and are usually pollinated by bees. Hybrids have been reported between the latter forms and *S. gracilis*, Southern Slender Ladies' Tresses, not illustrated. The seed capsule splits into 3 sections, releasing numerous irregular, dust-like seeds that become windborne in the slightest breeze.

This is a genus of about 25 species including at least 8 that are native to eastern North America. All species are characterized by 1 or 3 spiral rows of white flowers. These plants are usually not abundant in undisturbed habitats but some species have greatly increased in disturbed areas.

Nodding Ladies' Tresses is found in moist, acid soils in moist meadows, open woods, and bogs from Nova Scotia and Quebec to Wisconsin, south to Florida and Texas.

4 cm

7 mm

240

Oxalis europaea
Yellow Wood Sorrel*

273

Oxalidaceae
Wood Sorrel Family

*Also Lady's Sorrel

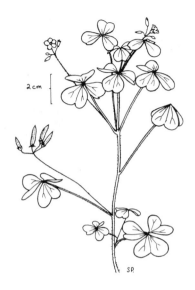

An annual or perennial with thick underground runners, of North America and Europe. The slender, pointed, seed capsule splits into 5 sections, sometimes explosively, hurling outward several light brown, ridged seeds. It is often an abundant weed in lawns, gardens, and fields. This species is recognized by some botanists as O. cor-

Flowers yellow, to 12 mm. wide (½ in.), on slender stalks, in terminal or axillary clusters of 2's to 6's; sepals 5, petals 5, stamens 10, pistil 1; leaves alternate, long stalked, with 3 heart-shaped leaflets, often purplish on underside; stems branched, hairy, often prostrate at base, to 37.5 cm. high (15 in.).

May–October

niculata. The plants of this genus are eaten by several species of songbirds and game birds.

This is probably the most variable of the 7–9 species of the genus in the Northeast. The several varieties are differentiated by the types of hairs on the stems, leaves, and flower stalks. Forms with orange-yellow and pale lemon colored flowers are occasionally observed. A related species, O. stricta, not illustrated, is very similar but has erect seed capsules usually with abruptly bent stalks. It occurs throughout most of the United States and southern Canada in the same types of habitats as O. europaea.

Yellow Wood Sorrel is found in cultivated and abandoned fields, along roadsides, and in waste areas from Quebec to North Dakota, south to Georgia, Arkansas, and Arizona.

✂ See O. acetosella.

Chelidonium majus
Celandine*

274

Papaveraceae
Poppy Family

*Also Swallowwort

A biennial thoroughly established in eastern North America but a native of Europe and Asia. The flowers do not produce nectar, but cross-pollination is usually effected by pollen-collecting bees. The seed capsule splits into 2 sections from the bottom upward, releasing several egg-shaped, glossy, black, pitted seeds. Each seed has a soft appendage that attracts ants as agents of dispersal.

The genus name is derived from the Greek

word for "swallow" and, by one account, refers to a belief held by ancient Greek scholars that the female swallow bathed the eyes of her young in the yellow-orange juice to give them improved vision. By another account the flowering period of the plant begins when the swallows arrive and ends when they depart. It is a genus of only one species.

Celandine is found in fields and untended gardens, along roadsides, and in waste areas, often near human habitations,

Flowers yellow, to 16 mm. wide (⅔ in.), in loose clusters at the ends of stems and branches; sepals 2, shedding when the flower opens, petals 4, stamens numerous, pistil 1; leaves alternate, deeply pinnately lobed, whitish on underside, to 20 cm. long (8 in.); stems branched, hairy, with yellow-orange juice, to 80 cm. high (32 in.).

May–August

from Quebec and Ontario, south to Georgia and Missouri.

✕ M. L. Fernald reported that he had eaten the young leaves of Celandine with oil and vinegar and found them pleasing. However, according to W. H. Lewis and M. P. F. Elvin-Lewis, loss of life in Europe has resulted from consuming the leaves after mistaking them for parsley. Caution is advised.

℞ The whole plant, collected while in flower and dried, and the fresh juice have been used in herbal medicine. The orange colored juice has been applied directly as a treatment for warts, corns, ringworm, and other skin diseases. The juice, mixed with milk, has been used as an eyewash and as a treatment for itch when combined with

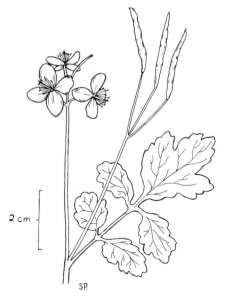

sulphur. A solution made by steeping the plant in hot water has been used in England as a mouthwash for toothache. An ointment made by boiling the plant in lard has been used for hemorrhoids.

Phytolacca americana Pokeweed*   **275**   *Phytolaccaceae* Pokeweed Family

Also Scoke, Garget, Pigeonberry

A perennial with a very large, deep tap-root and dangling clusters of mature fruits. The juicy ripe berries make up 2–10% of the diet of the Mourning Dove, Bluebird, Catbird, Mockingbird, and Cedar Waxwing, and are eaten by several other species of songbirds and mammals. Each berry contains 10 smooth, lens-shaped, glossy black seeds which are probably dispersed by these consumers.

This plant is sometimes a weed pest in gardens, where it may serve as an alternate host for cucumber mosaic disease. It can be destroyed by cutting the root just below the ground with a spade or hoe. The common name is said to be derived from an Algonquian Indian name for the plant which

means "a plant used for staining or dyeing". The juice from the mature berries has been used to color food and wine, as a pigment for paint, and as a writing fluid.

This is a tropical or semitropical genus with only one species in the Northeast. It is planted as a vegetable in southern Missouri, and it has been introduced into Europe where it is cultivated as a garden vegetable. In North America Pokeweed is found in pastures, fields, waste places, and woodland openings from Quebec and Ontario, south to Florida and Texas.

✕ The young shoots, to 15 cm. high (6 in.), can be prepared as asparagus or pickled, and the very young leaves can be cooked as greens. At least two changes of water are recommended in cooking (See the next par-

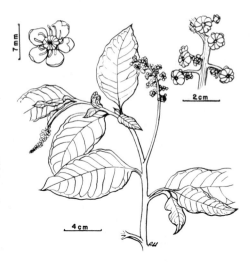

Flowers greenish-white tinged with pink, in stalked clusters, to 20 cm. long (8 in.), opposite the leaves; stamens 10, petals none, petal-like sepals 5; fruit a purple-black berry with a red stalk; leaves alternate, long stalked, entire, prominently veined, to 30 cm. long (1 ft.); stems smooth, thick, succulent, purple tinged, branching, to 3 m. high (10 ft.).

July–September

☠ In the nineteenth century the dried root was used to cause vomiting. The Pamunky Indians of Virginia used a tea made by boiling the berries to treat rheumatism. W. H. Lewis and M. P. F. Elvin-Lewis report that *P. americana* contains compounds that may have a harmful influence on blood cells, and that these compounds can be absorbed through skin abrasions. They are of the opinion that it should not be used as a potherb and that gloves should be used when removing the plant as a weed from gardens.

agraph). *Caution*: No part of the root or any portion of the stem that has a purple cast should be eaten. The root, the mature plant and the seeds are poisonous.

Plantago aristata
Bracted Plantain*

276

Plantaginaceae
Plantain Family

Also Buckhorn

Flowers greenish-white, crowded at end of leafless stalk, each subtended by a stiff bract twice as long as flower, or longer; flowers with 4 membranous petals and 4 hairy sepals; flower cluster to 10 cm. long (4 in.); leaves basal, very narrow, to 18 cm. long (7½ in.), pointed; flower stalk to 30 cm. high (1 ft.).

June–October

An annual or winter annual native to western North America. Pollination is by wind, and each flower produces 2 tiny, oval, brown seeds. These become sticky when wet and are dispersed by adhering to the coats of animals or by becoming attached to leaves that are subsequently blown in the wind.

The plants of this genus are eaten by a number of wildlife species, including the Ruffed Grouse, Cardinal, Cottontail Rabbit, Gray Squirrel, and White-tailed Deer. These animals no doubt help to spread the seeds on their coats and in their digestive tracts. All the plantains have been declared noxious weeds in the seed laws of 6 states. In order to eradicate extensive infestations in cultivated fields, two years of clean cultivation may be required.

Bracted Plantain was originally a western species growing from Illinois to Oregon, south to Louisiana and New Mexico. In the late nineteenth century it began spreading eastward. Today it is found in open fields, grasslands, dry meadows, and waste areas from Maine to British Columbia, south to Florida and Texas.

Plantago lanceolata English Plantain*

277

Plantaginaceae Plantain Family

*Also Ribgrass, Ripplegrass
Flowers greenish white, inconspicuous, in a dense, cylindrical cluster, to 8 cm. long (3 in.), at the tip of a leafless stalk; leaves basal, in a rosette, narrowly lance-shaped with entire margins, pointed at tip, with 3–5 prominent parallel veins, to 30 cm. long (1 ft.); flowering stem to 60 cm. high (2 ft.).

May–October

A perennial with a persistent thickened base and slender rootlets, introduced from Europe. Fertilization is mainly the result of cross-pollination by wind. It produces large quantities of pollen and is often a cause of hayfever, especially in early summer when flowering is at its peak. The flower clusters are visited by honeybees, which collect the very dry pollen after wetting it with regurgitated honey. The peak period of pollen release is between 7:00–10:00 a.m..

The seed capsule opens by a lid after splitting horizontally at or below the middle, to release 2 smooth, shiny brown, boat-shaped seeds. These are sticky when moist and are dispersed in the same manner as those of *P. aristata*. They are eaten by game birds and songbirds and are sometimes used to feed caged birds. The leaves are favored as a food by rabbits.

This species has been declared a noxious weed in the seed laws of 18 states. Heavy infestations in cultivated fields may require

4 cm

two years of clean cultivation to eradicate. Evidence from experimental studies suggests that the growth of *P. lanceolata* may be depressed when it occurs with *Trifolium repens* (White Clover).

English Plantain is a common weed of lawns, roadsides, pastures, and waste places throughout the United States and southern Canada.

℞ *P. lanceolata* and *P. major* are both mild astringents that have been used in fresh crushed form for bruises, cuts, scratches, and wounds. The same treatment has been used as an external application for hemorrhoids.

The use of these plants for healing apparently dates from ancient times. In *Romeo and Juliet*, one of Shakespeare's characters refers to the use of plantain leaves for a broken shin.

Plantago major
Common Plantain*

278

Plantaginaceae
Plantain Family

4 cm

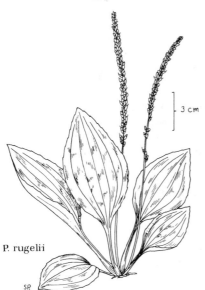

3 cm

P. rugelii

*Also Whiteman's-foot, Broadleaf Plantain
Flowers greenish, inconspicuous, in a
dense cluster, to 30 cm. long (1 ft.), on a
leafless stalk; leaves basal, in a rosette,
broadly oval-shaped, with long grooved pe-
tioles, with prominent ribs; flowering stem
to 50 cm. high (20 in.).

June–October

A perennial with a highly branched, fibrous root system, native to North America and Europe. Pollination is by wind, but pollen is shed in small quantities and this is not considered to be a hayfever plant. The pistil matures before the stamens, the style protruding before the corolla opens; thus self-pollination is avoided.

The seed capsule splits horizontally about the middle and opens like a lid to release up to 20 seeds. These become sticky when moist, and are dispersed in the same manner as those of *P. aristata*. The seeds and leaves are favored foods of rabbits and several other species of rodents.

This is a highly variable species with several very large and very small forms. It is native to some parts of northern North America but has been introduced from Europe into the Northeast. The name Whiteman's-foot was given by the American Indians because this plant seemed to grow everywhere the Europeans set their feet.

A related species *P. rugelii*, Rugels Plantain (Red-stemmed or Pale Plantain), is very similar (see illustration) but has slightly larger seed capsules, each of which contains 2–9 seeds and opens to below the middle by a lid. The leaves are thinner and have stalks that are reddish at the base. Rugels Plantain is a native of North America and is found in the same types of habitats as *P.*

major. Its range is restricted to eastern North America where it may be more abundant than *P. major.*

Common Plantain is found in lawns, along roadsides, and in waste areas throughout the United States and southern Canada.

✷ The very young leaves can be used in salads or cooked as greens. They must be picked young because they soon become tough and stringy.

℞ See *P. lanceolata.*

**Polygala sanguinea
Field Milkwort***

279

**Polygalaceae
Milkwort Family**

---

*Also Purple Milkwort
Flowers rose-purple, in very dense, cylindrical clusters, with flowers overlapping, at the ends of stems and branches; leaves alternate, narrow, to 4 cm. long (1½ in.); stems erect, branched, often bushy, to 40 cm. high (16 in.).

June–September

An annual with a root that has the odor of wintergreen when crushed. The flowers at the bottom of the cluster open first and continue toward the tip. The fruits are shed as soon as they mature and often also the subtending bracts, leaving the axis bare. Thus the stalk may seem to elongate as the flower cluster shrinks.

The name milkwort is derived from the genus name, which is Greek for "much milk" and refers to an early belief that some plants of this genus, if eaten by cows, would increase the flow of milk. The name should not be confused with milkweed (see *Asclepias*) which was named for its milky juice. The specific name is Latin for "blood" and inappropriately refers to the rose-purple color of the flowers.

Field Milkwort is found in open acid fields

2 cm

and sterile meadows and along roadsides from Nova Scotia to Minnesota, south to South Carolina, Louisiana, and Oklahoma.

℞ See *P. paucifolia.*

Polygonum aviculare **280** ***Polygonaceae***
Knotweed* **Smartweed Family**

**Also Doorweed, Knotgrass*
Flowers green with white or pink margins, tiny, in axillary clusters; leaves alternate, blue-green, to 3 cm. long (1½ in.), oblong or elliptical; stems usually prostrate, freely branched, mat forming, to 2 m. long (6½ ft.).

June–October

An annual with a thin taproot, of North America, Europe, and Asia. Fertilization may be the result of self-pollination, or the seeds may develop without fertilization. Each flower produces a single triangular, dark brown, longitudinally ridged seed. The plant itself may serve as an agent of dispersal, dropping seeds as it spreads over the ground. In addition, the seeds are eaten by upland game birds, songbirds, and small mammals that help spread the seeds. This use of the plant for food by birds is recognized in the specific name.

This is a highly variable species and botanists are not in agreement on its classification. In some systems it has been listed as *P. heterophyllum* or *P. monspeliense*. It is sometimes a weed in cultivated fields, driveways, paths, and lawns. Control methods include hoeing, hand pulling, and enriching the soil in order for more desirable plants to become established. It cannot survive in competition with other species.

Knotweed is found in sterile soil along roadsides and in open fields, sidewalk cracks, waste areas, salt marshes, and sandy beaches throughout the United States and southern Canada.

The roasted seeds have been used as a food and ground for use as flour. The flour is somewhat similar to buckwheat flour.

℞ An infusion of the flowering plant has astringent and diuretic properties and has been used as a substitute for quinine. It is reported to cause contact dermatitis in sensitive individuals.

Polygonum cuspidatum **281** ***Polygonaceae***
Japanese Knotweed* **Smartweed Family**

**Also Japanese Bamboo, Mexican Bamboo*

A perennial with a thick, creeping, underground rhizome, native to Japan. The flowers are either staminate or pistillate; thus an outside agent is required for pollination. The seed is tightly enclosed by the calyx of the pistillate flower. At maturity the calyx enlarges and becomes strongly 3-winged as an adaptation for dispersal by wind.

Introduced as an ornamental, this plant has escaped widely and is spreading rapidly. It is similar to true bamboo in its very rapid growth rate. As a result of this and its ability to spread quickly by rhizomes, it can take over a large area in a short period of time. Once it has become established it often persists in spite of efforts toward eradication.

Japanese Knotweed is found along moist roadsides, in neglected yards and gardens, and in waste places from Newfoundland to Minnesota, south to Maryland and West Virginia.

The young shoots, to 30 cm. high (1 ft.) can be cooked as asparagus or made into a

Flowers greenish-white, in numerous branching axillary clusters, to 15 cm. long (6 in.); fruit a triangular, black, shiny nutlet; leaves alternate, broadly oval, with a pointed tip and a flat or heart-shaped base, to 15 cm. long (6 in.); stems thick, woody at base, often with zig-zag branching, to 3 m. high (10 ft.).

July–September

puree. Slightly older stems can be used as a substitute for rhubarb in pie and a rhubarb-like jam. The young rhizome can be cooked as a vegetable. This plant is abundant wherever it occurs and is often a noxious weed. A recommended control measure is to serve it for dinner.

4 cm

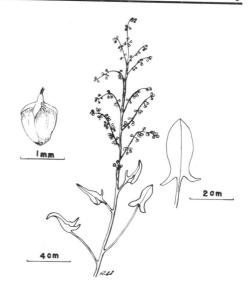

Rumex acetosella
Sheep Sorrel*

282

Polygonaceae
Smartweed Family

Also Red Sorrel, Sour Grass
Flowers inconspicuous, greenish at first, turning red later, in long branching, leafless, terminal clusters; leaves alternate, with 2 small basal lobes and a long pointed central lobe, sour to the taste; stems smooth, often branched, to 40 cm. high (16 in.).

June–October

A perennial with an extensive shallow creeping rhizome, introduced from Europe. Staminate and pistillate flowers are on different plants, and cross-pollination is by wind. The peak of airborne pollen is in June and July, but it is not considered an important hayfever plant outside of California and Oregon.

Each pistillate flower produces a single tiny, triangular, glossy brown, seed-like fruit. These and the leaves are eaten by the Ruffed Grouse, Redwing Blackbird, Hoary Redpoll, Eastern Grasshopper Sparrow, and other birds as well as rabbits and other small mammals, all of which probably contribute to seed dispersal.

Heavy infestations of this plant are usually an indication of acid soil conditions, but it also grows well on neutral or alkaline soils if they are low in nitrates. It can be eradicated in crop fields by clean cultivation and the addition of nitrogen fertilizer. This plant has been declared a noxious weed in the seed laws in 6 states.

Sheep Sorrel is found in fields and lawns, along roadsides, and in waste places throughout the United States and southern Canada.

The leaves are rich in vitamin C and can

1 mm

2 cm

4 cm

Open Fields

be added to salads, used to make a lemonade-like drink, or cooked as greens. Since it is quite tart, it is best cooked with milder greens. *Caution*: the sourness of the plant is the result of the presence of potassium oxalate, and excessive intake over a period of time can be harmful.

℞ The fresh plants have been used to reduce fevers and to treat kidney and urinary disorders. The juice of the fresh plants has been recommended for relief of the burning and itching caused by contact with stinging nettles.

Rumex crispus
Curly Dock*

283

Polygonaceae
Smartweed Family

Also Yellow Dock, Sour Dock

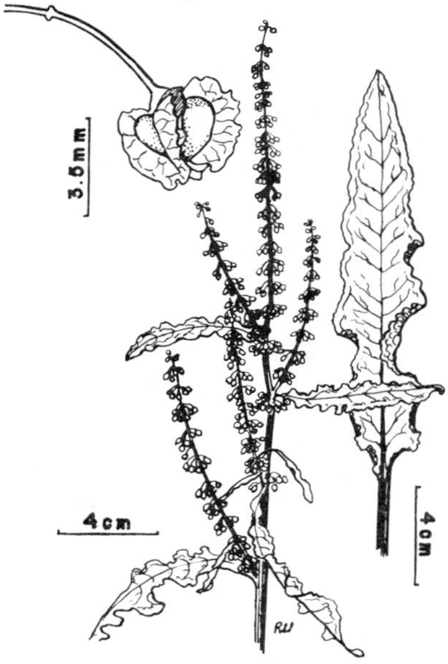

Flowers greenish, small, in dense whorls on terminal, long, leafless branches, becoming reddish brown in fruit; fruit 3-angled, glossy brown, tightly enclosed by 3 heart-shaped, nearly smooth margined wings; leaves alternate, smooth, lanceolate, rounded or heart-shaped at base, with wavy margins, largest ones to 30 cm. long (1 ft.); stems ribbed, usually unbranched below flowering tip, to 1.5 m. high (5 ft.).

May–September

water. Although not as important to wildlife as *R. acetosella*, the seeds and leaves of *R. crispus* are eaten by several species of birds and mammals. In feeding experiments the seeds remained viable after passing through the intestinal tracts of horses, cows, swine, and sheep. It has been declared a noxious weed in the seed laws of 10 states.

Curly Dock is a very common weed found in open fields, meadows, pastures, and waste areas and along roadsides throughout the United States and southern Canada.

A perennial with a long, branched taproot, introduced from Europe. As an adaptation to avoid self-pollination, the anthers mature and pollen is shed in each flower before the stigma becomes receptive. Pollination is by wind, but the plant is not considered an important hayfever plant outside of California and Oregon.

The seed-like fruits may be scattered in several ways. The membranous wings formed by the calyx gives the nutlet buoyancy that promotes dispersal by wind or

� The very young leaves can be added to salads or cooked as greens. Older leaves become bitter and must be cooked in several changes of water. The plant has a high protein content and is rich in vitamin A. The seeds can be ground and mixed with other kinds of flour. These uses also apply to the other species of Dock in our range.

℞ A solution made from the root, collected in late summer or autumn, is said to have astringent properties. It has been used for internal bleeding and hemorrhoids and as a laxative. As an ointment, it is considered effective for the relief of itching skin conditions, sores, swellings, and boils.

284

Portulaca oleracea
Purslane*

Portulacaceae
Purslane Family

**Also Pusley*

Flowers yellow, usually solitary, at the ends of stems and branches, in center of a cluster of leaves; petals usually 5 (4–6), sepals 2, stamens 8–10; leaves usually alternate, succulent, spatula-shaped, rounded at tip, to 3.8 cm. long (1½ in.); stems prostrate, smooth, branched, succulent, reddish, to 30 cm. long (1 ft.), often forming mats.

June–October

An annual with a substantial taproot, introduced from southern Europe. The flowers open only in bright morning sunshine and usually do not last more than 1–2 days. Fertilization is usually the result of cross-pollination by bees and butterflies. The top half of the seed capsule is shed like a lid, releasing numerous oval, rough, slightly glossy black seeds that are small enough to become windborne in a moderate breeze. A single plant may produce more than fifty thousand seeds.

It may become a serious weed pest in cultivated areas. The best control measure is repeated cultivation or hoeing while the plants are in the seedling stages. If severed plants are not removed, they will develop roots and continue to grow. The seeds are eaten by several species of songbirds, and both seeds and vegetation are eaten by small mammals such as rabbits, chipmunks, and kangaroo rats.

This is a large genus of mainly tropical and subtropical distribution. Of the 3–4 species that occur in the Northeast, *P. oleracea* is the most abundant. A related species, *P. grandiflora*, Rose Moss, not illustrated, a native of Argentina, is widely cultivated as a flower garden ornamental and sometimes escapes.

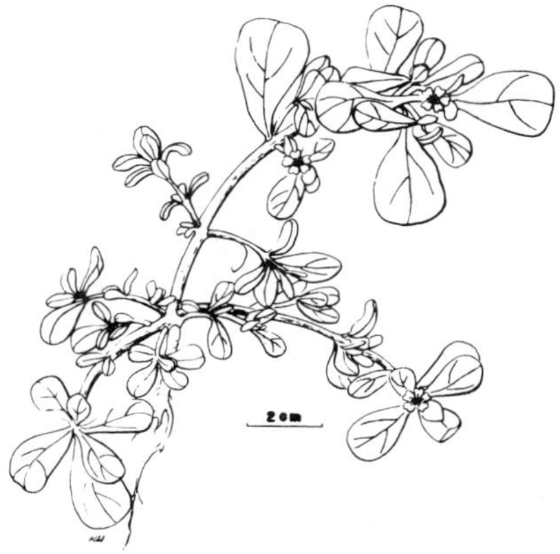

Purslane is found in cultivated fields, gardens, and lawns, along roadsides, and in waste spaces throughout southern Canada and the United States.

✂ Written records dating back to the sixteenth century describe the virtues of this plant as a potherb and a salad. It is nutritious, especially high in iron, and it loses very little bulk in cooking. This species regenerates rapidly, and a small patch will supply fresh greens until frost. It can be pickeled or dried and stored for year-round use. The seeds can be ground and used as flour.

℞ The fresh juice has been used for coughs and shortness of breath and applied externally for skin irritations and sores. The crushed seeds have been boiled in wine and given to children as a worm medicine. In ancient times the plant is said to have been spread on the floor around a bed to ward off evil spirits.

Lysimachia nummularia Moneywort*   285   *Primulaceae* Primrose Family

*Also Creeping Jenny

Flowers yellow, 5-lobed, with red dots, solitary, in leaf axils; leaves opposite, shiny green, almost round, with short stalks; stems prostrate, rooting at nodes, often forming mats, to 60 cm. long (2 ft.).

June–August

A perennial introduced from Europe. When sexual reproduction takes place, cross-pollination by insects is necessary for seed production since the flowers are self-incompatible. The fertile seeds are about 1 mm. long, 3-sided, dark brown or black with a surface roughened by minute ridges. However, more commonly the seeds are sterile and reproduction is asexual by the creeping stem.

The round coin-like leaves are responsible for the names Moneywort in North America and Herb Twopence in the British Isles. It grows well under cultivation and makes an attractive cover plant in areas that are subject to erosion. However, it spreads rapidly and may become a problem if unattended. Control measures include raking deeply to remove the stems and close mowing.

This plant is not difficult to distinguish from related species, since it is the only member of the genus with creeping stems. It has a high tolerance for water and can withstand complete immersion for several days at a time. It is easy to propagate by cuttings, and its trailing stems make it an attractive plant for indoor hanging baskets and planters.

Moneywort is found in moist fields, gardens, damp roadsides, lawns, and waste areas from Newfoundland to Minnesota, south to Georgia and Missouri, and on the Pacific Coast.

℞ In ancient times Moneywort was highly regarded in the treatment of wounds. The bruised fresh leaves were used as an application to wounds and persistent sores. The dried leaves were used in remedies for scurvy, internal bleeding, and whooping cough.

Anemone canadensis
Canada Anemone

286

*Flower white, 2–4 cm. wide (1½ in.), on a
long stalk, 5 unequal petal-like sepals; upper stem leaves usually a single pair, lower
ones in a whorl of 3, sessile, deeply parted;
basal leaves similar, with long petioles;
flowering stem to 60 cm. high (2 ft.).*

May–August

A perennial with a slender tough rhizome. The anthers release pollen over a period of 7–14 days and the flowers may be
visited by numerous insects during this time.
For this period the petaloid sepals close in
the evening and reopen the next morning
in sunny weather. Each of the numerous
pistils in the flower becomes a gray to brown,
single seeded fruit with a broad marginal
wing which aids in dispersal by wind.

Canada Anemone is found in roadside
ditches and damp meadows and on sandy
or gravelly shores, often in limestone areas,
from Quebec to British Columbia, south to
West Virginia and New Mexico.

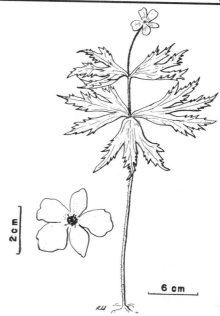

The *Anemones* are recorded as poisonous plants in several publications worldwide. Some tribes of American Indians are
said to have used the rhizome in a remedy
for headache and dizziness and the leaves
as a poultice for burns. Other tribes used
the root in a treatment for wounds. The
literature is not always clear as to the species utilized but all appear to contain a toxic
substance called protoanemonin.

Open Fields

Clematis virginiana
Virgin's Bower*

287

*Also Devil's Darning Needle, Old Man's
Beard

A woody perennial sometimes forming a
dense network of stems on other vegetation.
Some plants have staminate flowers only;
others have pistillate flowers and flowers
with both stamens and pistils. This arrange-
ment assures some cross-pollination but
does not eliminate self-pollination. The most
common insect visitors are bees, bee-like
flies, and brightly colored syrphus flies. Each
one-seeded fruit has a long feathery style
that serves as a parachute for dispersal by
wind.

This is a large genus of over 100 species,
widely distributed, mostly in temperate regions. Most of the large showy flowered garden varieties have been developed by
hybridization and selective breeding.

Virgin's Bower is found along woodland
borders, stream banks, moist roadsides, and
fence-rows from Nova Scotia to Manitoba,
south to Louisiana.

℞ In herbology, this plant is reported to

252

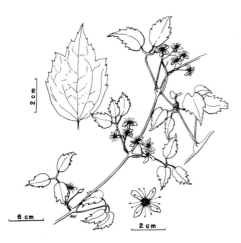

Flowers white, in stalked axillary clusters, petal-like sepals 4, stamens numerous; in autumn fruits characterized by dense growths of long, soft, gray hairs (thus Old Man's Beard); leaves opposite, pinnately compound, with 3 coarsely toothed leaflets, leaf stalks often curled around branches of host vegetation; stems climbing, to 3 m. long (10 ft.).

July–September

have tonic, diuretic, and diaphoretic properties. A remedy made from the flowers and leaves has been used as a treatment for headache.

The leaves contain compounds that may cause dermatitis if handled by sensitive individuals.

Ranunculus acris Tall Buttercup*

288

Ranunculaceae Crowfoot Family

*Also Butter Flower, Blister Plant

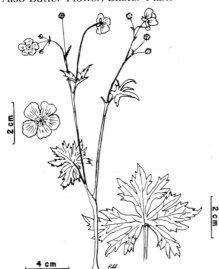

Flowers bright yellow, solitary or in clusters at the ends of branches and stems; petals usually 5, shiny, stamens numerous, pistils numerous; leaves alternate, palmately divided into 3–7 deep lobes, each of which is further dissected, upper ones smaller, 3-parted; stems slender, branched, usually hairy, most of leaves in lower half, to 1 m. high (40 in.).

May–September

The seeds and leaves of the plants in this genus are eaten in small quantities by several species of birds and mammals which probably contribute to seed dispersal. The seeds of *R. acris* have been shown to remain viable after passage through the digestive tracts of horses, cows, swine, and sheep.

As the result of its bitter juice, this species is usually not eaten by domestic livestock when other forage is available. It is thus often able to spread freely in pastures and meadows. The most effective control measure is to improve the drainage. In heavily infested fields, it can be eradicated by one year of clean cultivation.

Tall Buttercup is found in open fields, pastures, meadows, and waste areas and

A perennial with thick fibrous roots, introduced from Europe. The flowers produce large quantities of pollen, and each petal has a minute nectar scale at its base. It is self-incompatible and cross-pollinated by insects, but up to 1% of its seeds are produced without fertilization.

along roadsides, often in wet soil, from Labrador to Alaska, south to North Carolina,
Missouri, and Oregon. It is most abundant
in the Northeast.

☠ The very acrid juice of this plant contains a substance that causes blistering of

the mouth and intestinal tract if eaten by
humans or livestock. Cows that have eaten
the fresh plant produce unpalatable, sometimes reddish, milk. The dried plants as they
occur in hay do not exhibit these properties.

Ranunculus bulbosus
Bulbous Buttercup*

289

Ranunculaceae
Crowfoot Family

Also Yellow Weed, Blister Flower
*Flowers similar to R. acris but with sepals
bent back to the stem; basal leaves in a rosette, long stalked, 3-parted with divisions
further dissected, terminal segments
stalked; stem leaves alternate, scarce,
smaller, less dissected; stems hairy,
branched, to 60 cm. high (2 ft.).*

April–July

A perennial with a bulbous thickening at
the base of the stem. Introduced from Europe, this is the only species of the genus
arising from a bulb. The plants are self-incompatible, and cross-pollination is effected by insects. One study indicated that
over 60 different species of insects visited
the flowers of *R. repens*, *R. bulbosus*, and *R.
acris*. *R. bulbosus* is a morning plant releasing 49% of its pollen between 9:00 and
10:00 a.m. The numerous flattened, dark
brown, finely pitted, slightly hooked seeds
are similar to those of *R. acris* and are dispersed in the same manner.

This is a very large genus of northern and
Arctic regions. There are about 33 species
in the Northeast, several of which are circumboreal in distribution. *R. bulbosus* is one
of approximately seven species that have
been introduced from Europe. A related
species, *R. repens*, Creeping Buttercup, not

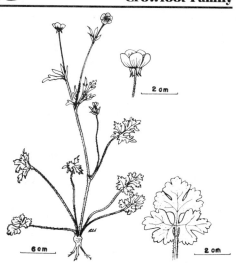

illustrated, has a double flowered variety that
is cultivated as an ornamental.

Bulbous Buttercup is found in open fields,
meadows, and lawns and along roadsides,
especially in limestone areas, from Newfoundland to Ontario, south to Georgia and
Louisiana.

✗ ☠ The bulb can be collected in Spring
and cooked in several changes of water as
an emergency food. It is edible but not tasty.
The vegetation has the same toxic properties as *R. acris*.

Duchesnea indica
Indian Strawberry*

290

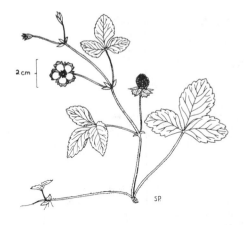

Also Mock Strawberry

Flowers yellow, on long stalks arising from nodes, petals 5, stamens numerous; flowers subtended by five 3-toothed bracts that are longer than the petals; fruit red and fleshy; leaves basal, 3-parted, with long petioles; slender runners spreading from underground stem, bearing alternate leaves, to 1 m. long (40 in.).

April–June

A perennial with a short rhizome, introduced from India. The fruit is strawberry-like in appearance, with numerous seeds on its surface. However, the fruit is non-juicy and tasteless. For those animals that are attracted by color rather than odor, the fruit is probably eaten by the same ones that eat strawberries.

This is a genus of only 2 species, both of which are small running plants of southern Asia. The genus is named for A. N. Duchesne, an eighteenth-century French botanist who wrote extensively about *Fragaria*, the strawberry genus. Duchesne also performed hybridization experiments and developed several varieties for growers. It is ironic that the species named for him has a tasteless and inedible fruit.

This species sometimes becomes established as a weed in lawns, gardens, and cultivated areas. It is fairly easy to control by close mowing and digging out individual plants. It is a useful soil stabilizer on exposed banks and road cuts. Its long graceful runners make it an attractive plant for hanging baskets or other containers.

Indian Strawberry is found in lawns and other grassy areas, along roadsides, and in moist waste places from New England to Iowa, south to Florida and the Gulf Coast, and on the Pacific Coast.

Fragaria vesca
Wood Strawberry*

291

Also Sow-teat Strawberry

A slender perennial of North America and Europe. The showy white petals and orange-yellow stamens attract a variety of insect pollinators. Many tiny oval seeds are attached to the shiny red surface of the mature fruit. The fruits and leaves are eaten by numerous songbirds, game birds, and small mammals. Seeds adapted for dispersal by birds typically do not have an odor, since birds have a weak sense of smell, but this has not deterred their consumption of the fragrant strawberry. Wood Strawberries are also a popular source of food for turtles such as the Wood Turtle.

There are two recognized varieties of this species. One has a round fruit and a more northerly distribution, and was introduced from Europe. The other has an oval shaped fruit and a range that extends farther south, and is native to the Northeast. The latter variety was at one time recognized as a separate species, *F. americana*. These varieties interbreed and intermediate forms are found throughout the range.

Vegetative reproduction by runners in this

Flowers white, in a cluster of 1–9, on a leafless stalk, often taller than the leaves; petals 5, stamens numerous; fruit red or white, with seeds on surface; leaves basal, 3 parted, leaflets toothed, terminal tooth usually longer than the two flanking teeth; leaves and flowering stem to 15 cm. high (6 in.).

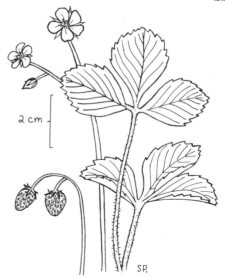

May–July

species is as important as propagation by seeds. Each plant may produce several slender runners that at intervals develop roots and a new plant. The result is that plants often occur in clumps or colonies.

A related species, *F. virginiana*, Field Strawberry, not illustrated, is similar but the flower stalk usually does not overtop the leaves, the terminal teeth of the leaflets are shorter than the flanking teeth, and the seeds are embedded in pits on the surface of the fruit.

Wood Strawberry is found along roadsides, in fields and pastures, and in open moist or dry woods from Newfoundland and Quebec to Manitoba, south to Virginia and Indiana, and westward in the Rocky Mountains.

✂ Wild food enthusiasts are in agreement that wild strawberries are superior, in flavor,

to the cultivated variety. However, they are much smaller and more difficult to pick. The white ones have a slightly different, but no less delicious, flavor. In order to have enough for a strawberry shortcake or a few jars of freezer jam, be prepared to spend the day searching through fields and woods.

The evergreen leaves are a last resort survival food, winter or summer. Fresh or dried they make a pleasant tea. The fresh leaves are rich in vitamin C.

Potentilla argentea
Silvery Cinquefoil

292

Rosaceae
Rose Family

A perennial with a well formed taproot, introduced from Europe. Each flower produces numerous yellow-brown, ribbed seeds. These are produced without fertilization; thus the new plant that grows from each seed is the exact genetic duplicate of the parent plant. When these plants occur as scattered weeds in lawns and fields,

hoeing and hand pulling may suffice as control methods. Heavy infestations in cultivated areas may require plowing and planting a smother crop.

This is a large and complex genus with at least 25 species in eastern North America. Some of these are restricted to specific geographic areas within our range. *P. argentea*

256

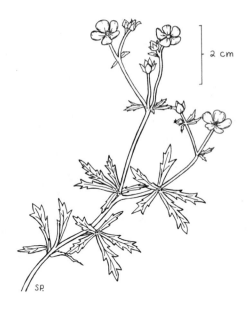

Flowers yellow, small, in clusters at ends of branches, petals 5, rounded, stamens about 20; leaves alternate, palmately divided into 5 narrow, deeply toothed leaflets, with silvery hair on underside; stems often sprawling, profusely branched, white woolly, to 50 cm. long (20 in.).

June–September

is one of about 7 species that have been introduced from Europe. A related species, *P. collina*, not illustrated, is very similar but has gray instead of silvery hairs on the underside of its leaves. The latter is of limited distribution and may be a hybrid of *P. argentea* and *P. verna*, not illustrated.

Silvery Cinquefoil is found in the dry soil of roadsides, open fields, and waste areas, particularly in limestone regions, from Newfoundland to Minnesota and Montana, south to Virginia and Illinois.

Potentilla norvegica
Rough Cinquefoil

293

Rosaceae
Rose Family

Flowers yellow, in clusters at the ends of branches, to 12 mm. wide (½ in.), petals 5, slightly shorter than the sepals; leaves alternate, with 3 broad, coarsely toothed leaflets, leaflets to 8 cm. long (3 in.); stems thick, bushy, hairy, to 90 cm. high (3 ft.).

June–October

seeds with longitudinally curved, branching ribs. These are formed without fertilization, and the new plants that grow from them are genetic duplicates of the parent plant. It is sometimes a weed in neglected gardens and fields but is effectively controlled by clean cultivation.

This is a highly variable species with a wide geographic distribution and diverse habitats in North America. It is probably represented in our flora by both native and introduced varieties. The native variety is classified as a distinct species, *P. monspeliensis*, by some botanists. The northern plants are smaller and non-hairy and have been designated as variety *Labradorica*.

An annual or biennial with a substantial taproot, of North America, Europe, and Asia. Each flower produces numerous light brown

Plants of the latter variety are sometimes observed in alpine areas of New England.

The plants of this genus, although abundant in many areas, are of relatively little value as wildlife food. There are more species, and they are of greater usefulness, in the western states. In the East, the leaves and seeds of these plants comprise a small part of the diets of the Ruffed Grouse, Woodcock, and Cottontail Rabbit.

Rough Cinquefoil is found along roadsides and in fields, pastures, and waste places from Greenland and Labrador to Alaska, south to North Carolina, Texas, and Arizona.

Potentilla recta
Rough-fruited Cinquefoil*   **294**   *Rosaceae*
Rose Family

*Also Sulphur Cinquefoil
Flowers pale yellow, numerous, in a branched, flat-topped terminal cluster; flowers to 2.5 cm. wide (1 in.), petals 5, notched, pistils and stamens numerous, sepals 5, subtended alternately by sepal-like bracts; leaves alternate, palmately compound, lower ones with 5–7 leaflets, upper ones smaller with 3 leaflets; stems stout, hairy, unbranched below the flower cluster, to 60 cm. high (2 ft.).

June–August

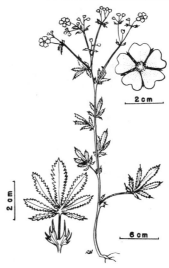

A perennial introduced from Europe and rapidly spreading in eastern North America. Each flower produces numerous dark to light brown seeds with sharp, curved, branching ribs. Since they develop without fertilization, the new plants that grow from the seeds will be genetic duplicates of the parent plant. The seeds have a slight marginal wing that may aid in dispersal by wind.

This plant has beautiful showy flowers, but it sometimes becomes a troublesome weed, especially in limestone regions. Its forage value is low, and it is usually not eaten by livestock. Consequently, the native species may be heavily grazed, eliminating the competition and allowing *P. recta* to spread freely. Scattered plants can be dug out or hand pulled, but in badly infested fields or meadows plowing and clean cultivation for a year may be necessary to bring this weed under control.

Rough-fruited Cinquefoil is found along dry roadsides and in fields, pastures, and waste areas from Newfoundland to Minnesota, south to North Carolina, Tennessee, and Arkansas.

Potentilla simplex
Old Field Cinquefoil*

295

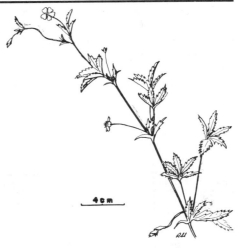

Also Common Cinquefoil
Flowers yellow, solitary, on long axillary
stalks, the first flower arises from second
leaf axil; leaves alternate, palmately com-
pound, with 5 toothed leaflets, teeth on up-
per ¾ of leaflet; stems slender, at first
erect, then arching and prostrate, rooting
at nodes, branched, to 1 m. long (40 in.).

April–June

A perennial with a short rhizome to 8 cm. long (3 in.). When roots form at the nodes, small tubers develop that will become the young rhizomes of the following year. The numerous seeds are yellow-brown with inconspicuous, wavy, broken ridges. This plant is capable of surviving on soil too acid and sterile for many other species, and when abundant may be an indicator of worn-out soil.

Strong infestations of this plant may be indicative of a need for reforestation. Less extensive invasions can be controlled by the addition of fertilizer and enriching the soil by plowing under a green manure crop.

A related species, *P. canadensis*, Dwarf Cinquefoil, not illustrated, is very similar but has leaflets with teeth only in the upper half,

the first flower from the axil of the first well developed leaf, and a rhizome no longer than 2 cm. (1 in.). Like *P. simplex*, it has a prostrate stem and is an indicator of impoverished soil.

Old Field Cinquefoil is found along dry roadsides and in fields, waste areas, and open woods from Newfoundland and Quebec to Minnesota, south to Georgia, Louisiana, and Texas.

Spiraea latifolia
Meadow-sweet

296

A shrubby perennial with clusters of slightly fragrant, miniature apple-like blossoms. Each flower usually produces 5 seedpods which split along one side, releasing several 3-angled, flattened, light brown seeds. These may spread to meadows and pastures, where it sometimes becomes a trou-

blesome weed in New England. Plants of this genus are browsed by and may make up to 5% of the diet of the White-tailed Deer.

This is a very large genus of deciduous shrubs native to the Northern Hemisphere. It includes some of the most popular ornamentals with a variety of flower colors

Flowers white or pale pink, in dense, branched, terminal clusters; petals 5, stamens numerous, pink-red; leaves alternate, oval, sharply toothed, to 7.5 cm. long (3 in.); stems reddish or purplish brown, smooth, to 1.2 m. high (4 ft.).

June–August

and flowering dates. *S. latifolia* is sometimes included in landscaping that emphasizes non-commercial native species. A related species, *S. alba*, not illustrated, is very similar but the flowering branchlets are hairy and it occurs in swamps and other wet areas.

Meadow-sweet is found in pastures, old fields, and meadows and along roadsides from Newfoundland and Quebec to Michigan, south to North Carolina and Missouri. ℞ The bark and leaves of a related species, *S. tomentosa*, Hardhack, not illustrated, were used by some North AmericanIndians to make an astringent for treating diarrhea.

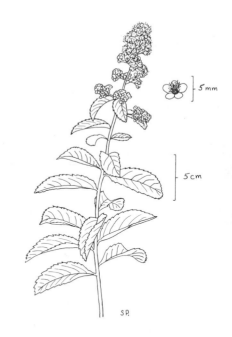

5 mm

5 cm

S.P.

Open Fields

Diodia teres
Buttonweed*

297

Rubiaceae
Madder Family

*Also Poorweed
Flowers white to light purple, funnel-shaped, in axils of leaves; corolla with 4 lobes, sepals 4, stigma 1; leaves opposite, stiff, narrow, with bristles at base, to 4 cm. long (1½ in.); stems hairy, trailing over the ground, branched, to 80 cm. long (32 in.).

June–October

An annual with a slender, shallow taproot. Each flower produces 2–3 hairy, light brown or grayish-brown seeds. These are eaten in moderate to small quantities by the Bobwhite Quail, Wild Turkey, and Greater and Attwater Prairie Chickens. However, the value of the plant as a wildlife food plant is not in proportion to its abundance.

This is a highly variable species with several varieties distinguished by the nature of the axillary bristles and the hairiness of the fruits, stems, and leaves. The name Poorweed recognizes the ability of this plant to survive on poor soils. A related species, *D.*

2 cm

virginiana, not illustrated, is very similar but has 2 stigmas and 2 sepals. It occurs on wet ground from New Jersey to Illinois, south to Florida and Texas.

Buttonweed is found in dry, open fields, along roadsides, and in waste spaces, usually in sandy soil, from southern New England to southern Wisconsin, south to Florida and Texas.

Galium boreale
Northern Bedstraw

298

Rubiaceae
Madder Family

3 cm

SP

A perennial with a substantial rootstock, spreading by runners. It usually grows in clumps with masses of bright white flowers that attract small insect pollinators. Each flower produces 2 black, finely wrinkled,

Flowers white, tiny, in dense branched clusters at the ends of stems and branches, corolla 4-lobed, stamens 4; leaves in whorls of 4's, 3-nerved, to 5 cm. long (2 in.); stems erect, branched, square, smooth, to 1 m. high (40 in.).

June–August

hairy seeds. It is usually not a troublesome weed because it is easily controlled by cultivation, mowing, or close grazing.

This is a highly variable species with several varieties based on differences in the size and hairiness of the leaves and fruits. Each variety is ordinarily associated with a geographic section of the range. Some forms extend the range into Europe and Asia, making the species circumboreal in distribution.

In North America, Northern Bedstraw is found in fields, meadows, roadsides, and shores often in rocky soil from Nova Scotia to Alaska, south to Kentucky and Missouri, and extending westward to the Pacific coast.

For other information on the genus and family see *G. aparine* and *G. verum*.

Galium mollugo
Wild Madder*

299

Rubiaceae
Madder Family

Also White Bedstraw, False Baby's-breath

A perennial with a horizontal rhizome, introduced from Europe. The dense clusters of minute flowers may serve to attract insect pollinators or the crowded overlapping flowers may pollinate one another by making physical contact. Each flower produces 2 gray to black, warty seeds. It sometimes occurs as a weed in meadows and cultivated areas but is usually easy to control by cultivation, mowing, or close grazing.

This family of plants is one of the largest and is common throughout the tropics and subtropics. *Galium* is the largest genus of

Flowers white, tiny, numerous, in branched terminal and upper axillary clusters; corolla 4-lobed, stamens 4; leaves in whorls of 6's or 8's, widest above the middle, to 4 cm. long (1½ in.); stems erect, or often with prostrate bases, sometimes forming mats, to 90 cm. high (3 ft.).

June–August

the family in North America. In addition to the approximately 28 species in the Northeast, there are many others in the West. Although *G. mollugo* is native to Europe and Asia, it has become thoroughly naturalized in North America and is spreading. Hybrid forms resulting from cross-pollination with *G. verum*, Yellow Bedstraw, have been reported.

Wild Madder is found in open fields, pastures, roadsides, and waste areas from Newfoundland to Ontario, south to Tennessee.

For other information on the genus and the family see *G. aparine* and *G. verum*.

4 cm

Galium verum
Yellow Bedstraw*

300

Rubiaceae
Madder Family

*Also Our Lady's Bedstraw, Cheese-rennet Flowers bright yellow, in dense branched, terminal and axillary clusters; corolla 4-lobed, stamens 4, sepals none; leaves in whorls of 6's or 9's, narrow, to 2.5 cm. long (1 in.); stems prostrate at base, then firm and erect, smooth, angled, to 90 cm. high (3 ft.).

June–August

A perennial with a somewhat woody base, introduced from Europe. Fertilization is similar to that in *G. mollugo*. Each flower produces 2 black, finely wrinkled, kidney-shaped seeds. It sometimes invades fields and lawns as a weed and can be controlled by cultivation, mowing, and close grazing.

The genus name is derived from a Greek word for milk and refers to the early use of *G. verum* for the curdling of milk to make cheese. It was mixed with calf rennet and according to some descriptions it made the cheese sweeter, tastier, and more whole-

4 cm

some. Yellow Bedstraw, like several other species in this genus, has a fragrant odor when dried. Consequently, in earlier times, these plants were used to stuff mattresses and pillows. According to legend, Yellow Bedstraw was present in the hay on which the mother of Jesus slept and which was used to fill the manger at Bethlehem; thus

the common name Our Lady's Bedstraw.

Yellow Bedstraw is found in open fields, pastures, and roadsides from Newfoundland to Ontario, south to West Virginia and Ohio.

For medicinal and other information on the genus and family see *G. aparine.*

Houstonia caerulea **Bluets***	**301**	*Rubiaceae* **Madder Family**

*Also Innocence, Quaker Ladies

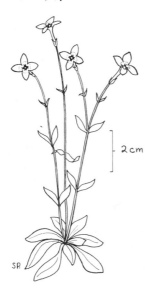

2 cm

SR.

A perennial growing in dense tufts from a thin creeping rhizome. Flowers are of 2 types: pistillate ones with non-functional stamens, and staminate ones with non-functional pistils. Cross-pollination is by small bees and the smaller butterflies such as the Clouded Sulphur, Meadow Fritillary,

Flowers pale blue with yellow centers, to 13 mm. wide (½ in.), tubular, 4-lobed, solitary, terminal; basal leaves in a rosette, spatula-shaped, to 13 mm. long (½ in.), stem leaves opposite, smaller; stems unbranched or sparingly so, smooth, to 20 cm. high (8 in.).

April–July

and Painted Lady. The seed capsule splits across the top, releasing several tiny seeds that are dispersed by wind.

The genus is named in honor of William Houston, an early eighteenth-century Scottish surgeon and botanist who studied and collected plants in Mexico and the West Indies. It is a North American genus with about 12 species in the East. One of the most abundant is *H. caerulea,* which is sometimes planted as a delicate, very attractive addition to rock gardens. A related species, *H. serpyllifolia,* Thyme-leaved Bluets, not illustrated, has very similar flowers but has roundish leaves and a prostrate branching stem. It occurs in the mountains from Pennsylvania to Georgia.

Bluets is found on moist soil in grassy meadows, lawns, fields, and open woods from Nova Scotia and Quebec to Wisconsin, south to Georgia and Arkansas.

Comandra umbellata
Bastard Toadflax

302

Flowers greenish-white, tiny, in flat-topped terminal clusters; petals none, sepals 5, stamens 5, attached to sepals by a tuft of hairs; leaves alternate, elliptical, pale green on underside, numerous, to 3.8 cm. long (1½ in.); stems branched, to 40 cm. high (16 in.).

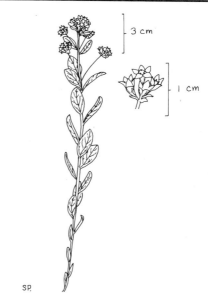

April–July

A perennial with a deep, horizontal rhizome. Each flower produces a single globular seed, and the plant usually withers by midsummer. Although it is green and carries on photosynthesis, its roots are partially parasitic on the roots of various other species. It absorbs water and mineral nutrients from the host plant but seldom causes death.

This is a genus of 6 species, one native to southeastern Europe, the others to North America. Of the 3 species native to the Northeast, *C. umbellata* is the most abundant. A related species, *C. richardsiana*, not illustrated, is very similar but has leaves that are not pale on the underside and a rhizome on the surface of the ground or shallowly buried. It occurs in woods from Labrador to Manitoba, south to Tennessee.

Bastard Toadflax is found in dry fields and thickets from Maine and Ontario to Wisconsin, south to Georgia and Alabama. ✖ The small, unripe fruits are oily and sweet but are usually present in quantities sufficient only for a nibble.

Linaria vulgaris
Butter-and-eggs*

303

*Also Toadflax

A perennial with a creeping rhizome and branching roots, introduced from Europe. The long spur of the flower serves as a repository for nectar. The orange blotch on the upper lip functions as a nectar guide for pollinating bumblebees that force open the lips of the blossom. Moths and butterflies are capable of pirating the nectar by inserting their long tongues between the lips without opening them and without effecting pollination.

The globular seed capsule opens by 2–3 pores or slits just below the top to release numerous circular, flattened, dark brown or black, winged seeds. These are windborne and widely dispersed over a period of time, as the dead stem persists into the winter months. When it becomes established as a weed in cultivated areas, control measures include cultivation until late in the season followed by a close-growing cover crop to smother the seedlings.

This is a very attractive plant, producing masses of flowers throughout summer and into autumn. It grows well as an ornamental for flower gardens, especially in sterile, exposed areas where less hardy species can-

2 cm

Flowers yellow, 2-lipped, upper lip with orange palate, with long spur; flowers numerous, in dense, elongate, terminal clusters; leaves alternate, crowded, very narrow, pale green; stems stiff, smooth, to 90 cm. high (3 ft.).

June–October

not survive. The genus name is derived from the Latin word for flax and refers to the similarity of the leaves of flax to those of *L. vulgaris*.

Butter-and-eggs is found in dry fields, along roadsides, and in waste areas from Newfoundland to British Columbia, south to Florida, Texas, and California.

℞ The dried plant has been used as a diuretic and a purgative. A poultice of the fresh leaves and an ointment made from the flowers are recommended in herbology for hemorrhoids, sores, and diseases of the skin.

Penstemon digitalis **Foxglove Beardtongue**	**304**	*Scrophulariaceae* **Figwort Family**

2 cm

6 cm

Flowers white or tinged with violet, with purple lines inside, in terminal and stalked axillary clusters; corolla tubular, 2-lipped, upper lip 2-lobed, lower lip 3-lobed; fertile stamens 4, sterile stamen 1, with a tuft of hairs at its tip; leaves opposite, sessile, lanceolate, toothed, to 15 cm. long (6 in.); stems smooth, tinged with purple, to 1.5 m. high (5 ft.).

May–July

A perennial with a short rhizome and a rosette of stalked basal leaves. In most of the members of this genus fertilization is the result of cross-pollination, often by bees and butterflies, and *P. digitalis* is probably not an exception. The oval seed capsule splits into 2 sections to release numerous finely pitted, gray, irregularly angled seeds. These are small enough to be windborne in a moderate breeze.

This plant sometimes becomes a weed in pastures and meadows where it is not highly valued as a forage plant. Scattered plants can be controlled by hoeing or close mowing, but in areas of heavy infestations plowing and cultivation for a year may be necessary for its eradication.

Most of the species of this genus are native to western North America. Many of the western species are very showy and are cultivated as ornamentals. It is not a significant wildlife food genus, but the seeds and leaves are eaten by a few species of western birds and mammals.

Foxglove Beardtongue was found originally mainly in the Mississippi Basin but has spread eastward in fields, meadows, and open woods, and now ranges from Maine and Quebec to Minnesota, south to Alabama and Texas.

Verbascum blattaria
Moth Mullein
305
Scrophulariaceae
Figwort Family

Flowers yellow or white, on slender stalks, scattered at tip of stem; corolla 5-lobed, about 2.5 cm. wide (1 in.), stamens 5, of unequal length, with orange anthers and purple hairs; stem leaves alternate, sessile, toothed or lobed; stems slender, usually unbranched, sticky-hairy in upper part, to 1 m. high (40 in.).

June–September

A biennial native to Europe and Asia. During the first year of growth a basal rosette of leaves and a deep taproot are produced. During the second year a tall flowering stem arises and pollination is probably by bees, butterflies, and moths. The flattened stigma opens only when it is receptive and closes when pollen has been deposited upon it.

The seed capsule splits into 2 sections, releasing numerous tiny, coarsely pitted seeds that are small enough to become windborne in a moderate breeze. In germination experiments seeds retained viability after burial in sand for 70 years. Other studies showed that they could pass unharmed through the intestinal tracts of swine and sheep.

Yellow and white flowered varieties may occur in the same field, but the yellow flowered form is usually the more abundant. Their only difference appears to be the color of the flowers. The hairy stamens suppos-

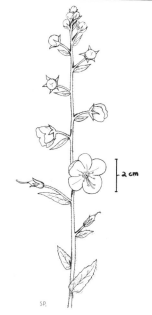

edly resemble the antennae of a moth; thus the name Moth Mullein. A related species, *V. phoeniceum*, Purple Mullein, not illustrated, is similar but has purple flowers. It occurs as a weed on Long Island, New York.

Moth Mullein is found along roadsides and in old fields, lawns, and waste areas throughout the United States and southern Canada.

266

Flowers yellow or white, in a branched, pyramid-shaped, terminal cluster; corolla 5-lobed, about 15 mm. wide (⅝ in.), stamens 5, unequal, white-woolly; stem leaves alternate, entire or shallowly toothed, sessile, white-woolly on underside; stems erect, white-woolly, to 1.5 m. high (5 ft.).

June–September

Similar to *V. blattaria*. This is a genus of about 250 species, mostly biennials, concentrated in the Mediterranean region of southeastern Europe. Five introduced species occur in eastern North America. The genus name is an ancient Roman word and its meaning is unclear. A few species are cultivated as border plants in flower gardens.

White Mullein is found along roadsides and in open fields and waste areas from New England and Ontario, south to Virginia and West Virginia. It is listed in the *Flora of West Virginia* but the authors report that they have not seen it in that state.

*Also Great Mullein, Flannel Plant
Flowers yellow, in a dense, club-shaped, terminal cluster; corolla with 5 unequal lobes, stamens 5, upper 3 short, white woolly, lower 2 much longer, smooth; stem leaves alternate, densely woolly, extending down the stem as a wing from point of attachment, decreasing in size upward, lower ones to 30 cm. long (12 in.); stems thick, densely woolly, to 2 m. high (6½ ft.).

June–September

A perennial introduced from Europe, producing a basal rosette of leaves and a deep taproot during the first year of growth. A tall flowering stem appears in the second year. The two types of stamens serve different functions; the short ones produce pollen for consumption, the long ones produce pollen for fertilization. Thus cross-pollination by pollen-collecting insects may be more common than self-pollination.

The seed capsule splits from the top into 2 sections, releasing numerous tiny, brown, coarsely pitted seeds. Seed dispersal studies have shown that most of the seeds are carried by wind 10–12 feet from the parent plant, but some are carried much further. Germination experiments have demonstrated that the seeds can pass undamaged through the intestinal tracts of cattle. They are common contaminants of commercial seeds.

This is a common weed on dry, gravelly, or rocky soils of roadsides, fields and waste areas throughout most of the United States and southern Canada.

✄ The dried leaves can be steeped in boiling water for a tea.

℞ The dried leaves and flowers have been used for a variety of ailments. A tea has been used as a remedy for coughs, hoarseness, bronchitis, and whooping cough. An ointment made by boiling the leaves in lard has been used in external applications for skin irritations and itching hemorrhoids. A nineteenth-century home remedy, learned from the American Indians, included smoking the dried roots, leaves, or flowers as a treatment for asthma and bronchitis.

Veronica arvensis — **308** — *Scrophulariaceae*
Corn Speedwell — **Figwort Family**

Flowers blue, tiny, in axils of progressively smaller alternate leaves on upper two thirds of plant; corolla 4-lobed, stamens 2; foliage leaves opposite, lower ones oval, short stalked, toothed, to 12 mm. long (½ in.); stems usually branched from base, to 25 cm. high (10 in.).

April–August

An annual or winter annual with a fibrous root system, introduced from Europe. Fertilization is the result of self-pollination or cross-pollination by insects. The deeply notched seed capsule splits into two sections, releasing several tiny, yellow, granular seeds with one side flat or slightly concave, the other side convex.

The genus is believed to have been named in honor of St. Veronica, who supposedly wiped the sweat from the face of Jesus on his way to Calvary. The specific name *arvensis* is a Latin word meaning "belonging to a field" and refers to its frequent status as a weed in grain fields. Heavy infestations can be controlled by harrowing in the spring and shallow plowing in the fall, the objective at both times being to destroy the seedlings.

Corn Speedwell is a weed of gardens, lawns, pastures, open fields, and waste areas from Newfoundland to Minnesota, south to Florida and Texas. It also occurs along the Pacific coast.

Veronica chamaedrys
Bird's-eye Speedwell*

309

Also Germander Speedwell

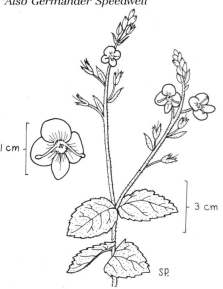

1 cm

3 cm

S.P.

A perennial introduced from Europe. Self-pollination is avoided by the maturation of the stigmas on a given plant before the pollen is shed. Plants growing in moist sheltered locations were observed to be visited most often by Syrphus flies, while those in dry exposed areas were most often visited

Flowers blue, in clusters to 15 cm. long (6 in.), from upper leaf axils; corolla 4-lobed, with dark blue lines and white center, to 1 cm. wide (½ in.); leaves opposite, sessile, oval, toothed, to 2.5 cm. long (1 in.); stems prostrate, rooting at nodes, hairy, with erect branches, to 40 cm. long (16 in.).

May–June

by Halictus bees. As a result of the requirement for cross-pollination, seed capsules are rarely observed.

This is a large genus mostly of the Northern Hemisphere and concentrated in Europe and Asia. There are 17–20 species in the Northeast, but only about 8 are native to North America. The specific name *chamaedrys* is derived from greek words that mean "oak tree on the ground" but its significance with regard to this species is unknown.

When in flower Bird's-eye Speedwell may form attractive, dense, blue mats in lawns, gardens, and open fields and along roadsides from Newfoundland and Quebec to Michigan, south to West Virginia, Ohio, and Illinois.

✖ The leaves have been used as a substitute for tea.

Veronica officinalis
Common Speedwell*

310

Also Gypsyweed

2 cm

4 mm

A perennial of North America, Europe, and Asia. The flowers are self-incompatible, and

Flowers blue or lavender, in long axillary clusters; corolla 4-lobed, with darker blue lines, lower lobe narrower than other 3, stamens 2; leaves opposite, short stalked, toothed, to 5 cm. long (2 in.); stems hairy, trailing, with erect flowering branches, rooting at lower nodes, to 30 cm. long (12 in.).

May–July

cross-pollination is probably accomplished by small bees and flies. The flattened, shallowly notched, glandular-hairy seed capsule splits into sections, releasing numerous finely wrinkled, lemon-yellow seeds. These are less than 1 mm. long and may become windborne in a moderate breeze.

The specific name means "of the shops" and probably indicates that it was at one time sold for a supposed medicinal property. The name Speedwell apparently refers to the ease with which this species and other members of the genus spread over the ground or into new areas. They are often weeds in meadows and pastures, but they can usually be eradicated by clean cultivation.

Common Speedwell is found on acid soil in abandoned fields, pastures, and open woods from Newfoundland and Quebec to Wisconsin, south to Georgia and Tennessee. ✖ The leaves have been used as a substitute for tea.

℞ The 1851 edition of *The Dispensatory of the United States of America* states, "*V. officinalis* has a bitterish, warm, somewhat astringent taste; has been considered diaphoretic, diuretic, expectorant etc.; and was formerly employed in pectoral and nephritic complaints, hemorrhages, and diseases of the skin, and in the treatment of wounds." In modern herbology it is recommended mainly for congestion of the respiratory tract.

| *Veronica serpyllifolia*
Thyme-leaved Speedwell | **311** | *Scrophulariaceae*
Figwort Family |

Flowers whitish or pale blue with darker stripes, occurring singly in axils of upper, reduced, alternate leaves; corolla 4-lobed, to 8 mm. wide (⅓ in.), lower lobe smallest, stamens 2; lower leaves opposite, oval or nearly round, sessile or short-stalked, to 12 mm. long (½ in.); stems creeping, with erect flowering branches, to 20 cm. high (8 in.).

May–July

1 cm

SP.

A mat-forming perennial of North America and Europe. The heart-shaped seed capsule is notched to about one-fourth its length. It splits into sections, releasing several light brown to orange, lightly pitted seeds that are small enough to become windborne in a slight breeze. Some of the European species of this genus are adapted for seed dispersal by ants, but apparently none of the species that occur in America are so adapted.

Both the specific name and the common name refer to the shape of the leaves. This small weed is usually not a serious problem in cropland because it yields readily to cultivation. However, it sometimes invades lawns, pastures, and other areas not ordinarily cultivated. For purists who cannot tolerate its presence in lawns, a spray of sodium chlorate or ammonium sulfate is effective for its eradication.

Thyme-leaved Speedwell is found in fields, pastures, and lawns, along roadsides, and in open woods from Newfoundland to Alaska, south to Georgia and California.

270

*Also Thorn Apple, Stramonium, James-
town Weed

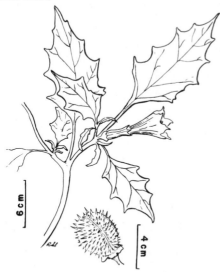

An annual introduced into North America
from Asia but probably of tropical origin.
Pollination is commonly by insects, but in
genetic studies man-induced self-pollination has produced viable seeds. The spiny
seed capsule splits into 4 sections, releasing
numerous dark brown or black, flattened,
kidney-shaped, pitted seeds.

There are two races in our area that differ
in the amount of pigmentation in the stems
and flowers: one has green stems and almost white flowers, the other has purple
stems and lavender flowers. In both, the
whole plant has a disagreeable odor.

Jimsonweed is found in cultivated fields,

*Flowers lavender to white, funnel-shaped,
to 12 cm. long (5 in.), in the forks of
branches; calyx tube green, with 5 teeth;
seedpod spiny; leaves alternate, pointed
with irregular sharp teeth, to 20 cm. long
(8 in.); stems thick, hollow, green or purplish, freely branched, to 1.5 m. high (5 ft.).*

July–October

abandoned feed lots, barnyards, and waste
areas from Nova Scotia to Minnesota and
Washington, south to Florida, Texas, and
California. It is most common in the South
and Southwest.

☠ All parts of this plant are poisonous, and
there are many cases on record of poisoning, especially among children, resulting
from ingestion, particularly of the seeds. The
settlers at Jamestown knew about the plant
and its properties; thus the common name
Jamestown Weed.

An early historian of the colony in 1705
described an event in which several soldiers
cooked and ate a quantity of young leaves
mistaking them for spinach. They became
deranged and had to be confined to keep
them from harming themselves. After eleven
days they recovered but did not remember
anything that had happened.

The whole plant, particularly the seeds,
contains the alkaloids atropin and hyos-
cyanin. The toxic symptoms resulting from
ingestion are dilated pupils, increased temperature, hallucinations, convulsions, un-
consciousness, and finally death. *Caution*:
Under no circumstances should this plant
be used in home remedies.

*Also Husk Tomato

A perennial with a deep, horizontal, fleshy
rhizome. Each berry contains numerous elliptical, granular, light brown seeds. If the
inflated calyx becomes detached from the

plant the wind may roll it and the enclosed
berry for some distance. However, the calyx
does not detach readily, so a more common
method of seed dispersal is by the birds and
mammals that eat the berries, including the

Flowers yellow with a brown center, funnel-shaped, with 5 shallow lobes, hanging from leaf axils; fruiting calyx green, inflated; fruit a yellow berry; leaves alternate, entire or irregularly toothed, rounded or heart-shaped at base; stems with sticky hairs, branched, erect or sprawling, to 90 cm. high (3 ft.).

July–September

Bobwhite Quail, Ring-necked Pheasant, Wild Turkey, Eastern Skunk, and Pine Mouse.

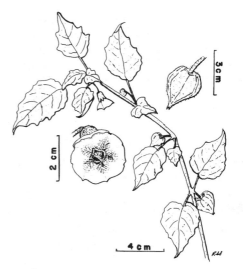

This is a large genus native mainly to Central and South America. *P. heterophylla* is probably the most common species of the genus in the Northeast. A related species, *P. virginiana*, Virginia Ground Cherry, not illustrated, is very similar but has narrowly oval leaves, a stem that is not sticky, and a red berry. Another related species is the cultivated Chinese Lantern-plant, *P. alkekengi*, not illustrated, with white flowers and an inflated bright orange calyx.

Clammy Ground Cherry is found in fields, gardens, and pastures, along roadsides, and in open woods from Nova Scotia and Quebec to Saskatchewan, south to Florida and Texas.

✗ The ripe fruits of all the species of this genus can be eaten fresh or made into pies, jam, or preserves. They are delicious, according to several experts on wild plant foods. Some species are cultivated and the fruits are occasionally available in food markets.

There have been reports of livestock poisoning from eating the vegetation and immature berries of *P. heterophylla*. Domestic animals usually do not eat this plant unless other forage is scarce. *P. alkekengi*, the Chinese Lantern-plant, is listed in a nineteenth-century German manual of poisonous plants.

Solanum carolinense
Horse Nettle*

314

Solanaceae
Nightshade Family

*Also Bull Nettle, Sand Briar
Flowers violet to white, in branched terminal or axillary clusters on prickly stalks; corolla star-shaped, with a prominent yellow cone in center formed by the fused anthers of 5 stamens; fruit a yellow berry; leaves alternate, to 12.5 cm. long (5 in.), with 2–5 coarse teeth or shallow pointed lobes, lower midrib spiny; stems branched, spiny, to 1 m. high (40 in.).

June–September

A perennial with a creeping underground rhizome, native to the southeastern United States. The flowers are visited by a variety of flying insects that may effect pollination.

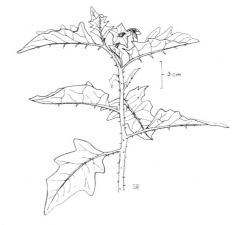

The berries contain numerous flattened, yellowish, glossy seeds and are eaten by large numbers of upland game birds, songbirds, and mammals. In feeding studies the seeds remained viable after passing through the intestines of horses, cattle, swine, and sheep. It is likely they show the same durability in the birds and mammals that normally consume the fruits.

It is often a troublesome weed in gardens, cultivated fields, and pastures but can be eradicated by close mowing and clean cultivation. The very prickly stems and leaves make gloves a necessity if it is to be handled. It has been declared a noxious weed in the seed laws of 6 eastern and midwestern states.

Horse Nettle is found in open fields, meadows, roadsides, and waste areas, especially in sandy soil, from Ontario to Minnesota, south to Florida and Texas. In the western states it ranges from Oregon and Idaho to California and Arizona.

☠ The dried berries and rhizomes are reported in herbology to have sedative and antispasmodic properties. They are said to have been used by southern Blacks in the treatment of epilepsy. Note: Poisoning in cattle, sheep, and deer have been reported as a result of consuming this plant. In 1963 a six-year-old boy in Pennsylvania died after eating what appears to have been the berries of *S. carolinense*.

Solanum dulcamara Bittersweet*	**315**	*Solonaceae* Nightshade Family

*Also Nightshade

Flowers violet or blue, in branched clusters on stalks from internodes or opposite leaves; petals 5, pointed, bent backward; stamens fused forming a yellow cone in center of flower; fruit a red berry; leaves alternate, oval, with a pointed tip, often with 1 or 2 basal lobes or leaflets; stems climbing or sprawling over other vegetation, becoming woody, to 3 m. long (10 ft.).

June–September

Since it yields readily to cultivation, it is not usually a troublesome weed in croplands. It is an alternate host for tomato mosaic disease and thus an undesirable weed near garden sites.

A woody perennial introduced from Europe. The flowers are visited by honeybees and bee-like flies that may effect pollination. The berries contain numerous flattened, finely pitted, yellowish seeds and are eaten by large numbers of upland game birds, songbirds, and mammals. These consumers serve as agents of dispersal.

This is the largest genus of a family consisting of more than 2000 mostly tropical American species. *S. dulcamara* is one of only 3 that are widespread in the Northeast.

Bittersweet is found along fence-rows, roadsides, and stream banks and in moist thickets and waste areas from Nova Scotia to Minnesota, south to North Carolina and Missouri. In the western states it occurs from Washington and Idaho to California.

☠ *The Dispensatory of the United States of America* (1851) states, "Dulcamara possesses feeble narcotic properties, with the power of increasing the secretions, particularly that of the kidneys and skin.... In overdoses it produces nausea, vomiting, faintness, vertigo and convulsive muscular movements." In herbology it is recommended for external use as a poultice for skin irritations, gout, herpes, and bruises. The plant contains a poisonous compound, solanine, which is most concentrated in the unripe berries. Cattle, horses, and sheep have been poisoned by eating the leaves and young shoots.

Solanum nigrum
Black Nightshade*

316

Solanaceae
Nightshade Family

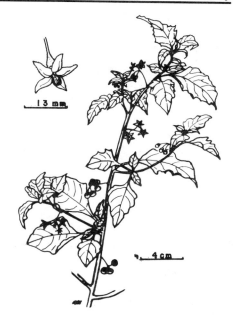

13 mm

4 cm

*Also Poison Berry, Garden Nightshade
Flowers white or pale blue, in branched clusters of 2's–10's, on stem between nodes; petals 5, pointed, stamens 5, fused, forming a yellow cone in center of flower; fruit a black berry; leaves alternate, long stalked, widest near base, tapering to a point, wavy-toothed; stems smooth, freely branched, to 60 cm. high (2 ft.).*

June–September

An annual with an extensive branched root system, introduced from Europe. The round, black berries contain numerous light brown, finely pitted seeds that are dispersed by the many birds and mammals that eat the fruits. Although widespread geographically, it is usually not very abundant. It may invade cultivated fields as a weed but is easily controlled by clean cultivation.

Black Nightshade is found in open fields, along roadsides, and on sandy beaches and dunes throughout the United States and southern Canada.

There are conflicting reports about the edibility of this species. In some varieties, apparently, the berries are edible while in others they are known to be poisonous. In 1909 a promoter advertised a new plant developed by Luther Burbank which the advertiser called "Wonderberry" and described as the greatest garden fruit ever introduced. This was subsequently identified as a western variety of Black Nightshade. Obviously the claims of the promoter were overly optimistic. (It should be noted, however, that some plants of this family are very important food plants: e.g., tomatoes, potatoes, green peppers, and chili peppers.) The best course to follow with regard to *S. nigrum* is not to eat the berries of local varieties unless recommended by someone who is familiar with the local flora.

It is listed as poisonous in manuals of poisonous plants. Numerous cases of poisoning of livestock have been reported from consumption of the leaves and berries.

℞ *The Dispensatory of the United States of America* (1851) states, regarding the leaves, "As a medicine they have been used in cancerous, scrofulous, and scorbutic diseases, and other painful ulcerous affections, being given internally, and applied at the same time to the parts affected in the form of a poultice, ointment, or decoction." In herbology, internal use of the plant is recommended only with supervision by a physician.

274

4 cm

Flowers greenish, tiny, in slender, branched, often drooping, axillary clusters; leaves opposite, stalked, lanceolate, sharp toothed, to 15 cm. long (6 in.); stems usually unbranched, densely covered with stinging hairs, to 2 m. high (6½ ft.).

June–September

the plants of the native variety, some botanists recognize 3 distinct species: *U. gracilis*, *U. procera*, and *U. viridis*.

Stinging Nettle is found in moist soil of roadsides, abandoned fields, waste spaces, woodland borders, and thickets throughout southern Canada and south to Georgia and Texas.

✕ The young shoots a few inches high and the incompletely developed terminal leaves can be cooked as greens. Cooking destroys the stinging quality. The leaves are said to be excellent additives for soups and stews. The rhizome and roots, although too tough to eat, can be boiled and the broth used for soup. Cooked fresh, the nettles are rich in iron, vitamin C, and vitamin A. Gloves should be worn in collecting these plants. If an accidental sting occurs, rub the area with the crushed stem of Jewelweed (Impatiens) to relieve the itching.

℞ In herbology the plant is described as astringent, tonic, and diuretic. It has been used for ailments of the urinary tract and as a treatment for rheumatism. An infusion of the seeds has been used for coughs and shortness of breath.

The stem contains fibers that were used during World War I to make a fabric used for tents and wagon covers. The fibers can be used to make linen for bedsheets, tablecloths, and clothing.

A perennial with a creeping rhizome, native to North America and Europe. Pollination is by wind, but it is not known to be a hayfever plant. The presence of nectaries in the flowers suggests that wind-pollination has evolved from an earlier insect-pollinated form. Each pistillate flower produces a single gray, egg-shaped, faintly pitted seed.

This is a highly variable species with both introduced and native varieties. The introduced variety has staminate and pistillate flowers on different plants. In the native variety the upper flower clusters are often staminate, the lower ones pistillate. Among

Phyla lanceolata
Fog-fruit

318

Flowers pale blue to pink, in dense cylindrical clusters, on long slender stalks from middle and upper leaf axils; leaves opposite, lanceolate, sharply toothed for over half of length, triangular at base, to 8.5 cm. long (3½ in.); stems sprawling, with erect flowering branches, rooting at nodes, to 60 cm. long (2 ft.).

May–October

2 cm

A perennial often becoming somewhat woody at the base of the stem. The flowers in the cluster open one ring at a time, and each produces 4 tiny seeds. This is not a significant food genus for wildlife, but the seeds are eaten by ducks, and in Florida by the Scrub Jay. The origin of the name Fog-fruit is obscure. N. C. Fassett refers to it as Frog-fruit, which is no easier to explain.

This is a genus of about 15 mostly tropical species. Of the 3–4 species that occur in the Northeast, *P. lanceolata* is the only one that is widespread. It is classified by some botanists as *Lippia lanceolata*. One species, *P. nodiflora*, not illustrated, is sometimes cultivated for ground cover in the southern part of the Northeast and in the southern United States.

Fog-fruit is found on flood plains and bottom lands and in roadside ditches from Ontario to Minnesota, south to Florida and California, and in brackish marshes along the Atlantic Coast.

Verbena hastata
Blue Vervain*

319

*Also Simpler's-joy, Wild Hyssop

A perennial with a short rhizome and fibrous roots. The flowers may be visited by a variety of insects, but pollination is effected mainly by a single species of bee (*Verbenajus verbenae*), which is a pollen collector that restricts its foraging to 3 species of *Verbena*. (The other two are *V. stricta* and *V. urticifolia*.)

Each flower produces 4 oblong, reddish-brown, triangular-convex nutlets. These are eaten by several species of songbirds and make up to 10% of the diet of the Swamp Sparrow. Germination studies have demonstrated that the seeds can pass undamaged through the intestinal tracts of cattle, and it is likely that some may successfully pass through the digestive tracts of the birds.

This is a plant of roadside ditches, moist open fields, marshes, and pond margins from Nova Scotia to British Columbia, south to Florida and California.

✗ California Indians roasted and ground the seeds into a slightly bitter flour. A way of reducing the bitterness is to wash the seeds several times in cold water.

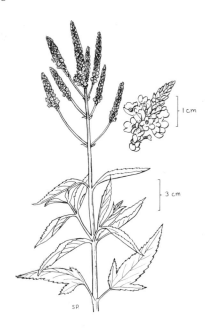

Flowers purplish-blue, small, 5 lobed, densely crowded in numerous, elongate, terminal clusters, flowering progresses from the bottom to top of cluster; leaves opposite, lanceolate, coarsely toothed, lower ones often 3-lobed; stems 4-angled, branched in upper half, hairy, to 1.5 m. high (5 ft.).

June–October

℞ Verbena is a Latin word for any sacred herb. How this came to be applied to the present genus is not clear. The Druids of ancient Britain are said to have considered it sacred because of the similarity of its leaves to those of some species of oak. The Crusaders reported it to be growing on Mt. Calvary and to have great curative powers. Some of its uses, according to herbology, are to relieve congestion in the throat and chest, to reduce fever, and to relieve pain in the bowels. Applied externally, it has been used to ease hemorrhoids and heal sores and wounds.

Verbena urticifolia
White Vervain*

320

Verbenaceae
Vervain Family

*Also Nettle-leaved Vervain

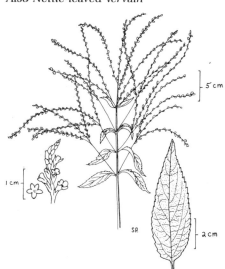

Flowers white, tiny, 5-lobed, scattered on numerous, slender, axillary and terminal branches; leaves opposite, lanceolate, coarsely toothed, to 20 cm. long (8 in.); stems 4-sided, grooved, usually hairy, branched, to 1.5 cm. high (5 ft.).

June–September

A perennial with a short rhizome and fibrous roots. Pollination is effected mainly by a single species of bee, *Verbenajas verbenae*, a pollen collector that restricts its foraging to 3 species of *Verbena* (see *V. hastata*). Each flower produces 4 reddish-brown, oval seeds. This species hybridizes with *V. hastata* when their ranges overlap.

The dead plant persists throughout the winter, the seeds providing a minor source of food for cardinals, juncos, and Song, Swamp, Tree, and White-crowned Sparrows. The summer vegetation is eaten in small quantities by the Cottontail Rabbit. It sometimes becomes a weed in cropland, and a year of clean cultivation may be necessary to eradicate heavy infestations.

White Vervain is found in moist fields, thickets, and woodland borders from Quebec and Ontario to Nebraska, south to Florida and Texas.

✕ ℞ See *V. hastata*.

Viola odorata — **321** — *Violaceae*
Sweet Violet* — Violet Family

*Also English Violet
Flowers usually deep violet, solitary, on leafless stalks, very fragrant, side petals usually with basal hairs; leaves basal, heart-shaped at base, roundish, equal to or longer than flower stalks, covered with fine hairs; stemless, to 15 cm. high (6 in.).

(all winter in sheltered places)

A perennial with a thick creeping rhizome and numerous leafy runners, introduced from Europe. It was introduced as an ornamental and is the cultivated violet of the florist, but it has escaped widely. Two kinds of flowers are produced: the showy violet ones that are insect pollinated, and self-pollinated ones that do not open. Seeds borne by both types of flowers are fertile but are probably more numerous in self-pollinated flowers.

The very sweet smelling flowers, to 2 cm. wide (¾ in.), have characteristics that make this species valuable to the florist. Double flowered and everblooming forms, as well as white varieties, are occasionally observed in nature. The flowers are used in the manufacture of commercial perfume.

1 cm

Sweet Violet has escaped cultivation and is commonly found along roadsides and around abandoned homesites and in open woods throughout much of the Northeast.

✕ ℞ See *V. canadensis*.

Viola papilionacea — **322** — *Violaceae*
Common Blue Violet* — Violet Family

*Also Meadow Violet
A prennial with a thick, horizontal, branched rhizome. Flowers are of two types: early ones visited and possibly cross-pollinated by bumblebees, and later ones produced on horizontal stalks that do not open and are self-pollinated. The seed capsule splits into 3 sections that shrink as they dry, forcibly hurling the seeds outward. As another means of dispersal the seeds have soft appendages that attract ants.

Closely related species of this genus hy-

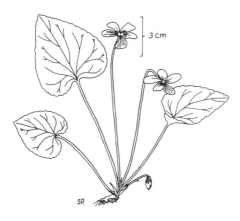

Flowers blue, solitary, on leafless stalks, side petals with hairs at base; leaves basal, smooth, heart-shaped, to 12.5 cm. wide (5 in.); stemless, leaves and flower stalks to 15 cm. high (6 in.).

April–June

rhizomes are reported to be a favored food of the Wild Turkey. The violet (unspecified) is the state flower of Illinois, New Jersey, Rhode Island, and Wisconsin.

This is probably the most common violet in the Northeast. White forms are occasionally observed and, rarely, forms in which pigment is lacking in seed capsules and seeds. A variety with grayish petals and a purple center called Confederate Violet occurs from North Carolina, Kentucky, and Arkansas south.

bridize freely, producing an array of intermediate forms that frustrate attempts at identification. Of the approximately 75 species in North America about 50 occur in the Northeast. Several species of birds and small mammals eat the seeds and leaves, and the

Common Blue Violet is found in fields and meadows, near human habitations, and along roadsides from Maine and Quebec to Minnesota, south to Georgia and Oklahoma.

Viola sororia
Woolly Blue Violet*

323

Violaceae
Violet Family

*Also Sister Violet

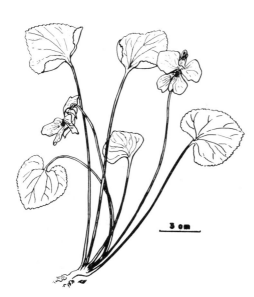

Flowers blue, on leafless stalks, about same height as leaves; corolla with side petals densely white-hairy at base; leaves basal, stalks and undersurface white-hairy, oval, to 10 cm. wide (4 in.); stemless, leaves and flowers to 18 cm. high (7½ in.).

March–June

A perennial with a thick, creeping rhizome. For pollination ecology and methods of seed dispersal see *V. papilionacea.*

It hybridizes freely with several other species, producing a variety of intermediate forms including a few that have been recognized by some botanists as distinct species. It resembles many forms in a sisterly way; thus the specific and common names. It can be distinguished from the very similar *V. papilionacea* by its hairy leaves and leaf stalks. An albino form with all white petals is occasionally observed.

This is a genus with 300–500 species of

worldwide distribution. There are about 50 species in the Northeast, mostly natives of North America. The cultivated Pansies, V. *tricolor* and V. *arvensis*, neither illustrated, were introduced from Europe as flower garden ornamentals but are sometimes observed in abandoned gardens and around old homesites.

Woolly Blue Violet is found in meadows and moist woods from Maine and Quebec to Minnesota, south to North Carolina, Missouri, and Oklahoma.

✖ ℞ See V. *canadensis*.

Plants of Wetlands

"Wetlands" is a broad term that refers to a wide variety of environmental conditions. Like forests and open fields, each wetland may represent a successional stage that will terminate in climax vegetation.

Plants that grow in water or in areas that are usually saturated with water are called Hydrophytes. Succession toward climax vegetation in a hydrophytic community is normally accompanied by a decrease in water depth with a corresponding increase in the amount of the plant body that protrudes above the saturated zone. Thus, the earliest stages of succession will consist of herbaceous plants attached to the bottom and totally submerged. As these submergents die and add their bodies to the accumulating sediments, the depth of the water will decrease and emergent plants will establish themselves. As the bottom sediments continue to build, a still later successional stage will follow, characterized by hydrophytic shrubs growing in wet soil or shallow water. Finally, as the substrate rises above the water table, deciduous trees will invade and become the climax vegetation. These successional sequences are most commonly associated with ponds and swamps.

In the glaciated northeast, ponds and lakes are characteristic features of the landscape. As the glacier melted back, ten to twelve thousand years ago, huge chunks of ice broke off the main body and were buried under glacial debris. As the ice melted, the debris caved in forming depressions called "kettle holes" that often filled with water. Around the margins of many of these water-filled basins, herbaceous plants such as grasses, sedges, and sphagnum moss established themselves in the moist soil. With good growing conditions these species formed dense bog mats which spread over the surface of the water and expanded toward the center of the basin. As the mat closed in toward the center of the pool, the basin became filled with peat originating from successive generations of herbaceous plants.

The earliest stage of succession in a bog is the bog mat, or bog meadow. It is dominated by grasses, sedges, sphagnum moss, and many other herbaceous bog species. This community will be invaded by shrubs, mainly of the heath family, which will eventually overtop and shade out the herbaceous plants. The final stage in some areas of the Northeast will be the establishment of evergreen species such as Black Spruce *(Picea mariana)*, White Pine *(Pinus strobus)*, White Cedar *(Chamaecyparis thyoides)*, and Tamarack *(Larix laricina)* which will become the climax species in bog succession.

Wetlands

281

Four widely accepted categories for wetland classification are marsh, swamp, fen, and bog. Marshes and swamps are defined as areas with mineral soil rooting substrates, but with different types of dominant vegetation: herbaceous plants in marshes, woody plants in swamps. Fens and bogs are similar in having peat as a rooting medium, but fens provide more mineral nutrients for plants than do bogs. Since marshes and swamps may have many species in common, as may fens and bogs, the main wetland categories used here are swamps and bogs. As indicated above, an important difference between the two is the rooting medium.

Wet meadows are areas where the water table is at, or near, the surface of the soil, keeping it saturated for much of the growing season. These areas are normally associated with agricultural land where frequent grazing or mowing prevents the establishment of trees and shrubs. Under these circumstances, emergent species ordinarily found in swamps are often observed.

The following species are sometimes observed in swamps, marshes or bogs, but they also occur in other types of habitats. These plants are described and illustrated in the section that describes the habitat where they are believed to be most common.

Asteraceae, the Aster Family
>   *Bidens frondosa*, Beggar Ticks
>   *Eupatorium maculatum*, Spotted Joe-Pye Weed
>   *Eupatorium perfoliatum*, Boneset
>   *Eupatorium purpureum*, Sweet Joe-Pye Weed
>   *Helenium autumnale*, Sneezeweed
>   *Helenium flexuosum*, Purple-headed Sneezeweed
>   *Senecio aureus*, Golden Ragwort

Campanulaceae, the Harebell Family
>   *Campanula aparinoides*, Marsh Bellflower

Cornaceae, the Dogwood Family
>   *Cornus canadensis*, Bunchberry

Ericaceae, the Heath Family
>   *Gaultheria hispidula*, Creeping Snowberry
>   *Gaultheria procumbens*, Wintergreen

Lamiaceae, the Mint Family
>   *Mentha piperita*, Peppermint
>   *Mentha spicata*, Spearmint

Liliaceae, the Lily Family
>   *Lilium superbum*, Turk's-cap Lily

Malvaceae, the Mallow Family
>   *Hibiscus palustris*, Rose Mallow

Onagraceae, the Evening Primrose Family
>   *Epilobium coloratum*, Purple-leaved Willow Herb

Orchidaceae, the Orchid Family
>   *Cypripedium acaule*, Mocassin Flower
>   *Cypripedium calceolus*, Yellow Lady's Slipper
>   *Cypripedium reginae*, Showy Lady's Slipper
>   *Habenaria clavellata*, Green Wood Orchis
>   *Spiranthes cernua*, Nodding Ladys' Tresses

Primulaceae, the Primrose Family
   Lysimachia ciliata, Fringed Loosestrife
   Trientalis borealis, Starflower
Ranunculaceae, the Crowfoot Family
   Coptis trifolia, Goldthread
Rosaceae, the Rose Family
   Dalibarda repens, Dewdrop
Scrophulariaceae, the Figwort Family
   Pedicularis lanceolata, Swamp Lousewort
Verbenaceae, the Vervain Family
   Verbena hastata, Blue Vervain
Violaceae, the Violet Family
   Viola renifolia, Kidney-leaved Violet

Alisma plantago-aquatica
Water Plantain*

324

Alismataceae
Water Plantain Family

*Also Mud-plantain, Mad-dog Weed

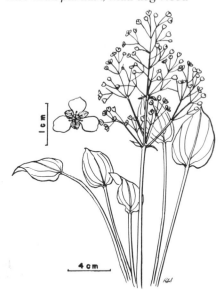

Flowers white or pinkish, tiny, petals 3, sepals 3, petals longer than sepals; leaves basal, long petioled, elliptical, heart-shaped or rounded at base; flowering stalk leafless, highly branched, to 90 cm. high (3 ft.).

June–September

honey-bees that visit the numerous flowers for both nectar and pollen.

Each flower produces 10–25 long-lived seeds which, under experimental conditions, have shown 94% germinability after storage for 5 years in water at 1–3°C. They are small and probably spread by sticking to the feet of waterfowl.

Water Plantain is found in marshes, lake margins, ponds, and ditches from Nova Scotia and Quebec to British Columbia and throughout the United States.

℞ The whole plant contains substances that have diuretic properties, and it has been used in treating kidney ailments. The bruised leaves are said to be beneficial when applied to bruises and swellings but may irritate or blister the skin. In the past the roots were regarded as a cure for hydrophobia (thus, Mad-dog Weed) but this has never been scientifically confirmed.

A perennial with a bulb-like fleshy rhizome. The individual flowers produce only minute quantities of pollen (.06 mg) and over 90% of the total pollen crop is released between 11:00 a.m. and noon. Pollination is by

---

Sagittaria latifolia
Broad-leaved Arrowhead*

325

Alismataceae
Water Plantain Family

*Also Wapato, Duck Potato

A perennial with a fibrous root system and several underground runners, each of which bears a tuber up to 5 cm. in diameter (2 in.). The flowers are visited by numerous insects which probably contribute to pollination. Each female flower produces several to many flattened seed-like fruits equipped with marginal wings which facilitate dispersal by wind or water. In longevity studies the seeds demonstrated a 93% germination rate after storage in water at 1–3°C for 5 years.

There is great variability in the leaves of this species from very narrow, sometimes almost grass-like, on plants in deeper water, to much broader ones on plants growing in mud or shallow water. A closely related species, *S. cuneata*, Wapato, not illustrated, is very similar but has smaller seeds. These two species are probably the most common members of the genus in the Northeast.

This is an important wildlife food genus. The seeds and tubers are eaten by at least 12 species of ducks, the Canada Goose and the Whistler Swan, all of which probably

Flowers white, on a leafless stalk, usually in whorls of 3's, upper flowers staminate or bisexual, on short stalks, lower ones pistillate, on longer stalks; stamens numerous, petals 3, sepals 3; leaves basal, arrow-shaped, highly variable, from very broad to very narrow, leaf stalks usually as long as the water is deep; flowering stalk to 1 m. high (40 in.).

July–September

contribute to seed dispersal. The tubers of *S. latifolia* are often too large to be eaten by ducks, but they are favored by muskrats.

Broad-leaved Arrowhead is found in ponds, swamps, and slowly flowing streams from Nova Scotia and Quebec to British Columbia, south to Florida, Mexico, and California.

✗ The tubers are edible and can be collected in the fall or early spring. They can be prepared in the same manner as potatoes but are said to have an unpleasant taste if eaten raw. Both eastern and western tribes

of American Indians used them for food. Lewis and Clark reported the tubers to be an item of trade among western Indians.

Wetlands

Angelica atropurpurea
Purple Angelica*

326

Apiaceae or Umbelliferae
Parsley Family

*Also Alexanders
Flowers white or greenish white, in spherical clusters, to 20 cm. (8 in.) in diameter; leaves alternate, long-stalked, sheathing stem at base, 2 or 3 times pinnately divided, upper leaves less divided and shorter stalked, leaflets toothed, veins ending at points of teeth; stems smooth, purple or purple streaked, hollow, to 2 m. high (6½ ft.).

June–September

A perennial with a thick, often branched taproot. The flowers are visited, and possibly pollinated, by bees, bee-like flies, and beetles. Each flowers produces 2 yellowish seeds that are flattened on one side and have narrow marginal wings that may aid in dispersal by wind. It may become a weed in wet meadows and pastures, but it can be eradicated by cutting the stem below the ground level and mowing the new shoots before they can produce seeds.

This is a genus of about 50 species mostly in the Northern Hemisphere. Of the 4 species native to eastern North America, *A. atropurpurea* is one of the most common, and it may be the most robust. A related

286

species, *A. venenosa*, Hairy Angelica, not illustrated, has fine white hairs on the stalks of the flower clusters and seeds that are heart-shaped at the base. It occurs in dry woods from Massachusetts to Minnesota, south to Florida and Mississippi.

Purple Angelica is found in swamps, marshes, and wet open ground from Labrador to Minnesota, south to West Virginia and Illinois.

✄ The young stems and leaf stalks can be peeled and used in salads. The strong taste can be removed by cooking in several changes of water. The roots and young shoots can be candied by boiling until tender then simmering in a sugar syrup. *Caution*: Do not confuse this plant with Water Hemlock (*Cicuta maculata*), which is deadly poisonous. In Water Hemlock the veins of the leaflets end in the notches between the teeth.

℞ The genus name is derived from the Latin word for "angel," and according to legend the medicinal properties of these plants were revealed to a monk by an angel during an outbreak of the plague. In herbology, the dried roots and seeds are recommended to relieve intestinal gas, stimulate the kidneys and sweat glands, initiate the menstrual flow, and to relieve indigestion.

Indian tribes of Arkansas mixed the root with smoking tobacco to give it flavor.

Cicuta bulbifera / Bulb-bearing Water Hemlock — 327 — Apiaceae or Umbelliferae / Parsley Family

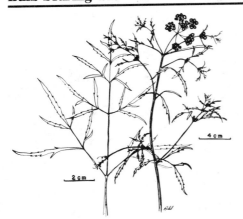

Flowers white, tiny, in flat-topped terminal clusters, to 5 cm. wide (2 in.); leaves alternate, 2 or 3 times pinnately compound, leaflets very narrow, sparsely toothed; axils of upper leaves and stems bearing clusters of small bulblets; stems slender, hollow, to 1 m. high (40 in.).

July–September

small bulblets that are produced in the axils of upper leaves and branches.

In this plant the veins of the leaflets do not end in the notches between the teeth as in most other species of the genus. However, *C. bulbifera* can be easily identified by its very narrow leaflets and axillary bulblets.

A prennial with a cluster of finger-like tuberous roots. The diaphragmatic stem base and odor are similar to *C. maculata*. Seeds usually do not mature, but the plant is propagated vegetatively by the numerous

Bulb-bearing Water Hemlock is found in swamps, marshes, and pond margins from Newfoundland to British Columbia, south to Virginia, Ohio, and Oregon.

For other information, see *C. maculata*.

Cicuta maculata
Water Hemlock*

328

Apiaceae or Umbelliferae
Parsley Family

*Also Spotted Cowbane, Musquash-root,
Beaver Poison
Flowers white, tiny, in flat-topped terminal
clusters, to 10 cm. wide (4 in.); leaves alter-
nate, 2 or 3 times pinnately compound,
leaflets oval-lanceolate with numerous
pointed teeth, veins of leaflets ending in
notches between teeth; stems thick, hollow,
freely branched, often with purple
splotches or lines, to 2 m. high (6½ ft.).

June–September

A perennial with a cluster of finger-like
tuberous roots that have the odor of parsnips. The base of the stem, to which the
roots are attached, is underground and enlarged. If this is split lengthwise it will reveal
a series of horizontal diaphragmatic parti-
tions, and the cut surface will exude a yellow, oily liquid with the odor of raw parsnips.
The flowers are visited by several species of
bees, butterflies, and wasps.

This is a genus of about 20 species in the
north temperate zone. Of the 3 species native to the Northeast, *C. maculata* is the most
widely distributed. It is represented in the
states south of Virginia by a variety with
larger leaflets. Three species are native to
the western states.

Water Hemlock is found in marshes,
swales, wet meadows, and roadside ditches
from Nova Scotia and Quebec to Manitoba,
south to Florida and Texas.

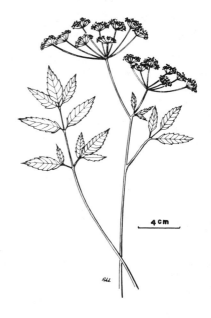

4 cm

☠ All the plants in this genus and all parts
of the plants are deadly poisonous. The poison is concentrated in the roots, and there
are many cases on record of fatal poisoning
of humans. Death may occur as quickly as
15 minutes after ingestion of the root and
is often preceded by violent convulsions. A
single bite is enough to kill a human, and
a root the size of a walnut will kill a cow.

Wetlands

Sium suave
Water Parsnip

329

Apiaceae or Umbelliferae
Parsley Family

A perennial with a fibrous root system.
The flowers are simple with unconcealed
nectar which attracts a wide variety of insects. Each flower produces 2 oval, dark
brown seeds with 5 prominent corky wings.
These are adapted for dispersal by wind or
water. It may become a weed in low areas
of pastures and meadows.

A form with slender round stems and
smaller flower clusters has been reported

for the southeastern states and identified
by some botanists as *S. floridana*, not illustrated. A poisonous plant sometimes confused with *S. suave* is *Oxypolis rigidor*,
Cowbane or Water Dropwort, not illustrated. It occurs in the same habitat, has a
similar growth form except its leaflets have
few teeth or are entire, and has about the
same geographic distribution.

Water Parsnip is found in the shallow bays

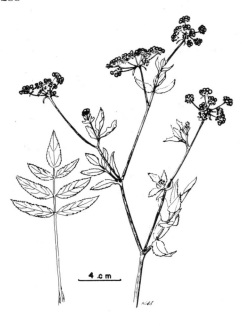

4 cm

Flowers white, tiny, in flat-topped clusters of umbels at the ends of stems and branches, to 12 cm. wide (5 in.); leaves alternate, pinnately compound, with 5–17 lanceolate, many toothed leaflets, upper leaves smaller and sometimes simple, underwater leaves finely dissected; stems thick, smooth, hollow, strongly ribbed, to 1.8 m. high (6 ft.).

July–September

of ponds and slow streams, swamps, swales, and wet meadows from Newfoundland to British Columbia, south to Florida, Louisiana, and California.

✖ The roots of Water Parsnip are edible but, since it grows in areas where several poisonous plants occur (see Water Hemlock), it should never be eaten unless collected by an expert on plant identification.

It is a suspect in the poisoning of a variety of livestock throughout its range.

Acorus calamus **Sweet Flag***	**330**	*Araceae* **Arum Family**

*Also Calamus

Flowers greenish-yellow, tiny, in a dense cone-like spadix projecting from the side of a leaf-like stem; leaves in a dense basal clump, long and narrow, ribbon-shaped, to 1.2 m. long (4 ft.); flower bearing stems leaflike, to 80 cm. high (32 in.).

May–August

A perennial with a thick, creeping, aromatic rhizome. Each ridged, pyramidal fruit includes 1–4 tiny seeds that were produced without fertilization; thus the new plants that grow from them are genetic duplicates of the parent plant. This plant sometimes spreads into drained areas that may become relatively dry in summer, but it does not ordinarily produce flowers or fruits in these areas.

This is a wetland genus with one other species native to central and eastern Asia. A form of *A. calamus* with longitudinal yellow stripes on its leaves is occasionally observed. Sweet Flag may be distinguished from Cattail (*Typha spp.*) and Blue Flag (*Iris*

8 cm

8 cm

SP.

spp.) by the off-center midvein of its leaves and their aromatic fragrance when bruised.

Sweet Flag is found in marshes and open swamps, along the margins of ponds, and in roadside ditches from Nova Scotia and Quebec to Oregon, south to Florida and Texas.

✖ The young inner shoots, while still less than 30 cm. high (1 ft.), give a distinctive flavor to salads. The rhizome can be peeled, boiled, and candied to make a confection treasured by some, but said to be most appreciated if taken in small quantities at a time. *Caution:* This plant should not be con-fused with Blue Flag, which is poisonous (see above).

℞ The most important herbal use of Sweet Flag is for disorders of the stomach. It has been used to stimulate the appetite and to relieve upset stomach and hyperacidity.

The rhizome contains asarone, which is chemically similar to mescaline. The Cree Indians of northern Canada chewed it as a strong stimulant and possibly a hallucinogen.

In former times, the rush-like leaves, because of their pleasant odor, were spread on the floors of churches on special days, and sometimes in private homes.

Calla palustris
Wild Calla*
331
Araceae
Arum Family

*Also Water Arum

Many tiny flowers on a spadix surrounded by an open, white, abruptly pointed spathe, on a leafless stalk; in fruit a globular cluster of red berries; leaves basal, heart-shaped, long petioled; stem lacking, leaves and flower stalk to 30 cm. high (12 in.).

May–July

A perennial with a long, thick, creeping rhizome. Pollination is by insects, and occasionally by pond snails. Each berry contains a few brown, barrel-shaped seeds with shallow pits in the upper half, and longitudinal ridges on the lower half. These may be dispersed by birds and mammals attracted to the red berries for food.

This is a genus of only one species found in North America, Europe, and Asia. It is found in bogs, swamps, or shallow water, ranging in North America from Newfoundland to Alaska, south to Virginia, Ohio, and Iowa.

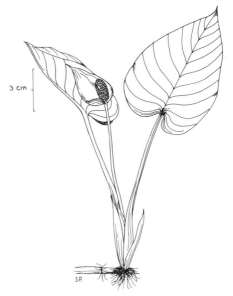

3 cm

S.R.

✖ The use of the rhizome of Wild Calla to make a bread-like food was first described by Linnaeus for Lapland. It apparently occurred there in such abundance that the rhizome could be collected by the cartload. The fresh rhizome contains needle-like crystals of calcium oxalate which, if taken into the mouth, pierce the mucous mem-branes and cause an extreme burning sensation. These are deactivated only by thorough drying. Both the seeds and the dried rhizome can be ground into a flour that is nutritious but distasteful. However, even if the flour was highly palatable, this plant does not occur in the Northeast in sufficient numbers ot justify collecting it as a food plant.

Peltandra virginica
Arrow Arum*

332

Araceae
Arum Family

*Also Tuckahoe

An emergent perennial with thick fibrous roots. At maturity the flower stalk bends until the spathe is under water or near the surface of the ground. Each berry contains 1–3 seeds surrounded by a gelatinous substance. The berries make up 10–25% of the diet of the Wood Duck and are also eaten by other marsh birds and shore birds.

This plant is found in swamps, pond margins, and shallow water from southern Quebec to Ontario and Michigan, south to Florida and Louisiana.

Staminate and pistillate flowers borne on a spadix enclosed in a green, tubular, pointed spathe; fruit consisting of a cluster of green or brown berries; leaves basal, with 3 prominent nerves and pinnate venation, broadly arrow-shaped with long petioles; stems lacking, leaves to 90 cm. high (3 ft.).

May–June

✗ Captain John Smith in 1626 wrote of a plant root used as food by the colonists that may have been Arrow Arum. The method of preparation he described was to dig a pit, place the roots in it, cover them with earth, and build a fire over them for 24 hours. After this the roots could be eaten but were very prickly to the throat unless sliced and dried. The prickly sensation he referred to resulted from the presence of needle-like crystals of calcium oxylate which cooking does not change but thorough drying does.

The berries may be cooked as dried peas or beans, and the dried root may be ground into a flour. The size and abundance of the roots of this plant makes it a good candidate for further research as a possible food source today.

Symplocarpus foetidus
Skunk Cabbage

333

Araceae
Arum Family

A perennial with a thick vertical rhizome bearing whorls of fleshy fibrous roots. Pollination is probably effected by carrion flies attracted to the flowers by the odor of decaying flesh. Cross-pollination is assured by the maturation of the anthers and stigmas at different times. Each flower produces a single large, finely wrinkled seed embedded in the tissue of the spadix.

The spathe enclosing the flower cluster develops and pollination occurs before the leaves unroll. When they first appear, the leaves are tightly coiled spikes beside the spathe. The mature leaves have a strong odor reminiscent of skunk. The seeds are eaten by the Wood Duck, Ruffed Grouse, Ring-necked Pheasant and the Bobwhite Quail.

Skunk Cabbage is found in swamps, in the margins of bogs, and in open woods from Nova Scotia and Quebec to Manitoba, south to Georgia and Illinois. It also occurs in eastern Asia.

✗ The leaves and rhizomes contain varying concentrations of calcium oxylate crys-

Many small pale purple to flesh-colored flowers crowded on a spadix enclosed by a green to purple-brown spathe, pointed and curved inward at the tip, to 15 cm. high (6 in.); leaves basal, short stalked, heart-shaped at base, prominently veined, to 60 cm. long (2 ft.).

February–April

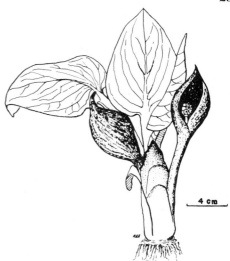

4 cm

tals which puncture the mucous membrane of the mouth and cause an intense burning sensation if these are eaten raw or cooked. The only way to dispel this property is by thorough drying. The young dried leaves can be added to other greens or stew dishes. The dried rhizome can be ground into a flour that has a taste suggestive of cocoa. *Caution:* This plant should not be confused with White Hellebore (*Veratrum viride*), which sometimes grows in the same habitat and is a violent poison.

℞ *The Dispensatory of the United States of America* (1851) states, "This root is a stimulant, antispasmodic and narcotic." It has been used to relieve the spasms of asthma, whooping cough, nasal catarrh, and bronchial irritation.

One tribe of American Indians is reported to have used Skunk Cabbage as a treatment for headache. The treatment consisted of binding a bunch of leaves, bruising them and inhaling the odor. This may be an instance of a remedy that cures by contrast because after a few deep sniffs almost any headache would pale by comparison.

Asclepias incarnata
Swamp Milkweed

334

Asclepiadaceae
Milkweed Family

Flowers pink to rose-purple, crown-shaped, in several terminal umbels; fruit a pod opening along one side at maturity, seeds numerous, each bearing a dense tuft of hairs; leaves opposite, numerous, lanceolate to oblong, tapering to the tip; stems erect, often branched in upper half, with milky juice, to 1.5 m. high (5 ft.).

June–August

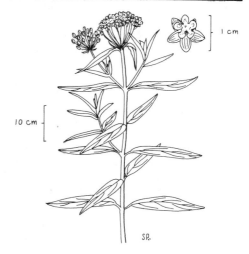

1 cm

10 cm

A perennial with a thick rhizome. Pollination is by insects, and the highly specialized structure of the flower in this genus is such that visiting insects sometimes get their legs caught, and the less hardy ones, being unable to free themselves, die there. Seed dispersal, with the aid of a parachute of hairs, is by wind. Most of the numerous

SR.

flowers do not produce mature seedpods.

Swamp Milkweed is found in wet meadows and along the margins of open swamps, ponds, lakes, and streams from Quebec to Manitoba and Wyoming, south to Florida, Louisiana, and New Mexico.

☠ In herbal medicine, the rhizome has been used to induce vomiting, as a diuretic, and in a treatment for intestinal worms. Almost all species of this genus contain toxic substances, called cardiac glycosides, that affect the heart muscles and if ingested uncooked can cause poisoning in livestock and man.

Aster puniceus **Purple-stemmed Aster**	**335**	*Asteraceae or Compositae* **Composite Family**

Flower heads with violet-blue rays, usually about 2.5 cm. wide (1 in.), on branching stalks from numerous upper leaf axils; leaves alternate, narrowly lanceolate, coarsely toothed, tapering at base and clasping the stem, to 15 cm. long (6 in.); stems purplish, rough-hairy, thick, to 2.4 m. high (8 ft.).

August–October

A perennial with a short, thick rhizome and dense fibrous roots. The flowers are self-incompatible and cross-pollination is probably accomplished by short-tongued bees. Each of the many flower heads produces 50–100 or more smooth or slightly hairy, seed-like fruits with parachutes of hairs that aid in dispersal by wind.

This is a very large and complex genus, and it sometimes requires a combination of characteristics to distinguish between related species. *A. puniceus* is a highly variable species with several varieties that differ in the hairiness of stems and leaves and in the nature of the flower heads. It hybridizes freely with one or more other species, and one hybrid has been recognized as a separate species by some botanists. Forms are occasionally observed with lilac, pink, or white ray flowers.

Purple-stemmed Aster is found in swamps, swales, damp thickets, and other moist areas from Newfoundland and Quebec to Manitoba, south to Georgia and Alabama.

Bidens cernua **Nodding Bur-marigold***	**336**	*Asteraceae or Compositae* **Aster Family**

*Also Stick-tights

An annual of North America, Europe, and Asia. Each flower head produces numerous seed-like fruits with 4 barbed awns that are very effective in dispersal by attaching readily to fur, feathers, and clothing. The fruits are eaten occasionally by ducks but are not an important source of food. It sometimes invades pastures and wet meadows as a weed but can usually be controlled by improving drainage.

This species is highly variable with regard

Flower heads with 6–8 yellow rays, numerous, subtended by 6–8 bracts, nodding with age; flower heads to 3 cm. wide (1¼ in.), rays sometimes absent; leaves opposite, narrow lanceolate, toothed, sessile, smooth, to 15 cm. long (6 in.); stems freely branched, sometimes horizontal on ground at base and rooting at nodes, to 1 m. high (40 in.).

July–October

to leaves and flower heads, and several varieties are recognized by some botanists. A related species, *B. laevis*, Smooth Bur-marigold, not illustrated, is very similar but has flower heads mostly over 3.7 cm. wide (1½ in.) and reddish subtending bracts. It is most abundant in wet areas along the Atlantic coast from Massachusetts to Florida.

Nodding Bur-marigold is found in swamps, pond margins, and roadside

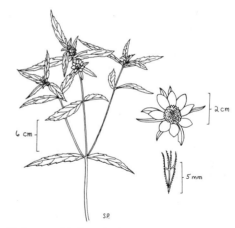

ditches and in brackish or saline areas from New Brunswick to British Columbia, south to North Carolina, and on the Pacific coast to California.

Mikania scandens
Climbing Hempweed*

337

Asteraceae or Compositae
Aster Family

*Also Climbing Boneset
Flower heads flesh colored to white, numerous, in round-topped, stalked, axillary clusters; leaves opposite, triangular, palmately veined, long stalked; stems twining, often forming mats on other vegetation, to 5 m. long (16½ ft.).

July–October

A perennial native of North America that has spread to tropical regions throughout the world. Each of the numerous flower heads produces 4 seed-like fruits which are equipped with plumes of hairs for dispersal by wind.

This is a North American genus of about 150 species restricted mainly to the tropical and sub-tropical regions. It was named in honor of Joseph C. Mikan, an eighteenth-century professor of botany at Prague University. The genus is unusual because the climbing habit is rare among North American members of the Aster Family. *M. scandens* is the only member of the genus in the Northeast.

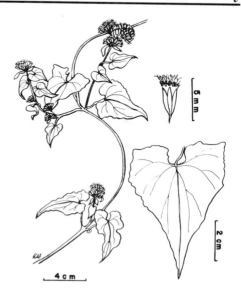

The flower heads are similar to those of Boneset, *Eupatorium perfoliatum*; thus the common name Climbing Boneset. It was

294

thought by early botanists to belong to that genus and was classified by Linnaeus as *Eupatorium scandens.*

Climbing Hempweed is found along the margins of swamps, on stream banks, and in thickets bordering salt marshes along the Atlantic Coast from Maine to Florida and Texas and locally inland to Ontario, Michigan, and Missouri.

Rudbeckia laciniata
Tall Coneflower*

338

Also Green-headed Coneflower

3 cm

SP

Ray flowers yellow, drooping, disk dome-shaped, greenish-yellow; flower heads several on the ends of stems and branches, to 10 cm. wide (4 in.); leaves alternate, basal and lower ones long petioled, pinnately divided with 5–7 irregularly lobed or toothed leaflets, upper ones smaller, 3–5 parted; stems smooth, branched, to 3 m. high (10 ft.).

July–September

25 species. It was named in honor of Olaf Rudbeck, a Swedish botanist of the seventeenth century and a teacher of Linnaeus. *R. laciniata* is the tallest member of the genus in the Northeast. A double-flowered form called Golden Glow is commonly cultivated in flower gardens. A form that usually does not exceed 1 m. in height (40 in.) and has 3-lobed leaves occurs in the mountains from Virginia southward.

A coarse perennial with a woody base. Although the flower heads are frequently visited by bees, some studies have indicated that at least some of the seeds are formed without fertilization. Only the disk flowers are fertile, and each flower head produces numerous black, 4-angled, longitudinally ribbed seeds. These are eaten by the Common Goldfinch and other birds.

This is a North American genus of about

Tall Coneflower is found in swamps, roadside ditches, low ground, and wet or moist woods from Quebec to Manitoba and Montana, south to Florida, Texas, and Arizona.

℞ In herbology, it is listed as a diuretic and a soothing tonic. It is said to have been used by the natives of New Mexico to make a tea for the treatment of gonorrhea or to initiate menstruation. For its effect on livestock, see *R. hirta.*

Impatiens biflora
Spotted Touch-me-not*

339

Balsaminaceae
Touch-me-not Family

Also Jewelweed, Snapweed, Lady's Earrings

Flowers orange with reddish-brown spots, with an inflated throat and a slender, curled basal spur; flowers dangling from slender axillary stalks; leaves alternate, long stalked, coarsely toothed, oval; stems succulent, usually branched, to 1.5 m. high (5 ft.).

June–September

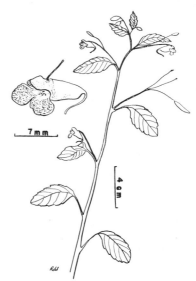

An annual with a dense cluster of fibrous roots. The stamens mature and are shed before the stigma becomes receptive to pollen, thus eliminating the possibility of self-pollination. Cross-pollination is effected mainly by honeybees, bumblebees, and the Ruby-throated Hummingbird. At maturity, the seed capsule is up to 3 cm. long (1¼ in.) and at the slightest touch bursts explosively. The capsule splits from the base into 5 sections which forcefully roll inward hurling the seeds for a distance of up to 2.4 m. (8 ft.). In addition the seeds are eaten by the Ruffed Grouse, Ring-necked Pheasant, and Bobwhite Quail, which may contribute further to their dispersal.

Of the 2 native and 3 introduced species of this genus in the Northeast, *I. biflora* is the most common. It is a highly variable species with several varieties based on flower color. In some systems of classification it is listed as *I. capensis*. A species often grown in green houses and flower gardens is *I. balsamina*, Garden Balsam, not illustrated. A related native species, *I. pallida*, Pale Touch-me-not, not illustrated, is very similar but has pale yellow flowers and is more common in limestone areas.

Spotted Touch-me-not is found along the margins of shady swamps, roadside ditches and wet woods from Newfoundland to Alaska, south to Florida and Alabama.

✖ The young stems and leaves of Pale and Spotted Touch-me-not can be cooked in 2 changes of water. The seeds are tasty, if tiny, snacks, tasting like butternuts.

℞ The water from cooking the plant or the fresh juice is said to prevent poison ivy rash if applied immediately after exposure. Several tribes of American Indians used the fresh plant to ease the itching caused by poison ivy rash and insect bites.

Myosotis scorpioides
Forget-me-not

340

Boraginaceae
Borage Family

A perennial with a shallow fibrous root system, introduced from Europe. The yellow ring in the center of the flower serves as an attractant and a nectar guide for insect pollinators. The contrasting yellow and blue colors of the flowers are a combination that has been found to be one of the most common among bee pollinated plants. Each flower produces 4 small, glossy, black or brown, narrowly winged or ridged seeds.

This genus has 5 introduced and 3 native species in the Northeast. The specific name

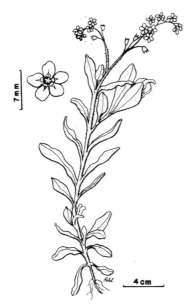

7 mm

4 cm

Flowers blue with yellow centers, broadly funnel-shaped, with 5 lobes; flowers on coiled branches at tip of stem, branches unroll as flowers open; leaves alternate, hairy, oblong, to 8 cm. long (3 in.); stems hairy, angled, prostrate at base with erect tips, often rooting at nodes, to 60 cm. long (2 ft.).

May–September

German folklore a knight and his damsel were walking beside a flooding stream when he fell and was swept into the raging current. In desperation he grasped a cluster of *Myosotis* and flung them to his beloved with the lament "forget me not!"

This species was introduced as an ornamental for moist gardens and has escaped widely. It often occurs in colorful masses in marshy areas during the summer months. It is the state flower of Alaska. A dwarf horticultural variety seldom exceeds 20 cm. in height (8 in.). *M. scorpioides* is classified as *M. palustris* by some botanists.

Forget-me-not is found in marshes, stream and pond margins, and wet meadows from Newfoundland to Ontario, south to Georgia and Louisiana, and along the Pacific Coast.

is a Latin word meaning "like a scorpion" and refers to the flower cluster, coiled like the tail of a scorpion. The name Forget-me-not is recognized in many countries, and the flowers are international symbols of remembrance and friendship. According to

---

Cardamine bulbosa **Spring Cress**	**341**	*Brassicaceae or Cruciferae* **Mustard Family**

A perennial with one or a cluster of short thick tubers. As the flowering season advances, the flower cluster elongates and the lowest seedpods mature first. At maturity the pods split into 2 sections, releasing several oval, greenish, slightly wrinkled seeds. It is a small inconspicuous plant that ordinarily does not become a troublesome weed.

This is a genus with at least 100 species

of the northern hemisphere. There are about 10 species in the Northeast, some of which are widely distributed also in Europe and Asia. *C. douglassii*, Purple Cress, not illustrated, is similar to *C. bulbosa* but has pink-purple petals, a hairy stem and begins flowering about 2 weeks earlier. It occurs in the same types of habitats from Ontario to Wisconsin, south to Tennessee and Missouri.

Spring Cress is found in wet meadows,

Flowers white, to 12 mm. wide (½ in.), in a terminal cluster that elongates in fruit; sepals 4, petals 4, stamens 6, 2 shorter than others; fruit a slender pod to 2.5 cm. long (1 in.); stem leaves alternate, oval to lanceolate, upper ones sessile; basal leaves long-stalked, roundish, to 3.8 cm. long (1½ in.); stems smooth, usually unbranched, to 60 cm. high (2 ft.).

April–June

stream borders, shallow water, and moist woods from Nova Scotia and Quebec to Minnesota, south to Florida and Texas.

✖ The young plants can be cooked as greens or added to salads. The grated tuberous underground parts and finely chopped young stems and leaves, when mixed with vinegar, make a mild substitute for horseradish. The older leaves become very bitter but the rootstock is edible year-round.

4 cm

Cardamine pratensis
Cuckoo Flower

342

Brassicaceae or *Cruciferae*
Mustard Family

Flowers white or pink, to 18 mm. wide (¾ in.), in dense terminal clusters, similar to those of C. bulbosa; fruit an erect slender pod, to 4 cm. long (1½ in.); leaves alternate, pinnately compound, with short petioles and narrow leaflets; basal and lower stem leaves with longer stalks and rounded leaflets; stems slender, erect, usually unbranched, smooth, to 50 cm. high (20 in.).

May–June

A perennial with a short rhizome, of North America, Europe, and Asia. When sexual reproduction occurs, the plant is self-incompatible and is cross-pollinated by insects. Forms that occur in the Arctic produce seeds without fertilization. Vegetative reproduction by short basal branches may be more common than reproduction by seeds. When mature seedpods are produced, they contain several oval, slightly winged, cinnamon-brown seeds.

This is the only species of the genus that

6 cm

2 cm

is cultivated as an ornamental. It is somewhat variable with pink and white flowered varieties. The pink form is native to Europe and Asia and extends into the Arctic region.

The white flowered variety is native to North America. A double flowered form is occasionally observed.

The pink form of Cuckoo Flower is found in lawns, meadows, and along roadsides from Newfoundland to Alaska, south to New Jersey and Ohio. The white flowered variety is found in swamps, bogs, and wet woods from Labrador to Mackenzie, south to Virginia and Illinois.

Nasturtium officinale — Watercress — **343** — *Brassicaceae or Cruciferae* Mustard Family

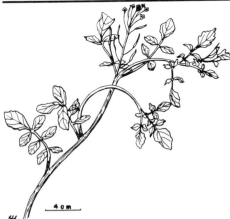

4 cm

Flowers white, small, in elongate clusters at the ends of stems and branches; petals 4, stamens 6; fruit a slender curved pod, to 2.5 cm. long (1 in.); leaves alternate, pinnately compound, with 3–9 roundish leaflets, succulent; stems prostrate, floating or submerged, succulent, rooting at nodes, to 1.5 m. long (5 ft.), or longer.

April–October

Radicula nasturtium-aquaticum. The present name was established by the Third International Botanical Congresss. The genus name is Latin for "twisted nose" or "convulsed nose" and refers to the pungent odor of the plant. It should not be confused with the flower garden ornamental Nasturtium, *Tropaeolum majus*, not illustrated. The common name was unfortunately applied to the latter because the odor of its leaves is similar to that of *Nasturtium officinale*.

A perennial with a freely rooting, brittle stem, introduced from Europe. Each of the numerous seedpods produces two rows of tiny, slightly glossy, dark brown seeds. The plant may remain green throughout the winter in some areas. It is eaten by ducks, muskrats, and deer, all of which may contribute to seed dispersal.

This is a good plant for trout streams because it provides food and shelter for small aquatic organisms, such as fresh water shrimp and water sowbugs, that are favored by trout for food. It is a valuable wildlife food plant for game farms, but it often grows in very dense tangled masses to the exclusion of native species.

In older systems of classification this species has been listed as *Rorippa nasturtium-aquaticum, Sisymbrium nasturtium,* and

Watercress is found in brooks, springs, roadside ditches, and other areas of cool shallow water throughout the United States and southern Canada.

It is cultivated in some areas for use in salads and sandwiches, and it can be cooked as greens. When it is collected in the wild, extreme care should be observed not to confuse it with the poisonous Water Hemlock. In addition, care should be taken to avoid plants growing in contaminated water. Water Cress is reported to be rich in Vitamin A, B, C, and B_2 as well as iron, copper, magnesium, and calcium.

Rorippa islandica
Marsh Yellow Cress

344

Flowers yellow, tiny, in long terminal or upper axillary clusters, to 25 cm. long (10 in.); stamens 6, 2 short, 4 long; petals 4, not longer than sepals; fruit a small oval pod with many seeds, standing out from stem; leaves alternate, lower ones pinnately dissected, to 15 cm. long (6 in.), upper ones smaller and less dissected or merely toothed; stems smooth or hairy, thick, often branched, to 1 m. high (40 in.).

May–October

A circumboreal annual or biennial with a thick taproot. Each flower has small nectar glands that presumably function or have functioned to attract insect pollinators. The seed capsule splits into 2 sections, releasing numerous irregularly shaped, finely pitted, gray-brown seeds that are small enough to be windborne in a moderate breeze.

This is a genus of about 40 species with worldwide distribution. There are 8 species in the Northeast, 4 of which are native to North America. Most species occur in moist to wet substrates. *R. islandica* is highly variable with several varieties based on leaf characters and distribution of hairs. One variety is widely distributed in Europe and Asia.

A related species, *R. sylvestris*, Creeping Yellow Cress, not illustrated, has similar leaves but has petals that are longer than the sepals and a straight narrow seed capsule.

Marsh Yellow Cress is found along the margins of ponds and lakes and on wet soil and mud from Labrador to Alaska, south to Florida, Texas, and California.

⚔ A related species *R. amphibia*, Amphibious Yellow Cress, not illustrated, is cooked as greens and used raw for salads in Europe. The young shoots of other species are probably equally palatable.

Butomus umbellatus
Flowering Rush

345

A perennial with a thick shallowly buried rhizome, introduced from Europe. In each flower the stamens mature before the pistils. As additional insurance against self-pollination the plants arising from one rhizome are self-incompatible and are receptive only to the pollen of plants from another rhizome. Each flower produces 6 seedpods that split along their inner sides, releasing numerous tiny, strongly ribbed seeds.

This is a genus of only one species widespread in Europe, Asia and northern India. The genus name is derived from Greek words that mean "ox cut" and possibly alludes to

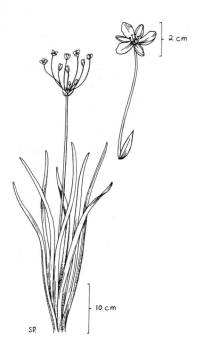

SP.

Flowers pink, very showy, to 2.5 cm. wide (1 in.), numerous, in a terminal cluster on a leafless stalk; flower divisions 6, stamens 9, pistils 6; leaves basal, erect or floating in water, narrow and ribbon-like; leaves and flowers stalk to 1.2 m. high (4 ft.).

June–August

the long sword-like leaves. A sterile form with long, thin, very limp leaves is occasionally observed. This species is sometimes planted along the edges of ponds and may be of value as a wildlife food plant.

Flowering Rush was introduced into North America along the marginal lowlands of the St. Lawrence River in the area of Montreal. It is spreading rapidly and is found in shallow to deep water on the shores of streams and ponds from Vermont, New York, and Quebec westward along the Great Lakes to Ohio and Michigan.

✄ The rhizomes are reportedly used for food in some parts of Europe and Asia. They are said to be roasted and eaten or ground into a flour for bread.

Campanula aparinoides
Marsh Bellflower

346

Campanulaceae
Harebell Family

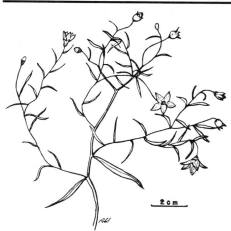

Flowers pale blue to white, to 12 mm. long (½ in.), on long terminal stalks; corolla funnel-shaped, with 5 lobes; leaves alternate, very narrow, to 5 cm. long (2 in.); stems slender, 3-angled, with fine bristles on angles, usually branched, weak and reclining, to 60 cm. long (2 ft.).

June–August

ous smooth, semi-glossy, yellowish or brown seeds. These are small enough to become airborne in a moderate breeze. The specific name alludes to the similarity of the rough stem to that of *Galium aparine*.

This species has two varieties, one in the southern and one in the northern part of its range. The southern phase has an almost white corolla and grades into the northern phase which has a light blue corolla. The blue flowered variety is recognized by some botanists as a separate species, *C. uliginosa*.

A perennial with a very slender horizontal rhizome. The flower is succeeded by a strongly ribbed seed capsule that splits into 3 sections from the base, releasing numer-

Many species of this large genus are cultivated for their showy flowers. Probably the most popular of these is *C. medium*, Canterbury Bells, not illustrated. It is a native of southern Europe, but is cultivated all over the world and has many horticultural varieties.

Marsh Bellflower is found in wet meadows, thickets, swales, and grassy swamps from New Brunswick to Saskatchewan, south to Georgia and Colorado.

| *Stellaria aquatica* Giant Chickweed | **347** | *Caryophyllaceae* Pink Family |

Flowers white, to 12 mm. wide (½ in.), long stalked, in upper leaf axils and in branched terminal clusters; petals 5, so deeply cleft there appears to be 10, styles usually 5, sepals 5, much shorter than petals; leaves opposite, sessile, heart-shaped at base, tapering to a pointed end, to 8.7 cm. long (3½ in.); stems angled, branched, with fine glandular hairs, often lying on the ground, to 80 cm. long (32 in.).

May–October

A perennial with tufted fibrous roots, introduced from Europe. The egg-shaped seed capsule opens at the tip by 10 teeth, releasing numerous tiny, brown, kidney-shaped, coarsely pebbled seeds. It is not as abundant as *S. media*, Common Chickweed, but probably resembles it enough to attract the same wildlife species as consumers.

This is a genus of 100 or more species of worldwide distribution but concentrated in temperate regions. Of the 14 species that occur in the Northeast, 5 are native to Europe and Asia. One of the latter, *S. holostea*, Easter Bells, not illustrated, has flowers to 2.5 cm. wide (1 in.) and is cultivated in flower gardens as a border ornamental and for ground cover. *S. aquatica* is the only species

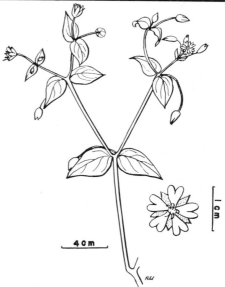

with 5 styles and because of this some botanists place it in a separate genus and classify it as *Myosoton aquaticum*.

Giant Chickweed is found in wet meadows, the edges of swamps and marshes, wet springy areas, and roadside ditches from Quebec and Ontario to Minnesota, south to North Carolina and Louisiana.

Wetlands

Ceratophyllum demersum
Hornwort*

348

Ceratophyllaceae
Hornwort Family

*Also Coontail

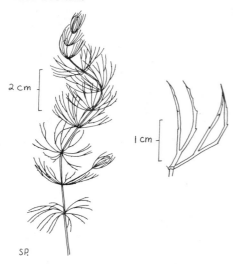

2 cm

1 cm

S.P.

Flowers small and inconspicuous, unisexual, scattered, in leaf axils, solitary; fruit a dark brown nutlet with a terminal spine and 2 basal spines; leaves in whorls of 5's–10's, 2 or 3 times palmately dissected, the fine divisions with teeth along one side, crowded at stem tip; stems submerged, freely branched, to 2 m. long (6½ ft.).

July–September

vigorous submerged aquatics. It is tolerant of fluctuating water levels and moderate light intensity. Under favorable conditions it may proliferate to the exclusion of other species and seriously clog waterways.

As the fall season advances, the stem tips may become shortened and thickened, then break off and sink to the bottom to serve as winter buds. It may be observed in the vegetative state all winter, even under a layer of ice. It is an attractive plant for fish bowls and home aquaria.

A related species, C. echinatum, not illustrated, is similar but the fruit has 3–5 lateral spines in addition to 2 basal ones, and it occurs only in the eastern half of the United States.

C. demersum is found in ponds and slowly moving streams, especially in hard water, ranging from southern Canada, throughout the United States, Mexico, and Central America. It is also widespread in Europe and Asia.

A submerged, evergreen perennial. Each staminate flower has 12–16 stalkless anthers which break loose and float to the surface at maturity. On the surface of the water the anthers split, releasing pollen that sinks and comes into contact with the stigmas of the pistillate flowers. The seeds and vegetation are eaten by several species of ducks, constituting up to 10% of the diet of the Gadwell, Mallard, Redhead, Ringneck, and Greater Scaup.

Hornwort is one of the most common and

Clethra alnifolia
Sweet Pepperbush

349

Clethraceae
White Alder Family

A woody perennial with large, pointed terminal buds. The fragrant clusters of flowers are visited by bees, small moths, and butterflies. The fruit is a small globular capsule that somewhat resembles a pepper-

corn and persists throughout the winter months. It is 3-celled and eventually splits into 3 sections, releasing numerous seeds.

This is a family with only one genus and about 30 species. Two species are native to

Flowers white, to 8 mm. wide (⅓ in.), in dense cylindrical clusters, to 15 cm. long (6 in.), at the ends of branches; leaves alternate, widest above middle, sharp pointed at tip, tapering at base, to 10 cm. long (4 in.); stems erect, woody, to 3 m. high (10 ft.).

July–September

the Northeast, both with limited geographic ranges. *C. acuminata*, Mountain Pepperbush, not illustrated, has elliptical, long-pointed leaves and hairy stamens. It occurs in mountainous woods from Pennsylvania to Georgia.

Sweet Pepperbush is found in swampy areas and in wet sand mostly along the Atlantic and Gulf coast from Maine to Florida and Texas.

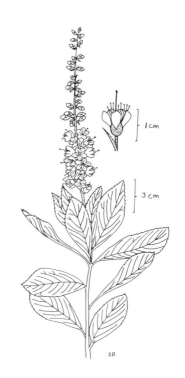

Wetlands

Cuscuta gronovii
Dodder*

350

Convolvulaceae
Morning-glory Family

Also Love-vine, Strangle-weed
Flowers white, tiny, in dense clusters; corolla bell-shaped, with 5 lobes, stamens attached between lobes; leafless; stems orange or yellow, tightly twined around stems and leaves of other plants.

July–October

An annual parasite with root-like structures that penetrate the conductive system of a host plant. Each flower produces 4 tiny, brownish, finely wrinkled, granular seeds. Some of the cells on the outer surface of the seed coat become enlarged and filled with air, providing a possible mechanism for dispersal by water or wind. The seeds germinate in the soil and the seedlings soon come into contact with host plants. Contact with the soil is then broken, and they become totally dependent on the hosts.

The 12 native and 3 introduced species of this genus in the Northeast are very sim-

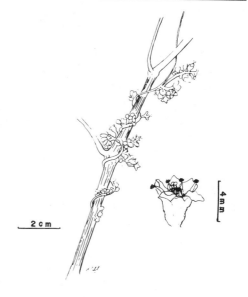

ilar and are distinguished from one another on rather technical characters. The introduced species are often very injurious to some crop plants, and all dodders have been declared noxious weeds in the seed laws of 42 states and by the Federal Seed Act.

Preventive or control measures for infestations of Dodder include early cutting of fence rows, ditches and weed fields to prevent seed production. Mowing infected areas and burning the residue is recommended for clover and alfalfa crops. Some of its common names, Strangle-weed, Hellweed, and Devil's Guts, suggest that plants of this genus have, from ancient times, had a bad reputation among farmers.

Cuscuta gronovii is the most common species. It is found on a great variety of host plants in swampy areas and wet fields from Nova Scotia to Manitoba and Montana, south to Florida and Texas.

℞ The whole plant is reported to be a rather harsh laxative. In some parts of the United States, it has been suspected of causing digestive disorders in horses and cattle.

Penthorum sedoides — Ditch Stonecrop — 351 — Crassulaceae — Orpine Family

Flowers greenish-yellow, in elongate, branching terminal clusters; petals none, sepals 5, stamens 10; pistils 5, united, forming a 5-horned seed capsule; leaves alternate, lanceolate, finely toothed; stems smooth, often branched, to 70 cm. high (28 in.).

July–September

seeds with longitudinal rows of spines. The seedpod persists throughout the winter, and the minute seeds are small enough to become windborne in a moderate breeze.

This is a small genus with one species in North America and three in eastern Asia. The genus name is derived from Greek words that mean "five mark" referring to the organization of the flower and seedpod into five parts. *P. sedoides* is somewhat similar to the Stonecrops (Sedum), thus the specific name *sedoides*. The Stonecrops differ in having succulent leaves.

A perennial with a thick, shallow, horizontal rhizome. The 5-lobed and 5-horned seed capsule opens as the 5 horns fall away, releasing numerous tiny, white or brown Ditch Stonecrop is found in swamps, roadside ditches, along stream banks, and in wet soil from New Brunswick and Quebec to Minnesota, south to Florida and Texas.

Drosera filiformis
Thread-leaved Sundew

352

Droseraceae
Sundew Family

Flowers light purple, to 12 mm. wide (½ in.), on a leafless stalk; sepals 5, petals 5, stamens 5; leaves basal, very narrow, to 30 cm. long (12 in.), covered with reddish glandular hairs, each with a sticky drop at its tip; flowering stalk to 25 cm. high (10 in.).

June–August

A perennial with a bulbous rootstock and leaves that develop by unrolling from the base. The leaves are covered with sticky hairs that trap and digest small insects. The flowering stalk is usually curved at the tip and bears 10–20 insect pollinated flowers that open 2 or 3 at a time on sunny days. The seed capsule splits into 3 sections, releasing numerous tiny, black, coarsely pitted seeds that may become airborne in a slight breeze.

This is a genus with at least 85 species of worldwide distribution but most abundant in Australia. All species are insectivorous and usually occur in habitats that are poor in mineral nutrients. The adaptation for trapping insects is believed to have evolved in response to low levels of nitrates in the

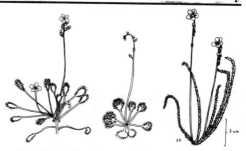

D. intermedia   D. rotundifolia   D. filiformis

substrate. Using radioactively tagged fruit flies, researchers have confirmed the supposition that *D. filiformis* absorbs considerable amounts of nitrogen from digested insect protein.

Thread-leaved Sundew is found in bogs and wet sand along the Atlantic coastal plain from Massachusetts to Florida and Louisiana.

✕ According to several European writers, the juice from the leaves of plants in this genus can be used as a substitute for rennet in curdling milk.

℞ See *D. rotundifolia*.

Wetlands

Drosera intermedia
Spatulate-leaved Sundew

353

Droseraceae
Sundew Family

Flowers white, several, in a terminal cluster, often on a leafless stalk, similar to D. filiformis; leaves in a basal rosette and sometimes, on lower few centimeters of stem, oval with widest part above middle, covered with reddish hairs, each with a sticky drop at its tip, leaf stalks without sticky hairs; flowering stem smooth, to 20 cm. high (8 in.).

June–August

An insectivorous annual or perennial of North America, Europe, and Asia. When a small insect becomes trapped in the sticky substance on the tips of the glandular hairs, the leaf folds over it like a closing hand, then unfolds, perhaps days later after the insect

is digested. The folding and unfolding process of the leaf is not rapid but takes place over a period of hours.

The flowering stem is often curved at the tip and bears 2–20 insect-pollinated flowers which open a few at a time on sunny days. The seed capsule splits into 3 sections, releasing numerous tiny, oblong, reddish-brown to black, pebbled seeds. These are small enough to become windborne in a slight breeze. This species occurs over much of the same range as, and sometimes hybridizes with, *D. rotundifolia*. In older systems of classification it is listed as *D. longifolia*.

Spatulate-leaved Sundew is found in

Sphagnum bogs and wet sand along the Atlantic and Gulf coast from Newfoundland to Texas and inland around the Great Lakes.

✂ See *D. filiformis*.
℞ See *D. rotundifolia*.

Drosera rotundifolia
Round-leaved Sundew

354

Droseraceae
Sundew Family

Flowers white, to 6 mm. wide (¼ in.), in a one-sided terminal cluster on a leafless stalk, similar to those of D. filiformis; leaves in a basal rosette, circular, on long stalks, leaf blades and stalks covered with glandular hairs, each with a drop of a sticky substance at its tip; flowering stalk smooth, to 22.5 cm. high (9 in.).

June–August

An insectivorous annual or perennial with a dense tuft of fibrous roots. It traps and digests insects in the same manner as described for *D. intermedia*. The sometimes branched, often nodding flower stalk bears 2–20 insect-pollinated flowers that open 1 or 2 at a time on sunny days. The seed capsule splits into 3 sections, releasing numerous tiny, light brown, spindle-shaped, winged seeds that are dispersed by wind.

This is a family of insectivorous plants with 2 genera native to North America; the other is *Dionaea*, Venus Fly-trap, not illustrated, of North and South Carolina. The name *Drosera* is derived from a Greek word that means "dewy" and refers to the dew-like viscous drops at the tips of the glandular hairs. *D. rotundifolia* is widespread in North America, Europe, and Asia. A dwarf form with only 1–3 flowers occurs south of North Carolina.

Round-leaved Sundew is found in bogs and peaty areas from Greenland and Labrador to Alaska, south to Florida and California.

✂ See *D. filiformis*.
℞ According to some herbologists this plant is useful in the treatment of whooping cough, bronchitis, asthma, old age, and arteriosclerosis. It has been used in the treatment of corns and warts. In Hindu medicine the seeds of one species are used in chest plasters for respiratory ailments. In Mexico, several species are used to relieve toothache. *Note*: The members of this genus are usually small plants and not abundant. To avoid their local extinction they should not be collected.

Vaccinium macrocarpon
Large Cranberry*

355

Ericaceae
Heath Family

**Also American Cranberry*

A perennial evergreen shrub with an extensive spreading root system. Pollination is probably at least partially effected by insects. The glossy red berry may be up to 18 mm. in diameter (¾ in.) and contains numerous light brown, finely pitted seeds. It persists through the winter and the seeds are spread by birds and mammals that use the berry for food.

A closely related species, *V. oxycoccus*, Small Cranberry, not illustrated, is very similar but the flower stalks are terminal and have bracts at or below the middle of the stalk, and pointed leaves rolled under along the margin. It is self-pollinated or partially cross-pollinated by bees, and hybridization with *V. macrocarpon* is avoided by an earlier flowering date (May–July). It has a more northerly distribution than *V. macrocarpon*,

Flowers pink, nodding, on long stalks from the axils of reduced leaves toward the base of the stem, stalk bearing 2 tiny bracts above the middle; corolla with 4 lobes bent backward, stamens protruding, forming a central cone; fruit a juicy red berry; leaves alternate, rounded at tip, not rolled under along the margin, to 17 mm. long (¾ in.); stems slender, trailing, very long, with erect branches.

June–August

extending to Greenland and Alaska, and it is native to Europe and Asia as well as North America.

Large Cranberry is found in sphagnum bogs and peaty soil and on wet sandy shores from Newfoundland to Manitoba, south to Virginia and Ohio, and in the mountains to North Carolina and Tennessee.

✖ This is the commercial cranberry which is cultivated in Nova Scotia, Cape Cod, New Jersey, Wisconsin, Oregon, and Washington. It grows wild throughout our range and is often harvested for local consumption. The wisdom of folklore advises that the best time

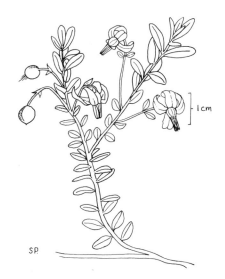

to harvest the berries is after the first frost. There is little need to elaborate further on the food uses of this plant, since the value of cranberry sauce to the Thanksgiving or Christmas turkey is well known to everyone.

Menyanthes trifoliata **Buckbean***	**356**	*Gentianaceae* **Gentian Family**

*Also Bogbean, Bog Myrtle

A perennial with a long fleshy rhizome, of North America, Europe, and Asia. There are two types of flowers: in some plants the style is much longer than the stamens, in others the stamens extend beyond the style. The plants are self-incompatible, and pollination is effected mainly by bees and bee-like flies. The thin walled, roundish seed capsules may persist for some time after the flowers wither before splitting irregularly to release numerous smooth, shiny brown seeds.

This is not a major wildlife food plant, but it constitutes at least part of the diet of the Wilson Snipe and several transitory migrating birds including the American Knot, Eastern Dowitcher, Semipalmated Sandpiper, and Hudsonian Godwit. Some of these birds migrate from the Arctic to South America and have probably contributed to the plant's widespread northern distribution.

This genus is circumpolar in distribution and includes only one species. Very robust forms of the plant can be observed in central Canada. According to folklore the rhizome is dug and eaten by deer in winter, hence the name Buckbean.

Buckbean is found in bogs and marshes and sometimes in the shallow water of pond margins from Labrador to Alaska, south to Virginia, Missouri, and California.

✖ The natives of Finland and Lappland dried and ground the thick rhizome into a flour. After it was washed several times to remove some of the bitterness, it was made into a nutritious but unpleasant tasting

308

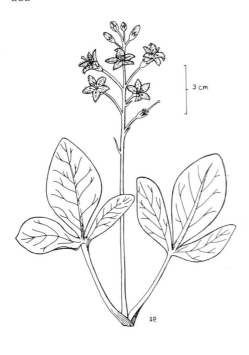

Flowers white or pinkish, funnel-shaped, 5-lobed, bearded on inside with white hairs; leaves basal, palmately compound with 3 oval leaflets, with long petioles forming sheaths around a thick rhizome; stems smooth, leafless, with 10–20 flowers at their tips, to 30 cm. high (1 ft.).

April–July

bread. As an emergency food the rhizome can be cooked in several changes of water. ℞ Bitter plants often get a high rating in herbal medicine, and Buckbean is no exception. *The Dispensatory of the United States of America* (1851) states, "With the ordinary properties of the bitter tonic *Menyanthes* unites a cathartic power, in large doses is apt to vomit." *Note*: This plant is usually not abundant in the Northeast. Its doubtful value as a food or medicinal plant does not justify its collection for these uses.

Myriophyllum spicatum
Water Milfoil

357

Haloragidaceae
Water Milfoil Family

Flowers inconspicuous, unisexual, on stem tip that emerges from the water, in axils of whorled bracts, upper flowers staminate, lower ones pistillate; stigmas 4, stamens 8; leaves in whorls of 4's or 5's, finely pinnately dissected; stems weak, submerged, rooting on bottom, to 1 m. long (40 in.), or longer.

July–September

of hayfever. It is not an important wildlife food plant, but it does provide up to 5% of the diet of several species of ducks including the Canvasback, Lesser Scaup, and Blue-winged Teal. The plants are also eaten occasionally by muskrats and moose. These animals probably contribute to seed dispersal. It also spreads by fragmentation of the stem and by axillary buds that are produced year round.

Myriophyllum is often sold in pet stores for home aquaria because of its attractive feathery leaves. There are about 9 species in the Northeast. They all have very similar

A perennial with a creeping rhizome, often rooting at lower nodes of the stem. Pollination is apparently by wind, but pollen production is low, so this plant is not a cause

leaves and are sometimes difficult to distinguish even with flowers and fruits. There is disagreement with regard to the classification of *M. spicatum*; in some systems it is listed as *M. exalbescens*. It is probably the most common species of the genus in the northern part of the region.

M. spicatum is a native of North America, Europe, and Asia. It is found in ponds and quiet streams, often in brackish or limestone areas, in water up to 4.5 m. deep (15 ft.). In North America it ranges from Newfoundland to Alaska, south to Maryland, Ohio, and California.

Anacharis canadensis
Water-weed*

358

Hydrocharitaceae
Frog's-bit Family

Wetlands

*Also Elodea

Flowers white, inconspicuous, to 10 mm. wide (⅖ in.), on long axillary stalks, to 15 cm. long (6 in.), that reach surface of water; sepals 3, petals 3, stamens 9, pistil with 3 branched stigmas; flowers unisexual, rare; lower leaves opposite; upper leaves in whorls of 3's, to 13 mm. long (½ in.), numerous, crowded toward tip; stems submerged, brittle, long and branched, often forming dense masses.

July–September

A perennial with horizontal runners rooted on the bottom. Pistillate and staminate flowers are on different plants and pollen reaches the stigmas by floating on the surface of the water. The seed capsule is oblong, about 8 mm. long (⅓ in.), and contains 1–5 spindle-shaped seeds. Staminate flowering plants are rare and vegetative reproduction by fragmentation is more common than reproduction by seeds.

It grows very rapidly and produces good support and shelter for aquatic insects. It is eaten by waterfowl but would probably be more important as a food plant if it produced fruits more abundantly. The staminate plant was introduced into European waters and by vegetative propagation has become a pest there in waterways. This species is a favorite in general biology laboratories for studying cell structure, protoplasmic streaming, and photosynthesis.

This is a genus of two native and one

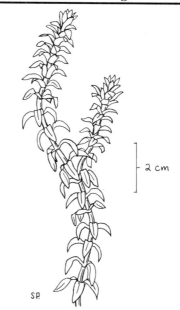

2 cm

SP.

introduced species, with *A. canadensis* the most abundant and widespread. A similar species, *A. nuttallii*, not illustrated, has narrower leaves and occurs in fresh or often brackish water of tidal river basins. *A. densa*, not illustrated, is a larger species with leaves to 3 cm. long (1.2 in.), in whorls of 4's or 6's. It was introduced from Argentina as an aquarium plant but has escaped and is spreading rapidly. *A. canadensis* is recognized by some botanists as *Elodea canadensis*.

Waterweed is commonly found in quiet

310

waters of ponds, lakes, and sluggish streams, especially in limestone regions, from Quebec to Saskatchewan and Washington, south to North Carolina, Oklahoma, and California.

Vallisneria americana Wild Celery*

359

Hydrocharitaceae Frog's-bit Family

*Also Water Celery, Tapegrass, Eelgrass

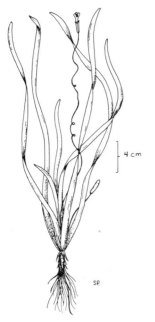

4 cm

S.P.

Pistillate flowers white, floating on surface of water, on long slender stalks, to 6 mm. wide (¼ in.); leaves in a basal tuft, long and ribbon-like, submerged, to 12 mm. wide (½ in.), to 2 m. long (6½ ft.), tips often floating on surface of water.

July–October

thers come into contact with one of the 3 spreading lobes of the stigma. After pollination the flower stalk coils and contracts, drawing the fruit underwater to mature.

The fruit is a berry-like cylinder up to 10 cm. long (4 in.) enclosing several seeds imbedded in a gelatinous substance. Reproduction is probably most often by winter buds produced at the ends of the stems. This is a very important food plant for waterfowl, marsh birds, shore birds, and muskrats. All parts of the plant are eaten, but the winter buds and rhizomes are especially favored.

It is easily propagated and is often transplanted as a wildlife management practice. It may comprise up to 50% of the diet of the Canvasback Duck, which is appropriately named *Aythya valisineria*. The plant also provides good food and shelter for aquatic insects and fish. It is an attractive indoor aquarium plant.

Wild Celery is found in quiet water to 2 m. deep (6½ ft.) from Nova Scotia and Quebec to North Dakota, south to Florida and Texas.

A perennial with a horizontal, shallowly buried stem, rooting at intervals. Staminate and pistillate flowers are produced on different plants. The staminate flowers develop underwater in dense many-flowered clusters. At maturity they break free and float to the surface, where they open, exposing 2 or 3 anthers. These float into the vicinity of the pistillate flower and the exposed an-

Hypericum mutilum
Dwarf St. John's-wort

360

Hypericaceae
St. John's-wort Family

Flowers yellow, tiny, in many branched terminal clusters; petals 5, stamens 5–12; leaves opposite, sessile, oval, rounded at base, to 2.5 cm. long (1 in.); stems diffusely branched in upper half, to 90 cm. high (3 ft.).

July–September

A perennial or sometimes an annual with decumbent stem bases. The 1-celled seed capsule splits to release numerous very tiny, wind-borne seeds. The tiny flowers gave this plant its common name. The specific name *mutilum* is a Latin word for "segmented" or "cut" and was applied by Linnaeus because, when he named the plant, he had only a segment rather than a whole plant.

The variety of this species that occurs in the Southeast is taller and has leaves that are narrower than the Northeastern variety. This species is normally not a weed pest in cultivated areas and gardens. It hybridizes with several other species, and intermediate forms are often observed.

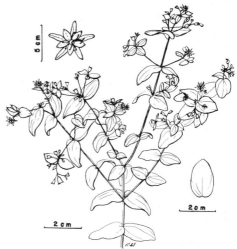

Dwarf St. John's-wort is found along the margins of swamps and marshes and other areas of low wet ground from Newfoundland to Manitoba, south to Florida and Texas. ✘ ℞ See *H. perforatum*.

Triadenum virginicum
Marsh St. John's-wort

361

Hypericaceae
St. John's-wort Family

Flowers pink, in upper axillary and terminal clusters; petals 5, stamens 9, in clusters of 3's, alternating with 3 conspicuous orange glands; leaves opposite, egg-shaped, rounded at base, sessile, with dark spots on underside, to 7.5 cm. long (3 in.); stems sometimes freely branched, to 60 cm. high (2 ft.).

July–August

A perennial with surface runners that root at intervals. In late summer and autumn it is characterized by deep red seed capsules that are about 20 mm. long (¾ in.). Each capsule has 3 cells and contains numerous tiny, black, cylindrical seeds with pointed ends.

There are only 4 species in this genus, all native to eastern North America. These are included in the genus *Hypericum* and the

Family Guttiferae by some botanists. They differ from the plants in the genus *Hypericum* in that the latter has yellow flowers and numerous stamens. The genus name *Triadenum* is derived from Greek words that mean "three glands" and refers to the orange glands that alternate with the 3 clusters of stamens.

A closely related species, *T. fraseri*, not illustrated, is very similar but has shorter sepals. It occurs from Newfoundland to Manitoba, south to West Virginia and Indiana.

Marsh St. John's-wort is found in swamps and bogs from Nova Scotia to Ontario, south to Florida and Mississippi. It is especially abundant along the Atlantic coast.

Iris versicolor
Large Blue Flag*

362

Iridaceae
Iris Family

Also Wild Iris, Poison Flag

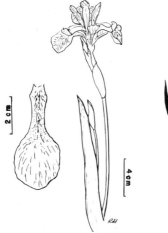

Flowers blue, 1 to several, terminal; flower parts 6, outer 3 arching with tips curving downward, inner 3 erect; leaves alternate, long, narrow, stem leaves shorter than basal ones; flowering stem usually longer than leaves, sometimes branched in upper part, to 90 cm. high (3 ft.).

May–July

capsule does not split along well defined sutures, and it may persist into the winter months before opening. The D-shaped, flattened, finely wrinkled, glossy, brown seeds are scattered by wind.

This is a genus of at least 200 species of the North Temperate Zone. There are about 10 species native to the Northeast and several others in the western and southern regions of North America. This is a favorite genus of many gardeners, and numerous horticultural forms and varieties have been developed. It has been hypothesized that, at some time in the distant past, *I. versicolor* originated as a fertile hybrid from a chance crossing of two other native species, *I. virginica*, Southern Blue Flag, and *I. setosa*, Beachhead Iris, latter two not illustrated.

A perennial with a thick creeping rhizome and many fibrous roots. The style is branched into 3 flattened, petal-like segments each covering a stamen. The pollinating insect, usually a bumblebee or honeybee, in order to reach the nectar must force its way under the style branch. In the process it comes into contact first with the stigma, leaving pollen picked up at the last flower, then with the anther where it becomes dusted with pollen that will be carried to the next flower.

The 3-celled, 3-lobed seed capsule may be up to 6 cm. long (2½ in.) and encloses numerous seeds in 2 rows in each cell. The

Large Blue Flag is found in marshes, swamps, wet meadows, roadside ditches, and the margins of ponds from Newfoundland to Manitoba, south to Virginia and Ohio. The leaves and rhizomes of *Iris* species contain a substance called irisin which has a violent affect on the intestinal tract, causing extreme vomiting and diarrhea. Great care should be taken not to confuse these with the leaves and underground parts of *Acorus calamus*, Sweet Flag, and *Typha spp.*, Cattails.

In contrast to Blue Flag, the foliage of Sweet Flag has a pleasant aroma when bruised. The rhizomes of Blue Flag have an unpleasant taste while those of Sweet Flag have a pleasant aroma and a spicy taste. Cattail foliage is often coarser than Blue Flag

and its rhizomes are odorless and have little taste.

Some individuals are very sensitive to Blue Flag and may develop severe dermatitis from handling the rhizomes or leaves.

Some North American Indian tribes used the boiled, crushed rhizome of Blue Flag as a poultice for burns, bruises, and sores.

Scheuchzeria palustris / Bog Arrow Grass — **363** — *Juncaginaceae* / Arrow Grass Family

Flowers greenish-yellow, to 6 mm. wide (¼ in.), few, subtended by bracts that decrease in size in upper flowers; perianth of 3 petals and 3 sepals, stamens 6, pistils 3; fruit of 3 segments, attached at base; leaves alternate, grass-like, forming expanded sheaths around stem at base, to 30 cm. long (12 in.); stems erect, unbranched, somewhat zig-zag, to 40 cm. high (16 in.).

May–July

A perennial with a jointed creeping rhizome. Each of the 3 seedpods produced by the flower splits along an inside seam, when mature, releasing 2 black, finely wrinkled seeds. These are to 5 mm. long (⅕ in.) and may provide a source of food for birds and small mammals.

The genus was named in honor of J. Scheuchzer, an eighteenth-century Swiss scientist. It is a genus with only one species which is circumpolar in distribution. Although it is often abundant in sphagnum bogs, it is an inconspicuous plant and easily overlooked. It is probably most noticeable in late summer and early fall when it is characterized by prominent triangular clusters of brown seedpods.

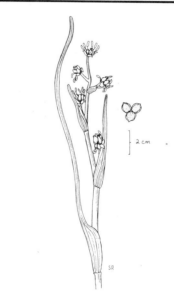

In North America, Bog Arrow Grass is found in wet sphagnum bogs and along peaty shores from Newfoundland to Alaska, south to Virginia, West Virginia, and California.

Lycopus americanus
Cut-leaved Water Horehound

364

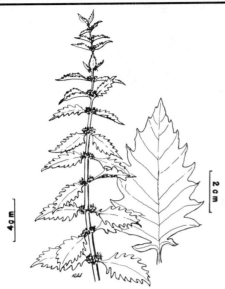

Flowers white, tiny, in dense axillary clusters; leaves opposite, oblong or lance-shaped in outline, lower ones pinnately lobed, upper ones merely toothed, not aromatic; stems square, slightly hairy, often branched in upper half, to 90 cm. high (3 ft.).

June–September

shoots can develop seeds. When it gets a late start, or under the pressure of mowing, it may produce flowers when only a few inches high.

A related species, *L. europaeus*, not illustrated, is very similar but has leaves that are oval in outline and are shallowly toothed on both the upper and lower part of the stem. It was introduced from Europe and has become established in wet sandy soil along the Atlantic Coast from Massachusetts to Mississippi.

Cut-leaved Water Horehound is found along the margins of swamps and ponds, in roadside ditches, swales, and other areas of low wet ground from Newfoundland to British Columbia, south to Florida, Texas, and California.

℞ It should be noted that the plants of this genus are in the same family but are not the horehound plants historically used in cough remedies. The traditional horehound is *Marrubium vulgare*, not illustrated, a native of Europe and Asia which has been introduced and is widespread in North America.

A perennial with fibrous roots and short runners. Pollination is mostly by small bees and bee-like flies. Each flower produces 4 tiny, triangular, brown, glandular-dotted seeds. The dead stem with its clusters of tubular calyxes persists into the winter months, dispersing the seeds as it sways in the wind.

This is the most common of the 7 species of the genus that occur in the Northeast. It sometimes becomes established as a weed in meadows and cultivated fields. Control measures include improvement of drainage and repeated close mowing before new

Lycopus virginicus
Bugle Weed*

365

*Also Water Horehound

A perennial with fibrous, sometimes tuberous, roots and usually several long runners. Pollination and seed dispersal as in *L. americanus*.

Bugle Weed is found in wooded swamps, roadside ditches, and low wet woods from Nova Scotia and Quebec to Minnesota, south to Georgia and Texas.

✗ At least two species of this genus, *L. un-*

Flowers similar to L. americanus. Leaves opposite, oval-lanceolate, sharply toothed; stems similar to L. americanus.

July–September

iflorus and *L. amplectens*, not illustrated, have tuberous roots. When it grows in open sand, the tubers of *L. amplectens* are large and abundant near the base of the plant. Collected in the autumn, these can be used fresh in salads, pickled or cooked as a vegetable. Tubers produced by other species of the genus, although smaller and more difficult to find, can be used in a similar way. ℞ The juice of this species is reported in *The Dispensatory of the United States of America* (1851) to be mildly narcotic, astringent, and sedative. It has been used for coughs, bleeding of the lungs, and tuberculosis.

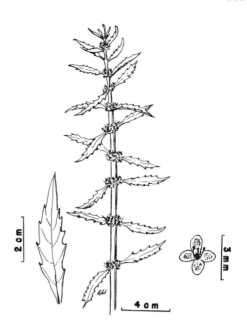

Scutellaria galericulata
Common Skullcap*

366

Lamiaceae or Labiatae
Mint Family

---

*Also Marsh Skullcap

Flowers blue, to 2.5 cm. long (1 in.), single in leaf axils, upturned, 2-lipped, upper lip entire or notched, lower lip 3-lobed; calyx 2-lipped, a conspicuous cap-shaped protuberance on upper lip; leaves opposite, lanceolate, with rounded teeth, sessile or short stalked, rounded at base, to 5 cm. long (2 in.); stems square, finely hairy along angles, usually branched to 1 m. high (40 in.).

June–September

A perennial with a slender creeping rhizome and thick underground runners. Cross-pollination is effected by small bees and butterflies, and each flower produces 4 tiny, brown, warty seeds. When the seeds are mature the upper lip of the calyx, with its protuberance, falls away exposing the seeds to dispersal as rain ballistics. Raindrops falling on the lower lip cause it to bend and rebound, hurling the seeds outward.

This is a very large genus with about 15

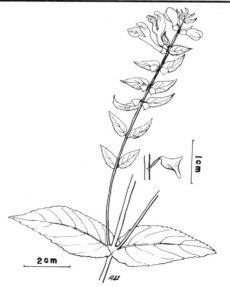

species native to the Northeast. *S. galericulata* is also widespread in Europe and Asia, and although very similar to the European plants, the North American plants are

316

sometimes classified as *S. epilobiifolia*. A related woodland species, *S. churchilliana*, not illustrated, has characteristics that are intermediate between *S. galericulata* and *S. lateriflora* and may be a hybrid of these species.

Common Skullcap is found in swamps, along pond margins, and in wet soil from Newfoundland and Quebec to Alaska, south to Delaware, West Virginia, Missouri, and California.

℞ See *S. lateriflora*.

Scutellaria lateriflora Mad-dog Skullcap	**367**	*Lamiaceae or Labiatae* Mint Family

Flowers blue, in long, one-sided, upper axillary clusters, to 10 cm. long (4 in.); corolla not curved upward, 2-lipped, upper lip notched or entire, lower lip 3-lobed; calyx 2-lipped, a conspicuous cap-shaped protuberance on the upper lip; leaves opposite, toothed, stalked, to 10 cm. long (4 in.); stems square, with soft hairs on the angles, slender, often freely branched, to 70 cm. high (28 in.).

June–September

The name Mad-dog is probably a reference to its former reputation as a treatment for rabies.

This species is easy to distinguish from *S. galericulata* by its long, one-sided clusters of small flowers and its distinctly stalked leaves. The specific name, which means "side flowers," refers to this characteristic. Pink flowered and white flowered forms are occasionally observed.

Mad-dog Skullcap is found in wooded swamps, marshes, and wet flood plains from Quebec to British Columbia, south to Florida, Louisiana, and Arizona.

℞ In herbology this plant and others in the genus have high reputations as nerve tonics, astringents, and muscle relaxants. They have been used to treat hydrophobia, hysteria, convulsions, St. Vitus' dance, hiccoughs, and other disorders of the nervous system.

A perennial with a slender branching rhizome and thread-like runners. Pollination is effected by small bees and butterflies. Each flower produces 4 tiny, pale, warty seeds which are dispersed as rain ballistics (see *S. galericulata*). The genus name is derived from a Latin word meaning "small dish" supposedly alluding to the protuberance on the calyx, which does not resemble a dish.

Stachys hispida
Rough Hedge-nettle

368

Lamiaceae or Labiatae
Mint Family

Flowers pale rose-pink, in terminal spaced whorls of 4's–8's; corolla strongly 2-lipped, upper lip forming a shallow hood enclosing the 4 stamens, lower lip 3-lobed; leaves opposite, usually hairy on upper surface, lanceolate, toothed, to 10 cm. long (4 in.); stems square, with stiff hairs on angles, unbranched, to 1 m. high (40 in.).

July–August

A perennial with an extensively creeping rhizome. Most of the members of this genus are pollinated by insects. The whorls of flowers are subtended by greatly reduced leaf-like bracts. Each flower produces 4 finely pitted, round-triangular, dull black seeds. In some members of this genus the seeds germinate in water and the seedlings are dispersed by water currents.

This is a genus of about 200 species mostly in the North Temperate Zone but with a few in the tropics and in northern regions. There is disagreement among botanists regarding the designation of species. Some botanists classify *S. hispida* as a variety of *S. tenuifolia*, Smooth Hedge-nettle. The latter does not have hairs on the angles of its stems. Another related species, *S. palustris*, Woundwort, not illustrated, is similar but is hairy on the flat surfaces of its stems as well as on the angles. It occurs in wet ground throughout eastern North America.

Rough Hedge-nettle is found in damp meadows, roadside ditches, and wet woods from New England and Ontario to Manitoba, south to South Carolina and Indiana. ✖ Several species of this genus, including *S. palustris* but apparently not *S. hispida*, produce white tubers at the end of the growing season that can be boiled and eaten or dried and ground into a flour.

Lemna minor
Lesser Duckweed*

369

Lemnaceae
Duckweed Family

*Also Duck's Meat

Flowers inconspicuous, in a marginal notch, consisting of 2 stamens and a pistil; true leaves and stems absent; plant body consists of an elliptical to circular, leaf-like structure floating on surface of water, to 5 mm. (⅕ in.) in diameter; a single root without vascular tissue is attached near center of underside of plant body.

June–August

Each flower usually produces a single

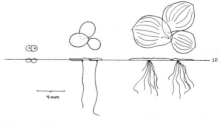

W. papulifera   L. minor   Spirodela

seed, but flowers are rarely observed. Reproduction is usually asexual by budding. A new plant grows from the margin of an existing one, the young plant separating from the parent plant or sometimes not separating until a colony of 2–8 is formed. Buds are produced in autumn that sink to the bottom and overwinter there. Dispersion is by water currents and on the feet of water birds.

This is a genus of 8 species, the 5 occurring in the Northeast all with an international distribution. *L. minor*, the most abundant species, is also widespread in Europe and Asia. It has a very rapid growth rate and can cover a pond in a short time.

It provides food and cover for a variety of small aquatic animals and is an important food plant for ducks and sometimes muskrats. A closely related species, *L. trisulca*, not illustrated, has an oblong plant body that remains attached to the parent plant by a long narrow stalk and grows in dense masses.

Lesser Duckweed is found in the quiet water of ponds and sluggish streams and on wet mud throughout the United States and southern Canada.

℞ In herbology this plant is listed as a demulcent and is recommended as a poultice for skin irritations.

Spirodela polyrhiza **Greater Duckweed***	**370**	*Lemnaceae* **Duckweed Family**

*Also Water-flaxseed

Flowers similar to those of L. minor; true leaves and stems absent, plant body a leaf-like structure floating on surface of water, oval, with 5–11 palmate veins, reddish or purplish on underside, to 8 mm. long (⅓ in.); attached to underside of plant body, toward one end, is a cluster of 2–16 roots, each with a central strand of vascular tissue.

June–August

As in *L. minor*, flowers are rarely observed and reproduction is mainly asexual, by budding. New plants grow from the margins of existing ones, remaining attached to the parent plants and forming small colonies.

Buds form in late autumn that sink to the bottom and overwinter there. Dispersion is by water currents and on the feet of water birds.

This is the largest of the duckweeds and it often grows in a mixture with *Lemna* and *Wolffia*. It is the only representative of the genus in our area and occurs also in tropical America and Europe. Although it is considered a poor food producer for fish, this plant is eaten by waterfowl and often by pheasants.

Greater Duckweed is found in ponds, marshes, and stream margins and in other quiet water throughout the United States and southern Canada.

Wolffia papulifera **Watermeal**	**371**	*Lemnaceae* **Duckweed Family**

Flowers microscopic, produced in a cavity on surface, bursting through upper layers of cells, stamen 1, pistil 1; true leaves and stems absent; plant body floating on surface of water, rounded on underside, with a slight peak on upper surface, to 1 mm. wide (1/25 in.); rootless.

June–August

Flowers are rarely produced and reproduction is mainly asexual by budding. New plants grow from the margins of existing plants, soon becoming detached from the parent plant. They become heavy with stored starch in autumn and sink to the bottom where they overwinter, rising to the surface in spring as the starch content diminishes. Dispersion is by water currents and on the feet of water birds.

This family includes the least complex of the flowering plants, and *Wolffia* has the distinction of being the smallest flowering plant. The genus was named in honor of J. F. Wolff,

an eighteenth-century German botanist who studied and wrote about *Lemna*. There are 3 almost indistinguishable species native to the Northeast. These are usually associated with *Lemna* and *Spirodela* and are eaten with them by waterfowl.

The Watermeals are found in quiet water and sometimes buried to 10 cm. (4 in.) deep under the debris on the bottom of dried-up ponds throughout the eastern United States and southern Canada.

Pinguicula vulgaris
Common Butterwort

372

Lentibulariaceae
Bladderwort Family

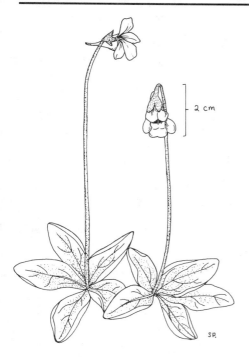

2 cm

SP.

Flowers violet, solitary, on a leafless stalk, 2-lipped, upper lip 2-lobed, lower lip 3-lobed, with a slender spur; leaves in a basal rosette, yellowish-green, shiny, sticky, with rolled-in margins, to 5 cm. long (2 in.); flowering stalk to 15 cm. high (6 in.).

June–July

are small enough to become windborne in a moderate breeze.

The genus name is derived from Latin words meaning "little fat" and refers to the greasy look of the leaves of *P. vulgaris*. Most species of this genus occur in the mountainous regions of Central and South America. Three closely related and similar species, *P. primulifolia*, Primrose-leaved Butterwort, *P. caerulea*, Violet Butterwort, and *P. pumila*, Dwarf Butterwort, none of these illustrated, occur in the southern states, the latter two extending northward to North Carolina.

Common Butterwort is found on wet rocks and in moist meadows, shores, and bogs, often in limestone areas, from Greenland and Labrador to Alaska, south to New England, Michigan, and Washington. It also occurs in northern Europe and Asia.

✂ This plant is reported to have been used by Laplanders as a substitute for rennet in curdling milk. The species occurring in the southeastern United States probably have the same properties.

A perennial with 5–6 elliptical leaves on the ground. Small insects that alight on these cannot get loose and are digested by secreted enzymes. The seed capsule splits irregularly into 2–4 sections, releasing numerous pitted, light brown seeds. These

Utricularia vulgaris
Greater Bladderwort

373

Lentibulariaceae
Bladderwort Family

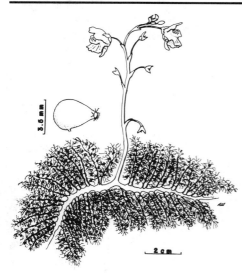

Flowers yellow, in a cluster of up to 20, scattered at the tip of a leafless stalk; corolla 2-lipped, with a spur shorter than the lower lip; leaves alternate, submerged, finely dissected, with many tiny bladders; stems long and slender, often becoming detached from rooting base and floating just beneath surface of water, to 2 m. long (6½ ft.).

May–September

highly valued as wildlife food. They may be eaten sparingly by moose, muskrats, and ducks; thus seed dispersal by animals is probably minimal.

An outstanding feature of the bladderworts is the presence of the structures that give them their name and provide the mechanisms for their carnivorousness. A large plant of *U. vulgaris* may have over a thousand bladders. Each bladder normally has concave sides and a ring of hairs around an opening at one end. When a small aquatic organism, such as a water flea, touches one of these hairs the bladder inflates suddenly and, like a syringe, sucks in water along with the hapless organism. A door then closes over the opening and the organism is digested.

A carnivorous perennial that survives the winter as a bud consisting of a dense ball of small leaves. When a bee alights on the flower it comes into contact first with the 2-lobed stigma, depositing pollen it may have acquired at another flower of the same species. The lobes of the stigma immediately come together like the pages of a book, enclosing the pollen and shielding the stigma from pollen of its own flower.

The seed capsule splits into 2–4 sections, releasing numerous tiny, black, 4 or 5 angled, slightly winged seeds. These are dispersed by water currents, or they may become airborne in a slight to moderate breeze. The plants in this genus are not

Greater Bladderwort is the most widespread of about 14 species of this genus that occur in eastern North America. It is found in shallow or deep quiet water of ponds, lakes, and sluggish streams from Newfoundland to Alaska, south to Florida and Texas. It also occurs in several western states.

Smilacina trifolia
Three-leaved Solomon's Seal

374

Liliaceae
Lily Family

A perennial with an extensively spreading, slender rhizome and soft whitish surface runners. The berries contain 1–2 brown, oval seeds with faint longitudinal ridges. The seeds are dispersed by the birds and mammals that eat the berries. This is the smallest

member of the genus in the Northeast. It is also widespread in northern Asia.

This species somewhat resembles False Lily-of-the-valley (*Maianthemum canadense*) but the two are easy to distinguish: *S. trifolia* has 6 petals and 6 stamens, in

Flowers white, 3–8, in an unbranched terminal cluster, to 5 cm. long (2 in.); fruit a dark red berry; leaves alternate, usually 3, tapering at both ends, clasping the stem, to 12.5 cm. long (5 in.); stems slender, unbranched, soft, to 25 cm. high (10 in.).

May–July

False Lily-of-the-valley there are 4 of each; the leaf bases of three-leaved Solomon's Seal are tapering and clasping, in False Lily-of-the-valley they are heart-shaped.

Three-leaved Solomon's Seal is found in bogs and wet woods from Newfoundland and Labrador to Mackenzie and British Columbia, south to New Jersey, Ohio, and Illinois.

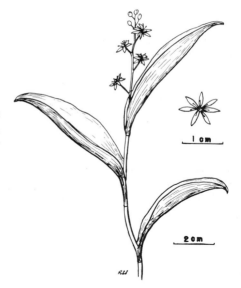

Wetlands

Veratrum viride
False Hellebore*

375

Liliaceae
Lily Family

Also White Hellebore, Indian Poke, Itchweed

Flowers green, to 2.5 cm. wide (1 in.), numerous, hairy, in a branched terminal cluster; flower divisions 6, stamens 6; leaves alternate, with prominent ribs, often clasping the stem, to 30 cm. long (12 in.); stems erect, unbranched, to 2 m. high (6½ ft.).

June–July

A perennial with a short rhizome and coarse fibrous roots. The seed capsule splits into 3 sections, releasing numerous brown, flattened, winged seeds. These are small enough to become windborne in a slight to moderate breeze. It is not a serious weed in cultivated fields because it is ordinarily eradicated by normal drainage procedures.

This is a genus of the Northern Hemisphere with 3 species native to the Northeast. *V. viride* is the most widespread member of the genus in eastern North America. A related species, *V. woodii*, Wood's False Hellebore, not illustrated, has purplish,

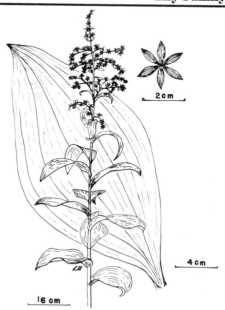

hairless flowers and narrow upper leaves. It occurs in upland woods from Ohio to Iowa, south to Missouri and Oklahoma.

False Hellebore is found in swamps and wet woods from New Brunswick and Quebec to Alaska, south to Georgia in the East and to Oregon in the West.

☙ The early American settlers used a solution made by boiling the roots as an external application for herpes infections and to kill head lice. However, all parts of this plant are poisonous and eating it has caused death in livestock and chickens. It contains compounds called cardiac glycosides that influence the heartbeat, and it should never be used in home remedies.

| *Zigadenus glaucus* **White Camass*** | **376** | *Liliaceae* **Lily Family** |

*Also Death Camass

4 cm

sp.

A perennial arising from a bulb with a coarsely fibrous outer covering. The 3-lobed, 3-beaked seed capsule is surrounded by the persistent withered flower parts and contains several relatively large, flattened seeds. The genus name is derived from Greek words that mean "a yoke of glands," referring to the 2-lobed glands on the petals of some species.

This is a genus of 18–20 species, mostly

Flowers white, on stalks from axils of reduced bract-like leaves, in a terminal cluster; flower with 6 divisions, each with a 2-lobed gland between base and middle, underside tinged with green or purple; leaves mostly basal, narrow and grass-like, to 40 cm. long (16 in.); stems thick, erect, to 80 cm. high (32 in.).

July–September

native to North America but with at least one in Asia. It is more common in western North America than in the East. Although several of the species that occur in the Northeast are most abundant along the Atlantic coastal plain, *Z. glaucus* has a comparatively wide inland distribution. In older systems of classification it was listed as *Z. chlorantha*.

White Camass is found in alkaline bogs, swamps, and wet meadows and on sandy shores from Quebec to Minnesota, south to North Carolina and Illinois.

☙ Death Camass is a name sometimes applied to any plant in this genus. All species contain the toxic alkaloid zygacine, and deaths among sheep, cattle, and horses have resulted from eating the plants. The food reserves stored in the bulb allow it to be among the first shoots to appear in spring, and the young plants are the most poisonous. Death to as many as 2000 sheep have been reported in single instances of poisoning. These plants retain their toxic properties when dried and may cause poisoning if they are in hay.

Lobelia cardinalis
Cardinal-flower

377

Flowers scarlet, 2-lipped, lower lip with 3 prominent lobes, stamens united into a tube projecting between the 2 upper lobes; flowers in a terminal cluster to 40 cm. long (16 in.); leaves alternate, lance-shaped, toothed, with prominent veins; stems coarse, usually unbranched, to 1.5 m. high (5 ft.).

July–September

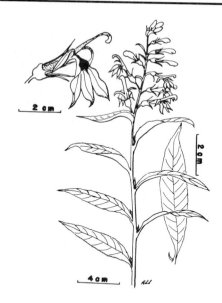

A perennial with a shallow horizontal rhizome. Pollination is effected mainly by the Ruby Throated Hummingbird, or rarely by bumblebees. The 2-celled seed capsule opens at the top, and the numerous tiny brown seeds are shaken into the wind as the stem sways with the breezes.

This is the most colorful of the lobelias in the Northeast. A single plant is an eye-catching attraction, but a dense cluster of these plants in flower is a spectacular sight. The flower supposedly has the shape and color of a churchman's or cardinal's hat, or miter, hence the specific and the common names. They do not refer to the red color of the cardinal bird.

This plant is usually difficult to find and very often occurs as a few isolated plants. It has declined in abundance as the number and sizes of wetlands have declined in eastern North America. It should not be collected and should be protected at all times. A related species with bright red flowers, *L. splendens*, not illustrated, has very narrow leaves and occurs in southwestern United States.

Cardinal-flower is found in wooded swamps, pond and stream margins, and wet woods from New Brunswick and Quebec to Minnesota, south to Florida and Texas.

℞ Although it is reputed to have medicinal properties as a nerve tonic and for intestinal worms, it should not be collected for reasons given above. In addition, *L. cardinalis* is closely related to, and probably contains the same compounds as, *L. inflata*. The latter contains toxic compounds that in overdose may cause death in humans. Neither plant should be experimented with in home remedies.

Lobelia siphilitica
Great Lobelia

378

A perennial with short basal sprouts that appear in the fall. The seeds are similar to those of *L. cardinalis* and are dispersed in the same manner. This is the largest of the blue-flowered lobelias.

This species is represented west of the Mississippi River by a variety that is not as

tall and has smaller leaves and fewer flowers. It often occurs in the same type of habitat, but it is not as widely distributed as *L. cardinalis*. The specific name refers to the reputed use of the root by the Mohawk Indians as a treatment for syphilis. In some systems of classification this genus is in-

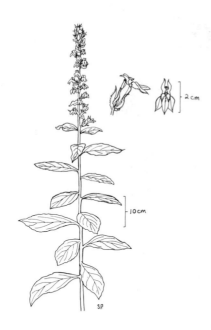

Flowers blue, 2-lipped, lower lip 3-lobed, upper lip 2-lobed, in an elongate, dense terminal cluster, to 30 cm. long (12 in.); leaves alternate, tapering at both ends, irregularly toothed, often widest toward tip; stems coarse, unbranched, to 1.5 m. high (5 ft.).

August–September

cluded in the family Campanulaceae, the Harebell Family.

Great Lobelia is found in swamps, along stream and pond margins, and in open wet woods from Maine to Manitoba, south to North Carolina, Alabama, and Texas.

☠ Although reputedly used with success by Mohawk Indians as a treatment for syphilis, English physicians of the early 1800's discarded it as ineffective. As with other lobelias, it contains toxic compounds that in overdose may be fatal to humans. It should not be experimented with in home remedies.

Decodon verticillatus
Swamp Loosestrife*

379

Lythraceae
Loosestrife Family

*Also Water Willow, Water Oleander

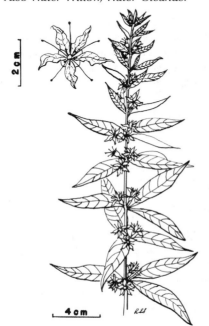

Flowers pink to rose, clustered in axils of upper leaves; petals 5, constricted at base; stamens 10, 5 protruding, 5 shorter; leaves opposite or in whorls of 3's or 4's, lanceolate; stems somewhat woody, arching, and often rooting where tip touches ground, with corky thickenings on submerged parts, to 2.5 m. long (8 ft.).

July–August

A perennial often forming dense thickets. Three types of flowers are produced with regard to relative lengths of styles and stamens. This is usually interpreted as an adaptation to avoid self-pollination. The 3–5 celled seed capsule splits to release numerous slightly shiny, finely pitted, brown seeds. The seedpods usually persist on the bare stems far into the winter months.

This is a native genus with only one species. Two well established varieties are recognized. One variety has smooth leaves and is most common inland. The other variety has hairs on the undersides of the leaves and occurs mostly along the Atlantic coast and in the Mississippi Valley.

Swamp Loosestrife is found in swamps, along the margins of ponds and streams, and in shallow water from Nova Scotia to Minnesota, south to Florida and Louisiana.

Lythrum salicaria
Purple Loosestrife*

380

Lythraceae
Loosestrife Family

Also Spiked Loosestrife
Flowers pink to red-purple, clustered in axils of small leaf-like bracts on upper part of plant, petals 5–6; leaves opposite or in whorls of 3's, lanceolate, sessile; stems erect, thick, often hairy, to 1.5 m. high (5 ft.).

July–September

A perennial with a cluster of fleshy roots, introduced from Europe. FLowers are of 3 types with regard to relative lengths of stamens and style. Two types of pollen are produced: green and yellow. Some flowers include long stamens with metallic green anthers and shorter ones with yellow anthers. The species is self-incompatible, and cross-pollination is effected by bumblebees, honeybees, and butterflies. The 2-celled seed capsule splits along two seams to release numerous glossy, light brown seeds that are small enough to be dispersed by air currents, in water, or in mud on the feet of waterfowl.

This is an aggressive species invading marshes, wet meadows, pond margins, and flood plains often to the exclusion of native species. It ranges from Newfoundland and Quebec to Minnesota, south to Virginia and Missouri, and it is spreading rapidly.

℞ The roots and leaves have been used in the treatment of dysentery and leukorrhea. Applied externally, they have been used for preserving sight and treating sore or bruised eyes. A solution made from the plant is said to be beneficial as a gargle for sore throat and, made into a salve, good for sores and other skin irritations.

Potamogeton amplifolius
Broad-leaved Pondweed*

381

Najadaceae
Pondweed Family

Also Muskie Weed, Bass Weed

A perennial with a long, slender, underwater rhizome. The flower clusters extend above the surface of the water, and pollination is by wind. However, none of the plants of this genus is known to be a cause of hayfever. The oval seeds are reddish-brown, and each has a prominent beak and three rounded ridges on one side.

This is an entirely aquatic genus of about

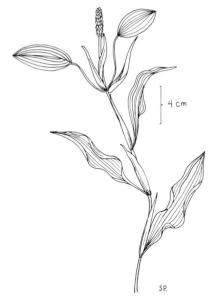

4 cm

S.P.

Flowers inconspicuous, in dense terminal clusters; leaves alternate, of two types, floating and submerged: floating leaves rounded or heart-shaped at base, oval, to 10 cm. long (4 in.), petioles to 20 cm. long (8 in.), 30–55 nerved; submerged leaves folded longitudinally and arched, sickle-like, tapering at both ends, to 20 cm. long (8 in.) and 7.5 cm. wide (3 in.); stipules slender, pointed, to 12.5 cm. long (5 in.); stems slender, unbranched, may attain a length of several feet.

July–September

coarsest of the pondweeds. Often occurring in hard water, it is especially useful as a food plant for ducks. It also provides food and shelter for many species of aquatic insects which are eaten by fish. This species hybridizes with several other species of pondweeds.

Broad-leaved Pondweed is found in somewhat deep water of ponds, lakes, and slowly moving streams from Newfoundland to British Columbia, south to Georgia and Alabama, and on the Pacific coast to California.

✖ Several species of this genus have thickened rhizomes or tubers. These can be eaten raw or cooked as potatoes. They are reported by some wild food fans to be the best potato-like food to be found outside of cultivation.

80 species with at least 30 in northeastern North America. It is a very important wildlife food genus, particularly for waterfowl, comprising more than 50% of the diet of some species. All parts of the plants are eaten including leaves, seeds, and in some pondweeds, tubers.These plants are also attractive to, and often heavily eaten by marsh birds, shore birds, muskrats, beaver, and moose.

P. amplifolius is one of the largest and

Potamogeton crispus
Curly Pondweed

382

Najadaceae
Pondweed Family

A perennial with a slender underwater rhizome, introduced from Europe. As with most other species of this genus, the flower clusters extend above the surface of the water, and pollination is by wind. The seeds

are flattened with 3 rounded ridges on one side and are shaped somewhat like an arrowhead. Winter buds may be more important than seeds in the propagation of this species. These are bur-like, to 2.5 cm.

Flowers inconspicuous, in short, dense clusters, to 1 cm. long (½ in.); leaves alternate, floating leaves none, leaf margin finely toothed and wavy, 3–5 nerved, to 8 cm. long (3.2 in.), and 12 mm. wide (½ in.), sessile; stems flattened, branched, to 80 cm. long (32 in.).

May–September

long (1 in.), and form in the bases of leaves.

This plant may become an aggressive weed in polluted water. Its winter buds and seeds are eaten by ducks, and the plants provide food, shelter, and shade for several types of fish. It was introduced into the Northeast before 1814 and has become widespread. It is easier to identify than most of the other plants in this genus by its wavy margined, fine toothed leaves.

Curly Pondweed is found in ponds, lakes, sluggish streams, and sometimes in muddy, polluted, or brackish water from Quebec and Ontario to Minnesota, south to Virginia and Missouri.

✖ See *P. amplifolius*.

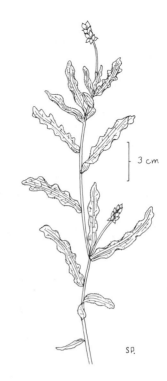

3 cm

S.P.

Potamogeton diversifolius
Variable Pondweed

383

Najadaceae
Pondweed Family

Flowers inconspicuous, in numerous, stalked or sessile axillary clusters, to 12 mm. long (½ in.); leaves of two types, floating and submerged: submerged leaves very narrow, numerous, to 5 cm. long (2 in.), floating leaves elliptical, to 3 cm. long (1.2 in.); stems bushy, branched, to 50 cm. long (20 in.).

June–September

A perennial with a long underwater rhizome. The seeds are greenish, almost circular and flattened, with a prominent ridge along one side. Winter buds are usually formed in the upper leaf axils in late summer and autumn. As with other species of this genus, the seeds and winter buds are favorite foods for ducks, and in some situations are especially favored by the Pintail.

A related species, *P. spirillus*, not illustrated, is very similar, and a technical manual will be needed to distinguish it from *P.*

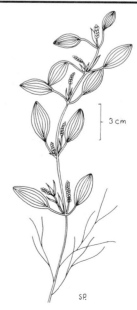

3 cm

S.P.

diversifolius. The range of *P. spirillus* extends northward to Newfoundland and Quebec and does not extend southward beyond Virginia. In an older system of classification, *P. diversifolius* was listed as *P. hybridus.*

Variable Pondweed is found in shallow water of ponds, lakes, and slowly moving streams from Maine to Wisconsin and Montana, south to Florida, Mexico, and California. It is most common in the southern and midwestern states and also occurs in the West Indies.

✖ See *P. amplifolius*.

Potamogeton epihydrus
Ribbonleaf Pondweed*

384

Najadaceae
Pondweed Family

*Also Nuttall's Pondweed

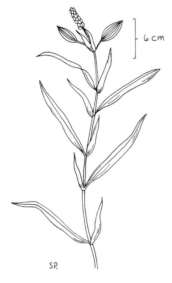

6 cm

S.P.

A perennial with an extensively spreading underground rhizome. The flower clusters extend above the surface of the water, and pollination is by wind. The seeds are nearly circular, brownish, flattened with a ridge along one side. The widths of the sub-

Flowers inconspicuous, in numerous, dense, stalked clusters, to 4 cm. long (1.6 in); leaves of two types: submerged leaves alternate, ribbon-like, to 20 cm. long (8 in.), and 1 cm. wide (.4 in); floating leaves usually opposite, numerous, elliptical, to 8 cm. long (3.2 in.); stems flattened, not extensively branched, to 2 m. long (6-½ ft.).

July–September

merged leaves are somewhat variable and may be associated with the richness of the substrate in which the plants are anchored.

This is a very important wildlife food genus, especially for waterfowl. In several species, short branches called winter buds are produced in the axils of leaves. These are shed and fall to the bottom in autumn, then grow into new plants in spring. The winter buds and seeds are favorite foods for many species of ducks. Consumption and subsequent elimination of the intact seeds has contributed to the spread of some species.

Ribbonleaf Pondweed is found in the quiet water of ponds, lakes, and slowly flowing streams from Newfoundland and Quebec to southern Alaska, south to Georgia, Missouri, and California.

Potamogeton natans Floating Pondweed — 385 — *Najadaceae* Pondweed Family

Flowers inconspicuous in a dense, cylindrical, terminal cluster, to 5 cm. long (2 in.), on a thick stalk to 15 cm. long (6 in.); leaves are of 2 types: submerged leaves very slender, almost circular in cross-section, to 20 cm. long (8 in.); floating leaves elliptical, to 10 cm. long (4 in.), rounded at base and tip, with long flexible petioles; stems unbranched or only slightly so, to 2 m. long (6½ ft.).

July–September

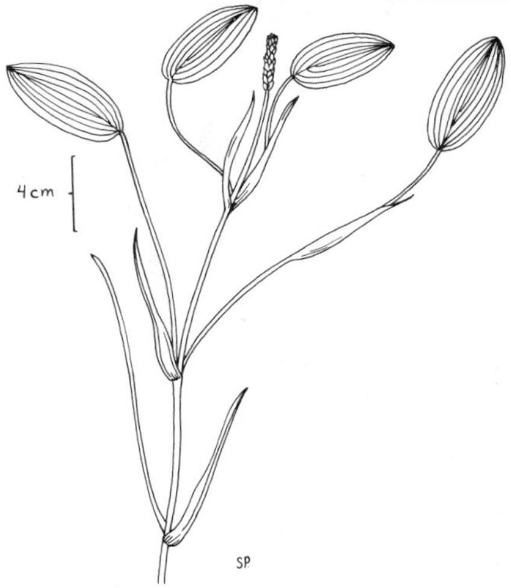

A perennial with a long, red-spotted rhizome of North America, Europe, and Asia. The flower cluster extends above the surface of the water, and pollination is by wind. The stalk of the flower cluster contracts at maturity and the seed-like fruits are drawn under the water. The seeds are roundish in outline, slightly flattened with a coarsely wrinkled surface and 3 ridges on one side. They contain a spongy tissue that increases their buoyancy and facilitates dispersal by water.

The seeds of this species are retained on the plant until late in the season, and this greatly enhances the species' value as a food for ducks. The rootstocks also are eaten by ducks and muskrats. The plants of this genus are important in providing food and shelter for fish. The leaves of some species are eaten by Bluegills. The plants improve the environment for fish by softening the water and adding oxygen.

Floating Pondweed is tolerant of acid conditions and grows best in ponds and slowly flowing streams to 1.5 m. (5 ft.) in depth. It is widespread in Europe and Asia, and is found in North America from Newfoundland to Alaska, south to Pennsylvania, Illinois, and California.

✂ See *P. amplifolius*.

Potamogeton nodosus Long-leaved Pondweed — 386 — *Najadaceae* Pondweed Family

A perennial of deep or shallow water with a pale, sometimes purple spotted, rhizome. The flower clusters extend above the surface of the water, and pollination is by wind. The seed-like fruits are egg-shaped, reddish, with a short beak and 3 rough ridges on one side.

In older systems of classification, *P. nodosus* has been listed as *P. americanus*, *P. fluitans*, and *P. rotundatus*. This is an especially useful food plant for waterfowl, but it is only fair as food and shelter for fish. However, it supports a variety of insects that are eaten by fish.

Most of the species in this genus are important wildlife food plants. Consumers include waterfowl, shore birds, marsh birds, muskrats, beaver, deer, and moose. These consumers serve sometimes as agents of seed dispersal. In fact, the seeds of one spe-

330

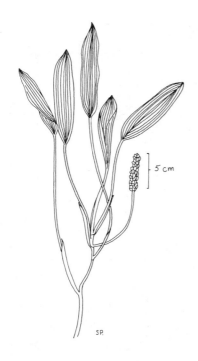

5 cm

Flowers inconspicuous, in dense clusters to 7 cm. long (2.8 in.), on long stalks usually thicker than the stem; leaves of 2 types: submerged leaves alternate, to 30 cm. long (1 ft.) and 3.5 cm. wide (1-½ in.), tapering at both ends, with long petioles; floating leaves elliptical, to 13 cm. long (5.2 in.), with long petioles; stems branched, to 2 m. long (6½ ft.).

August–September

cies were found to germinate only after being exposed to conditions that simulated passage through the digestive tract of a bird. Seeds may also be dispersed by water, and several species produce hardy winter buds that sink to the bottom where they eventually grow into new plants.

Long-leaved Pondweed is found in ponds and streams from New Brunswick and Quebec to British Columbia, south to Georgia, Texas, and California.

✖ See *P. amplifolius*.

Potamogeton zosteriformis **Flatstem Pondweed***	**387**	*Najadaceae* **Pondweed Family**

**Also Eel-grass Pondweed*

3 cm

Flowers inconspicuous, in dense terminal clusters, to 2.5 cm. long (1 in.); leaves narrow, to 20 cm. long (8 in.) and 5 mm. wide (⅕ in.), with 3 large nerves and many fine ones; no floating leaves; stems flattened, winged, freely branched.

June–August

A perennial with a short rhizome and sometimes rooting from the lower nodes of the stem. The flower clusters extend above the surface of the water on stalks to 9 cm. long (3½ in.), and pollination is by wind. Each flower produces 4 elliptical, seed-like fruits with narrow, slightly toothed, wings and finely wrinkled surfaces.

Reproduction is also by short branches in the upper axils that fall to the bottom and function as winter buds. These and the seeds are sometimes eaten by ducks, but this species is outranked by some other species of the genus as a wildlife food plant. It frequently occurs in deep water in association with one or more of the larger species of pondweed. This species has been classified by some botanists as *P. zosterfolius* and *P. compressus*.

Flatstem Pondweed is found in lakes, ponds, and slowly flowing streams from New Brunswick and Quebec to British Columbia, south to Virginia, Indiana, and California.

Brasenia schreberi
Water-shield

388

Nymphaeaceae
Water Lily Family

Flowers dull purple, to 2 cm. wide (¾ in.), solitary, on long axillary stalks that extend above surface of water; sepals 3, petals 3, stamens 12–18, pistils 4–8; leaves alternate, floating, elliptical, to 10 cm. long (4 in.) with long stalks attached to center of leaf undersurface; stems long and slender coated with a layer of slime, to 2 m. long (6½ ft.).

June–August

A perennial aquatic with a long horizontal rhizome shallowly buried in bottom mud. Most of the plants in this family, probably including *B. schreberi*, are insect-pollinated. Each of the several pistils in the flower produces 1–3 relatively large, brown, warty, egg-shaped seeds that eventually open by lids.

This species is important in some areas as a source of food for ducks including the Mallard, Redhead, Pintail, Ring-necked, Wood, and others. They eat the seeds and probably contribute to the dispersal of the plant. It also provides good shade and shelter for fish. In wildlife management practice, it has been successfully transplanted into numerous ponds in the Southeast.

Water-shield is found in the quiet water of shallow lakes, ponds, and sluggish streams

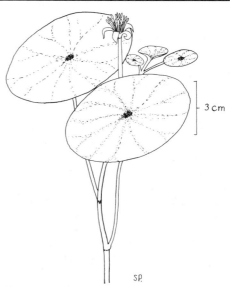

from Nova Scotia and Quebec to Minnesota, south to Florida and Texas. It also occurs in several Pacific coast states, Cuba, and Europe.

⚔ The very young leaves, before they are fully expanded, can be used in salads or cooked as greens. The small tuberous roots were used for food by Indian tribes of California.

Wetlands

332

Cabomba caroliniana
Fanwort*

389

Nymphaeaceae
Water Lily Family

*Also Fish-grass

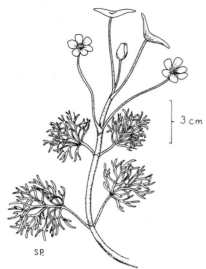

3 cm

S.P.

An aquatic perennial with a horizontal rhizome shallowly buried in bottom mud. The very large, football-shaped pollen grains are probably transferred from anthers to stigmas by insects. After flowering the flower stalk bends downward and the 3-seeded pods are released into the water. This is a valuable food and shelter plant for fish but

Flowers white to yellowish, to 12 mm. wide (½ in.), solitary, on long stalks that extend above surface of water; flower divisions 6, stamens 3–6, pistils 3–6; submerged leaves opposite, stalked, finely palmately dissected, a few upper leaves alternate, on long stalks, floating, narrowly elliptical, attached in center, to 20 mm. long (¾ in.); stems long and slender, to 2 m. long (6½ ft.).

June–August

is only slightly used as food by waterfowl.

This is a small genus of North and South America with a single species native to the Northeast. The genus name is an aboriginal word of Guiana. In southern areas of eastern North America it grows so prolifically that it may clog ponds and waterways. This is a popular plant for indoor aquaria because it is relatively odor free and, since it is eaten by the fish, there is usually less debris.

The widespread use of fanwort as an aquarium plant is probably responsible for its spread into New England and New York. It is found as an attached submergent in ponds, lakes, swamps, and sluggish streams from Massachusetts and New York to Michigan and Ohio, south to Florida and Texas.

Nelumbo lutea
American Lotus*

390

Nymphaeaceae
Water Lily Family

*Also Water Chinquapin, Pond-nuts, Wonkapin, Lotus Lily

A perennial with a tuber-bearing underwater rhizome that may reach a length of 15 m. (50 ft.). The fruiting receptacle is woody, about 10 cm. wide (4 in.), and inversely coneshaped. The flat surface of the receptacle has numerous small pits, each of which contains a single seeded fruit about 1 cm. in diameter (½ in.). In late autumn the pits expand and many of the fruits fall into the water. The seeds may germinate on the surface of the water and be dispersed as floating seedlings.

The seeds of some species of this genus

have remarkable longevity. A Japanese botanist recovered viable seeds similar to those of *N. nucifera*, not illustrated, within a layer of peat on a dry lake bed in Manchuria. Some of these were dated by the Carbon-14 method and were found to be between 830 and 1250 years old.

American Lotus is sometimes very abundant, forming dense colonies of several acres in extent in some parts of its range. In those areas it may be of importance as a wildlife food plant. The seeds are eaten to a limited extent by marsh birds, waterfowl, and songbirds, and the rhizomes are eaten by beavers.

There are 3 species in the genus, includ-

*Flowers pale yellow, to 25 cm. wide (10 in.),
sepals and petals similar, usually 25–30;
flowers on long stalks, high above water;
leaves circular, to 60 cm. wide (2 ft.), cup
or bowl-shaped, petiole attached to center
of leaf, held high above water.*

July–September

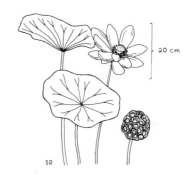

ing one in the West Indies and one in southeastern Asia and Australia. The genus name
is the Ceylonese name for the species occurring in that country (*N. nucifera*). This
species has pink flowers and has been introduced into North America as an ornamental. It has escaped and is occasionally
seen growing wild. The woody receptacle
of the native species is often used in dried
flower arrangements.

American Lotus is found in the quiet water
of ponds, lakes, estuaries, and sluggish
streams from southern New England and
Ontario to Minnesota, south to Florida and
Texas.

✖ This plant has long been known as a
food plant of the American Indians. Its range
has probably been expanded as a result of
introductions into new areas by Indians. The
seeds may be eaten raw or cooked, and have
a taste somewhat like those of the Chinquapin (*Castanea pumila*), thus the names
Water Chinquapin and Pond-nuts. The
young unrolling leaves and leaf stalks can
be cooked as greens. The tubers are very
starchy and can be prepared the same ways
as sweet potatoes. *Note*: This plant is often
rare in the northern part of its range, and
collecting it for food may result in local
extinction.

**Nuphar variegatum
Spatterdock***

391

**Nymphaeaceae
Water Lily Family**

*Also Bullhead Lily, Yellow Pond-lily, Cow-
lily*

*Flowers yellow, to 7.5 cm. wide (3 in.), on
long stalks, floating on the surface of the
water, usually with 5–6 fleshy, petal-like sepals; petals small, often not visible; stigma
disk-like, stamens numerous; leaves floating, rounded at tip, basal lobes often overlapping, to 30 cm. long (12 in.), on stalks to
3.5 m. long (11½ ft.).*

May–October

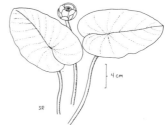

A perennial with a thick creeping underwater rhizome. Cross-pollination is effected
generally by bees of the genus *Halictus*. Relatively large quantities of pollen grains are
produced but few become airborne. The
surplus pollen becomes incorporated into
the sediments and is used as an environmental indicator in the analysis of postglacial deposits.

The berry-like fruit includes numerous

334

finely pitted, greenish seeds. These are eaten by several species of waterfowl and marsh birds, comprising up to 25% of the diet of the Ring-necked Duck in some areas. Since a proportion of the seeds pass through their intestinal tracts undamaged, these migratory birds have had an important influence on the plant's distribution. This plant also contributes up to 10% of the diets of beaver and porcupines in the Northeast.

This is a genus of about 25 species in the Northern Hemisphere with possibly 7 species in the Northeast. There is disagreement among botanists with regard to the classification within this genus. *N. variegatum* has been variously classified as *N. americana, N. advena, N. lutea,* and *Nymphaea advena.*

A related species, *N. advena,* Southern Pond Lily, not illustrated, is very similar but has leaves usually held above the water. It ranges from southern New England to Wisconsin, south to Florida and Texas.

Spatterdock is found in ponds, shallow lakes, and the still water of slowly flowing streams from Newfoundland to British Columbia, south to New Jersey, Ohio, and Iowa. ✕ The seeds can be boiled or roasted and eaten whole, or ground into a flour. The rhizomes can be boiled like potatoes, but may require several changes of water. There is a long history of usage of this plant for food by American Indians of both the East and the West Coast.

**Nymphaea odorata
Fragrant Water Lily***   **392**   *Nymphaeaceae*
Water Lily Family

*Also Sweet-scented Water Lily, Pond Lily, White Water Lily

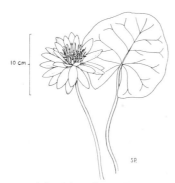

A perennial with a long, branched, underwater rhizome. The stigma matures before the anthers, reducing the possibility of self-pollination. Cross-pollination is effected mainly by bees and beetles, although the flower is visited by a variety of insects. Large quantities of spiny pollen grains are produced and, like those of *Nuphar,* they are used as environmental indicators in the analysis of postglacial sediments.

Flowers white, floating, very fragrant, petals numerous, tapering at the tip, decreasing in size toward the center; flowers opening in the morning, closing in the afternoon, to 15 cm. wide (6 in.); leaves floating, almost round, usually purplish on underside, to 25 cm. wide (10 in.), on long non-striped petioles.

June–September

After flowering the flower stalk contracts and the globular, many seeded fruit matures under water. Each seed has a fleshy appendage, called an aril, that gives it added buoyancy. As the fruit decays, the seeds float to the surface and are dispersed by water currents. The seeds and other parts of the plant are eaten by marsh birds, waterfowl, muskrats, beaver, porcupines, moose, and deer.

A related species, *N. tuberosa,* Tuberous Water Lily, not illustrated, is very similar but has petals with broad rounded tips, leaves with green undersides, and petioles with 4–5 purple streaks. It is most abundant in the upper Mississippi Valley and around the Great Lakes. The specific and common names refer to small tuberous branches that break loose and develop into new plants. In older systems of classification *N. odorata* and *N. tuberosa* are listed as *Castalia odorata* and *C. tuberosa*.

Fragrant Water Lily is found in quiet ponds, shallow lakes, bog pools, and slowly flowing streams from Newfoundland to Manitoba, south to Florida and Texas. ✖ The seeds are rich in protein, oil, and starch. They can be boiled and eaten, or dried and ground into a flour. The unopened flower buds and young unrolling leaves can be used as potherbs. The rhizomes and tubers are very rich in starch and can be prepared as potatoes.

Species of this genus are reported to be widely used as food by natives of Egypt and the western coast of Africa. However, Euell Gibbons stated that no matter how he prepared them, the tubers of the North American species were impossibly bitter to eat. ℞ The rhizomes and tubers contain tannin, a bitter substance. As with many bitter plants, the water lilies have been adopted in folk medicine. They have been recommended as a remedy for stomach problems, as a gargle for sore throat, and as an eyewash. The effectiveness of these remedies is questionable.

Wetlands

Epilobium coloratum
Purple-leaved Willow Herb   **393**   *Onagraceae*
Evening Primrose Family

Flowers pink, numerous, tiny, at the ends of branches, petals 4; seedpods to 5 cm. long (2 in.), numerous, each seed with a tuft of light brown hairs; leaves narrowly lanceolate, toothed, often purplish; stems freely branched, bushy, sometimes purplish, to 1 m. high (40 in.).

July–September

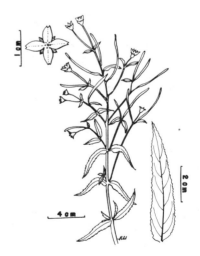

A perennial with a dense cluster of roots at the base of the stem. The long seed capsule opens by 4 valves to release numerous tiny, rough, light to dark brown seeds, each equipped with a tuft of silky, brownish hairs that aid in dispersal by wind. At the end of summer it develops a rosette of sessile leaves that remains green throughout the winter.

Purple-leaved Willow Herb is found in open marshes, roadside ditches, springy areas, and wet meadows from Quebec to Minnesota, south to Georgia and Arkansas.

For further information on ecology, food, and medicinal uses see *E. angustifolium*.

Arethusa bulbosa
Arethusa*

394

Orchidaceae
Orchid Family

*Also Dragon's Mouth

3 cm

S.P.

A perennial with a solid, white or greenish bulb. The flower has a very delicate odor but offers no pollen or nectar for visiting insects. The yellow pollen-like hairs on the lower lip attract early queen bumblebees

Flowers pink, solitary, petals and petal-like sepals erect or bending forward; lower lip white or pale, with purple spots and a crest of yellow hairs; a single grass-like leaf, developing from upper bract after flowering season; stem leafless, smooth, with 2 or 3 rounded bracts near base, to 30 cm. (12 in.).

May–June

that effect pollination before they learn to avoid flowers that offer no rewards. The seed capsule splits into 3 sections, releasing innumerable dust-like seeds that are dispersed by wind.

The orchid family is the largest family of flowering plants. It is probably best known commercially for that special occasion orchid corsage, and for vanilla flavoring that is made from the seeds of several species of the genus Vanilla. The genus Arethusa has only two species, the other one native to Japan. The genus is named in honor of the wood nymph by that name in Greek mythology.

Arethusa is found in sphagnum bogs and peaty swales from Newfoundland to Ontario and Minnesota, south to Delaware and Indiana, and in the mountains to South Carolina. It is very rare south of the Canadian border and may be becoming extinct.

Calopogon pulchellus
Grass Pink*

395

Orchidaceae
Orchid Family

*Also Swamp Pink

A perennial with a bulb-like tuber. Although this plant is insect pollinated it offers neither pollen or nectar for the pollinator.Inexperienced bees are attracted to the crest of yellow hairs on the erect lip of the flower, apparently mistaking it for a cluster of yellow anthers. However, when the bee lands on the front surface of the lip, it bends forward bringing the back of the

bee first into contact with the stigma, then with the single stamen. The process is repeated at a second flower, and pollen masses from the first are deposited on the stigma of the second. After a few such visits the bee may learn to recognize the deception and avoid the yellow hairs. Fortunately, there is an abundance of inexperienced bees during the flowering period.

The seedpod opens by 3 valves to release

Flowers pink to deep rose, in a loose terminal cluster of 2-several; flowers showy,
to 5 cm. wide (2 in), upper lip with crest of
yellow hairs; 1 leaf, long and grass-like,
sheathing the stem; flowering stem to 40
cm. high (16 in.).

June–July

numerous dust-like seeds that are dispersed by air currents. The seeds of the
members of this family are typically irregular in shape and very small, with no co-
tyledons (embryonic leaves), no endosperm
(stored nutrients), and an embryo consisting of relatively few undifferentiated cells.

Grass Pink is found primarily in peat bogs
from Newfoundland to Minnesota, south to
Florida and Texas.

℞ All species of native orchids are protected by law in some states. The plants of
this family are usually not abundant and
their greatest value is the beauty they impart to the habitat in which they occur. They
should be protected and enjoyed.

| Cypripedium reginae | **396** | Orchidaeceae |
| Showy Lady's Slipper | | Orchid Family |

Flowers white, lip tinged with pink, 1–3 at
tip of stem; leaves alternate, hairy, to 18
cm. long (7 in.) and 10 cm. wide (4 in.), with
prominent parallel veins; stems with stiff
hairs, unbranched to 1 m. high (40 in.).

May–July

A perennial with a leafy stem of ridged,
overlapping leaves and coarse fibrous roots.
Like other members of the family this species produces many minute seeds, each of
which has a barely differentiated embryo
and no stored food.

The specific name means "of the queen"
and many people are of the opinion that
this is the most beautiful of the Lady's Slip-
pers. The hairs on the stems and leaves contain a substance that may cause an allergic
reaction similar to that caused by Poison
Ivy in sensitive individuals.

It is illegal to pick any member of this
genus in New York and some other states

338

of the Northeast. It is the state flower of Minnesota.

Showy Lady's Slipper is found in swamps, bogs, and wet woods from Newfoundland to Manitoba, south to Georgia and Missouri.

For more ecological information see *C. acule.*

℞ See *C. calceolus.*

Habenaria clavellata
Green Woodland Orchid*

397

Orchidaceae
Orchid Family

*Also Club-spur Orchid

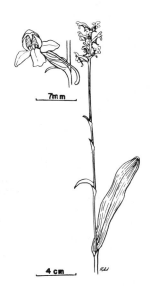

7mm

4 cm

Flowers greenish white or greenish yellow, in an elongate terminal cluster; each flower twisted to one side, with a long spur; fully developed leaf 1, basal, forming sheath around stem, other leaves much smaller or scale-like; stems slender, smooth, to 40 cm. high (16 in.).

July–August

mainly by moths, and the numerous dust-like seeds are spread by wind. This slender plant has delicate but not showy flowers. Since most of its seeds do not grow into new plants, this species is usually not abundant and to find it growing in its natural habitat is a special treat.

This is a very large genus of worldwide distribution. There are about 18 species found in bogs and wet woods native to the Northeast. The genus name is derived from a Latin word meaning "strap" or "rein" and refers to the shape of the spur of some species. The plants of this genus are sometimes referred to as the Rein Orchids.

The Green Woodland Orchid is found in acid bogs and wet woods from Newfoundland to Minnesota, south to Florida and Louisiana.

A perennial with a cluster of oblong, slender, tuberous roots originating from the base of the stem. Cross-pollination is effected

Habenaria dilatata
Tall White Orchid*

398

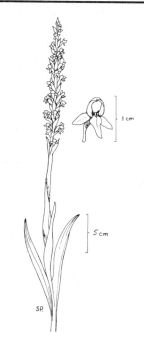

Also Leafy White Orchid, Bog Candle, Scent-bottle

Flowers white or greenish yellow, in a crowded or open elongate terminal cluster; lower lip of flower not fringed, spur about the same length as lip; leaves alternate, narrow, decreasing in size to bracts on upper part of stem; stems fleshy or slender, to 1 m. high (40 in.).

June–September

A perennial with thick fleshy roots extending from the base of the stem. Pollination and seed dispersal as in *H. clavellata*. This species is taller than most other species of the genus, and it has spicy fragrant flowers. These characteristics have contributed to two of its common names.

A related species, *H. hyperborea*, Northern Green Orchid, not illustrated, is similar but has green or yellow-green flowers and larger leaves. Its range does not extend as far south as *H. dilatata*.

Tall White Orchid is found in bogs and wet woods from Greenland to Alaska, south to Pennsylvania and Indiana, and westward to Colorado and California.

For further ecological information on this genus see *H. clavellata*.

Listera australis
Southern Twayblade

399

Flowers greenish-red, in a long terminal cluster, to 10 cm. long (4 in.); lower lip deeply cleft, to 10 mm. long (½ in.), about 4 times longer than petals; leaves opposite, a single pair about middle of stem, sessile, oval, pointed, to 4 cm. long (1½ in.); stems erect, slender, to 25 cm. high (10 in.).

June–July

A small herbaceous plant with slender but fleshy roots. The flower produces nectar but has an unpleasant odor and is pollinated by wasps and flies. The many dust-like seeds are spread by air currents. The genus was named in honor of Martin Lister, a seventeenth-century English naturalist. The name twayblade refers to the pairs of leaves in the middle of the stem.

This is a genus with 5 species native to the Northeast. These inconspicuous plants are rather widespread but are not common. *L. cordata*, Heart-leaved Twayblade, not illustrated, is similar to *L. australis* but is smaller and has a lip about 5 mm. long (⅕ in.), appoximately 2 times longer than the petals. It occurs in the same type of habitat and ranges from Greenland to Alaska, south to North Carolina and West Virginia.

Southern Twayblade is found in bogs and damp mossy woods from Quebec and Ontario, south to Florida and Louisiana.

Pogonia ophioglossoides
Rose Pogonia*

400

Orchidaceae
Orchid Family

*Also Snake Mouth, Beard Flower

2 cm

SP

A perennial with a short rhizome and long, horizontal, shallow roots. The roots sometimes produce buds that give rise to new shoots. Pollination is effected by bees that are lured to the fragrant flower by the yellow pollen-like hairs on the lip. The seed capsule splits into 3 sections, releasing a great number of dust-like seeds that are dispersed by air currents.

Flowers rose-pink, usually solitary, at tip of stem, lip bearded with yellow hairs; leaf 1, sheathing the stem at about midpoint, narrowly lance-shaped, to 10 cm. long (4 in.), a leaf-like bract subtending flower; stems slender, to 40 cm. high (16 in.).

June–July

This species flowers at about the same time as *Calopogon* and they frequently occur together. It is relatively uniform in appearance but occasionally a white flowered form is observed. Long stalked basal leaves often accompany new shoots that arise from the roots. Like many of the native orchids this one is widespread but usually not abundant. It should be admired in its natural setting but not picked.

The genus name is derived from a Greek word meaning "beard" and refers to the 3 rows of yellow or brown tipped fleshy hairs on the flat surface of the fringed lip. There are only 2 species, the other one located in eastern Asia. These species were dispersed at a time when North America and Asia were probably connected by a land bridge.

Rose Pogonia is found in bogs, peaty glades, and wet shores from Newfoundland to Minnesota, south to Florida and Texas.

Polygonum coccineum
Swamp Smartweed

401

A perennial with a thick, creeping, woody rhizome. The flowers are of two types. On some plants the styles are long and the anthers are small and non-functional. On others the filaments are long with fertile anthers and the pistil is short and sterile. Thus in order for fertilization to occur, cross-pollination must take place. The seeds are spread by the numerous species of waterfowl that use them for food.

This species exists in several forms related to its occurrence in water, on dry soil, or in an in-between environment. Its bright pink to red flowers make it one of the most showy native species of the genus. It is a hardy plant that sometimes invades ponds

Flowers pink to scarlet, in 1 or 2 dense, cylindric, terminal clusters, to 15 cm. long (6 in.); fruit a lens-shaped, dark brown nutlet; leaves alternate, lanceolate, long stalked in aquatic forms; stems erect, floating or submerged, hairy at the nodes and on the stalks of the flower clusters of terrestrial forms, nodes often enlarged, to 2 m. long (6½ ft.).

June–October

and impoundments as a vigorous weed. In most circumstances it is a good food and shelter plant for fish. The seeds are important as a food for shore birds, game birds, and songbirds in addition to waterfowl.

The great variability in this species has led to disagreement among botanists with regard to its classification. The aquatic forms are distinguishable from the dry soil forms, but there is continuous variation between these extremes associated with intermediate environmental conditions. Among the names that have been recognized for this species by various botanists are *P. muhlenbergii*, *P. amphibium*, and *Persicaria muhlenbergii*.

Water Smartweed is found in shallow or deep water, sluggish streams, swamps, ditches, and wet fields from Nova Scotia and Quebec to Alberta and Washington, south to North Carolina, Texas, and California.

Wetlands

Polygonum sagittatum
Arrow-leaved Tear-thumb*

402

*Also Arrow-vine

An annual with a slender stem and widely spaced leaves. Each flower produces a single, 3-sided, slightly glossy, brown or black nutlet. The seeds and leaves of the plants of this genus are eaten by large numbers of

game birds, songbirds, and small mammals. These consumers probably also serve as agents of seed dispersal.

A closely related species, *P. arifolium*, Halberd-leaved Tear-thumb, illustrated, is similar but is a perennial, has leaves to 25 cm.

342

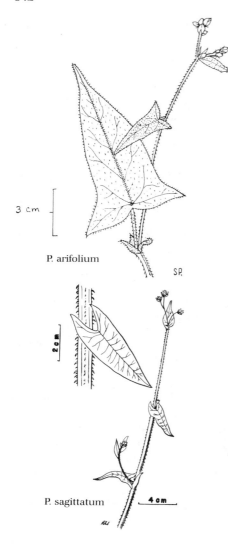

3 cm

P. arifolium

S.P.

2 cm

P. sagittatum

4 cm

R.U.

Flowers pink to whitish, in few flowered clusters on long axillary or terminal stalks, petal-like sepals 5, stamens usually 8; leaves alternate, narrow, arrow-head shaped, with basal lobes pointing backward, to 10 cm. long (4 in.); stems 4-sided with small backward pointing prickles on each of the 4 angles, sprawling on other plants, to 2 m. long (6½ ft.).

July–September

long (10 in.) with basal lobes flaring horizontally; the flowers have 4 petal-like sepals and 6 stamens and produce lens-shaped seeds. It occurs in the same types of habitats as *P. sagittatum* but has a less extensive geographic distribution. These species are easy to distinguish from other members of the genus in the Northeast by their prickly stems. In an earlier system of classification, they were listed as *Tracaulon sagittatum* and *T. arifolium*.

This is a genus of 150 species or more, widely distributed throughout the world. The genus name is derived from Greek words meaning "many knees," referring to the enlarged nodes of many species. A characteristic of the genus is the presence of stipules in the form of membranous sheaths that surround the stem at each node.

Arrow-leaved Tear-thumb is found in swamps, swales, roadside ditches, wet meadows, and (along the Atlantic coast) in brackish tidal marshes from Newfoundland and Quebec to Saskatchewan, south to Georgia and Texas.

Pontederia cordata
Pickerelweed*

403

Pontederiaceae
Pickerelweed Family

*Also Tuckahoe

A perennial with a thick creeping rhizome. Pickerelweed has an interesting association with a species of small bee (*Halictoides novae-angliae*). The emergence of the adult form of the bee coincides with the flowering of the plant, and it is believed that this bee does not seek pollen or nectar from any other plant. What makes this association unusual is that many other plants rich in pollen and nectar are in flower at the same time, and Pickerelweed has other insect visitors.

The ovary is three celled, but only one cell is fertile, and it produces a single brownish, irregularly longitudinally-ridged

Flowers blue, densely clustered at the top of the stem, flower cluster subtended by a clasping, sheath-like bract; flower with an upper and lower lip, each with 3 lobes, upper lip usually with a yellow spot; basal and cauline leaves similar, cauline leaf 1, heart-shaped at base, entire, tapering to blunt point, petiole sheathing base of stem; stems erect, thick, to 1 m. high (40 in.).

July–September

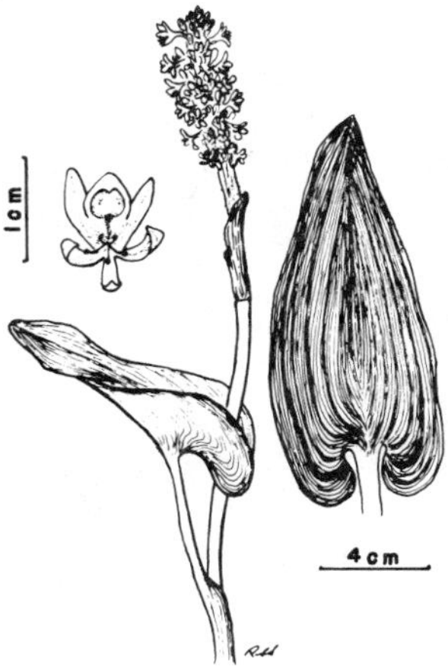

seed. These are known to be a source of food for Black Ducks, Wood Ducks, and Muskrats. Whitetail Deer of the Adirondack Region are reported to visit lake shores to feed on the plant.

This attractive plant is often found growing in dense stands in marshes, on the margins of ponds and lakes, and in shallow water from Nova Scotia to Ontario and Minnesota, south to South Carolina and Texas.

✂ The fruits can be eaten raw, or roasted and ground into a flour. The young unrolling leaves can be added to salads or cooked as greens.

Zosterella dubia **Water Stargrass** **404** ***Pontederiaceae*** **Pickerel-weed Family**

Flowers pale yellow, solitary, at tip of a long slender, axillary tube that reaches the surface of the water; flower with 6 narrow divisions, stamens 3, all alike; leaves very narrow, grass-like, to 15 cm. long (6 in.); stems usually long and slender, submerged or prostrate on mud, rooting at nodes, to 90 cm. long (3 ft.), or more.

June–September

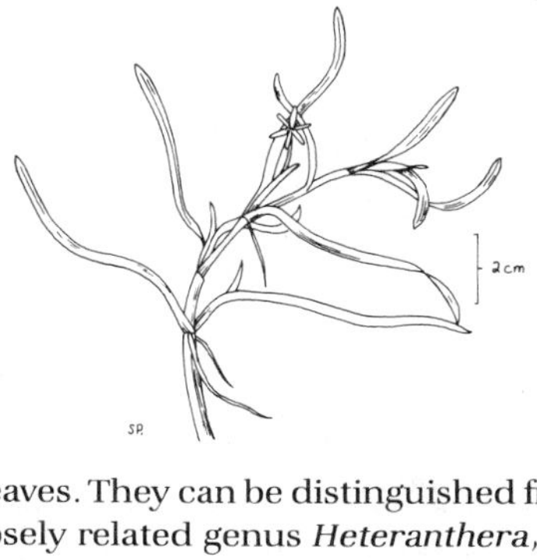

An aquatic perennial with a creeping stem and translucent leaves. The flowers often do not open and are self-pollinated in the buds. The fruit is a 1-celled, few to several seeded, capsule that splits along irregular lines, or not at all. This species is often eaten by waterfowl and it provides good food and shelter for fish.

This is a genus of 2 species, both native to North America. They are very similar to some species of *Potamogeton* but can be distinguished by the lack of a midvein in their leaves. They can be distinguished from the closely related genus *Heteranthera*, the Mud Plantains, not illustrated, by their narrow leaves and stamens all alike rather than of 2 types. *Z. dubia* is classified by some botanists as *Heteranthera dubia*.

Water Stargrass is found in the quiet water of ponds, lakes, and slowly flowing streams, especially in limestone areas, from Quebec to North Dakota, south to Florida and Texas.

Lysimachia ciliata
Fringed Loosestrife

405

Primulaceae
Primrose Family

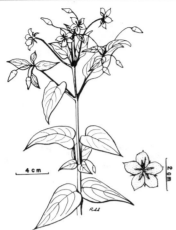

A perennial with a slender horizontal rhizome. The roundish seed capsule splits into 5 sections to release numerous angular, dark brown seeds that are small enough to be windborne. This species is easily distinguished from other plants of this genus by the toothed, sometimes fringed appearance of the petal lobes and the fringe of hairs along the leaf stalk.

Flowers yellow, often nodding, on long axillary stalks, with 5 pointed or toothed petals, stamens 5; leaves opposite, oval or lanceolate, with sharp pointed tips, rounded or heart-shaped at base, petioles fringed with hairs; stems square in cross section, smooth, usually branched, to 1.2 m. high (4 ft.).

June–August

The genus may have been named in honor of Lysimachus, King of Thrace, who is said to have used a loosestrife plant to pacify a maddened bull that was chasing him. It is a genus of about 140 species that are widely distributed in temperate and subtropical regions. A native of the Northeast, *L. ciliata* has been introduced and is widespread in Europe. In some systems of classification it is listed as *Steironema ciliatum*.

Fringed Loosestrife is found along the margins of marshes, open swamps, streambanks, and other exposed low areas from Nova Scotia to the Yukon, south, in the East to Florida and Alabama, and in the West to New Mexico and Arizona.

Lysimachia terrestris
Swamp Candles*

406

Primulaceae
Primrose Family

*Also Yellow Loosestrife

A perennial with a creeping rhizome that, at intervals, produces roots and stems. The flowers are visited by bumblebees and honeybees apparently for the purpose of collecting pollen. The globular seed capsule splits from the top into 5 sections, releasing several dull brown, irregular, warty seeds. It also reproduces vegetatively by bulblets produced in upper leaf axils.

The rather inappropriate specific name was applied by Linnaeus who mistook this plant for a terrestrial species of an epiphytic genus. Cross-pollination of *L. terrestris* with

Flowers yellow, to 12 mm. wide (½ in.), numerous, in a terminal cluster, to 25 cm. long (10 in.); petals 5, dotted or streaked with dark lines, sepals 5; leaves opposite, lanceolate, pointed at both ends, longest ones at upper middle part of stem, to 10 cm. long (4 in.), decreasing in size upward and toward base; axils of upper leaves with purplish bulblets late in season; stems smooth, slender, to 1 m. high (40 in.).

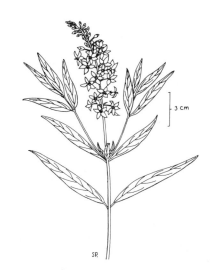

June–August

L. quadrifolia has resulted in a fertile hybrid, *L. producta*, not illustrated, that has intermediate characteristics and occurs from Quebec to South Carolina.

Swamp Candles is found in marshes, open swamps, wet meadows, and other low ground from Newfoundland to Minnesota, south to Tennessee and Georgia. It has been introduced locally in cranberry bogs in Pacific coast states.

Caltha palustris
Marsh Marigold*

407

Ranunculaceae
Buttercup Family

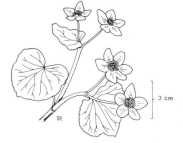

*Also Cowslip, King-Cup, May-Blob, Palsywort, Verrucaria

Flowers bright yellow, petal-like sepals 5–9, stamens numerous, pistils 4–12; leaves alternate, long stalked, heart-shaped at base with rounded tips, shiny green; stems hollow, succulent, often decumbent, to 60 cm. long (2 ft.).

April–May

A perennial with a spongy root and fibrous rootlets. Some plants of this species produce only staminate flowers, others produce both staminate and bisexual flowers, and still others produce only bisexual flowers. It is self-incompatible and cross-pollination is effected mainly by yellow syrphus flies and bees.

During a time when the air is humid, such as during a rain, the seedpods open to release numerous seeds which are rather spongy on one end with an irregularly ribbed and finely pitted surface. These are apparently adapted for dispersal by water.

These plants often occur in masses and are spectacular when in flower. They may be successfully cultivated and make a beautiful addition to water gardens. They are normally found in marshes, wet meadows, swamps, and wet woods ranging from as far north, in a reduced form, as the subarctic, south to South Carolina, Tennessee, and Nebraska.

✖ The young leaves, before or during flowering, can be eaten as a potherb if thoroughly cooked in 2 or 3 changes of boiling water. The flower buds cooked and pickled are said to be passable substitutes for capers. *Caution*: The fresh plant contains the toxic substance protoanemonin, and it should not be eaten without cooking. Cooking or drying deactivates the poison.

☠ According to folklore, Marsh Marigold is effective in curing warts (or verrucae); thus the common name Verrucaria. An infusion of the flowers has been used in the treatment of various types of seizures or fits. This may account for the common name Palsywort. However, in view of its poisonous qualities this plant should not be experimented with in home remedies.

| *Ranunculus septentrionalis*
 Swamp Buttercup | **408** | *Ranunculaceae*
 Crowfoot Family |

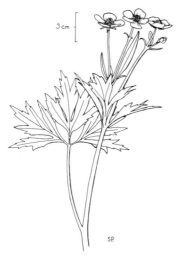

3 cm

s.p.

Flowers yellow, to 2.5 cm. wide (1 in.), solitary, on long terminal stalks, in few-flowered clusters; stamens numerous, petals 5–8, sepals 5; stem leaves alternate, similar to basal leaves, 3-parted with each division further divided, the primary divisions usually stalked; stems usually hairy, often sprawling, hollow, to 90 cm. long (3 ft.).

April–July

has been reported to be favored by moose as a food plant in Isle Royal National Park.

This is a highly variable species with several varieties recognized by the nature and density of the hairs on the stems. It is closely related to *R. hispidus*, Hispid Buttercup, not illustrated, and some botanists have classified it as a variety of that species. Additional study is needed in order to classify the status of these species.

Swamp Buttercup is found in shady swamps and wet woods from Labrador and Quebec to Manitoba, south to Georgia and Texas.

☠ For toxic properties see *R. acris*.

A perennial with thick fibrous roots and prostrate stems that root at the nodes. Pollination is mostly by bee-like flies and small bees. The fruiting receptacle is globular and contains up to 60 somewhat flattened, finely wrinkled, long beaked nutlets. This species

Thalictrum polygamum
Tall Meadow Rue*

409

*Also Muskrat-weed

Flowers white with globular clusters of stamens, not drooping, anthers white, flower clusters with many branches; leaves alternate, sessile, lower ones more than once ternately divided, leaflets commonly with 3 lobes; stems smooth or slightly hairy, to 2.4 m. high (8 ft.).

June–August

A perennial with a tuft of fibrous roots. Plants often have staminate and pistillate flowers, but the pistillate flowers usually have a few stamens. The staminate flowers are typical of wind-pollinated species, but they have a delicate fragrance and are visited by bees and butterflies that may contribute to pollination. Each pistillate flower produces a cluster of black, spindle-shaped, coarsely ridged seeds.

This is a genus of more than 100 species mostly of the North Temperate Zone, with at least 11 native to eastern North America. A few horticultural varieties with rose-purple stamens have been developed as ornamentals from Asiatic and European species.

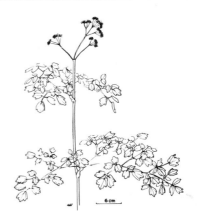

T. polygamum is sometimes confused with Blue Cohosh (*Caulophyllum thalictroides*) but the latter is smaller and has greenish-yellow flowers with 6 petals and 6 sepals.

Tall Meadow Rue is found in swamps, roadside ditches, and wet meadows from Newfoundland and Labrador to Ontario, south to Georgia and Tennessee.

Wetlands

Trollius laxus
Globeflower

410

A perennial with a persistent stem base and thick fibrous roots. The flower has numerous sterile stamens that have nectar glands at their bases. These may serve to attract insect pollinators. The flower produces several to many seed capsules each containing numerous seeds.

This is a rare species that has been placed on the endangered list in several eastern states and is likely soon to be included on the national list. Although its center of distribution seems to be in central New York, its abundance there is decreasing as the result of habitat distruction. Every effort needs to be made to protect this species if it is to be saved from extinction.

This is a genus with about 15 species of the Northern Hemisphere. The name Glob-

Flowers greenish-yellow, solitary, terminal, to 3.7 cm. wide (1½ in.); petals none, petal-like sepals 5–7, stamens numerous; stem leaves alternate, basal leaves long stalked, palmately dissected into 5–7 segments, upper leaves sessile; stems smooth, erect, to 50 cm. high (20 in.).

April–May

eflower refers to the shape of the flower in some species but not that of *T. laxus*. The latter is the only species native to North America, and it is represented in the western states by a white flowered variety that is less rare. A related species, *T. europaeus*, not illustrated, is often cultivated as an ornamental and sometimes escapes.

Globeflower is found in swamps, wet meadows, swales, and open wet woods from southern New England to Michigan, south to Delaware, Pennsylvania, and Ohio.

Potentilla palustris
Marsh Cinquefoil*

411

Rosaceae
Rose Family

*Also Purple Cinquefoil

A perennial with a long shallow rhizome, native to North America, Europe, and Asia. Each flower produces numerous tiny, smooth, greenish seeds on an expanded receptacle often enclosed by the persistent calyx. It is easy to distinguish this species from other members of the genus. This is the only species in the Northeast that occurs in wetlands and has purple flowers.

Flowers red-purple, to 2.5 cm. wide (1 in.), petals 5, about one-half as long as the purple sepals; leaves alternate, pinnately compound, with 5–7 toothed leaflets, long petioles sheathing the stem; stems thick, reddish, erect, from a prostrate, woody base, with few or no branches, to 60 cm. long (2 ft.).

June–August

This is a genus of about 300 species restricted to the North Temperate and Arctic Zones. The genus name is derived from a Latin word meaning "powerful" and was originally applied to *P. anserina* in recognition of supposed medicinal properties. *P. palustris* is one of several species that has a circumboreal distribution. It is highly variable with several recognized varieties and forms in the Northeast. In an earlier system of classification, it was listed as *Comarum palustre*.

Marsh Cinquefoil is found in swamps,

bogs, wet meadows, and along the margins of ponds and streams from Greenland to Alaska, south to New Jersey, Ohio, and Illinois, and in the West to California.

Rosa palustris / Swamp Rose — **412** — *Rosaceae* / Rose Family

Flowers crimson pink, to 7.5 cm. wide (3 in.), solitary or in small terminal clusters; petals 5, stamens very numerous; leaves alternate, pinnately compound, usually with 7 leaflets, base of petiole bordered by a very narrow stipule; stems highly branched, bearing scattered, thick, hooked spines, to 2 m. high (6½ ft.).

June–August

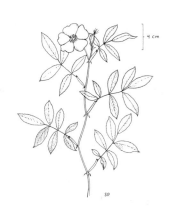

A shrubby perennial with a long runner-like rhizome. The flower produces large quantities of pollen which is collected by bees and sometimes eaten by beetles. Very little pollen gets into the air, and it is not a hayfever plant. Those individuals suffering from "rose colds" in early summer when roses are in flower are probably reacting to grass which is in flower at the same time and is wind-pollinated.

As the season progresses the flowers of this genus are replaced by bright red, globular fruits or "hips", which contain numerous seeds. These remain on the plants throughout the winter months and are an important source of food for several species of game birds, songbirds, small mammals and hoofed browsers. Wild roses often grow in thickets that provide good protective cover and nesting sites for birds.

Swamp Rose is found in swamps, wet thickets, and low ground from New Brunswick and Quebec to Minnesota, south to Florida and Texas.

✖ Rose petals can be eaten as a nibble, used in salads, or candied. The rose hips are rich in vitamin C and can be made into jam or jelly. One of the best species for this purpose is *R. rugosa*, the Japanese Rose or the Salt Spray Rose of the dunes of Cape Cod, not illustrated. It was introduced from eastern Asia and is commonly cultivated as a hedge shrub.

Wetlands

Sarracenia purpurea
Pitcher Plant*

413

Also Sidesaddle-flower, Huntsman's Cup, Indian-Cup

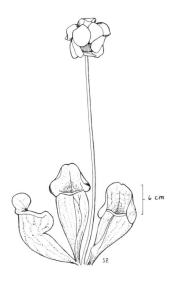

6 cm

S.P.

A perennial evergreen with a horizontal rootstock. Pollination is by insects, probably carrion flies. The 5-chambered seedpod splits to release numerous tiny, brownish seeds, each with a wing on one margin that aids in dispersal by wind.

The hollow leaves or pitchers are efficient insect traps. The hood above the pitcher trap has stiff hairs that point downward. An insect alighting there finds it easy to move downward but almost impossible to move away from the opening. The result is that it usually falls into the water, where it is apparently digested by enzymes secreted by the leaf. A species of the Southeast, *S. flava*, not illustrated, has been shown to produce within its pitcher the alkaloid coniine, which has the odor of rotting meat. This substance not only lures the insects into the trap but also paralyzes them and makes them easier for the plant to digest.

The carnivorous habit in these species

Flowers dark red, large and showy, nodding, solitary, on long stalks, to 35 cm. high (14 in.); petals 5, falling a few days after they appear; style expanding to form a 5-pointed, umbrella shaped cover for the 5 stigmas, 1 beneath each point; the flower stalk with sepals and pistil often persists into the next flowering season; leaves basal, hollow, pitcher-shaped, with a wing down one side and an erect, arching, bristly hood; pitcher-like leaves purpleveined, usually containing water, to 30 cm. long (12 in.).

May–June

has evolved apparently in response to substrates low in nitrates. An interesting note is that some insect species have evolved a resistance to the digestive enzymews and are normal inhabitants of the Pitcher Plant during their larval stages.

S. purpurea is the provincial flower of Newfoundland. Unlike related species of the Southeast, it can withstand being frozen for extended periods during winter. It is commonly found in bogs or peaty substrates ranging north-westward to the Mackenzie District of the Northwest territories and south to Florida and Louisiana. In the Southeast it is restricted chiefly to bogs of the coastal plain. It has been introduced by man and is well established at several sites in West Virginia.

℞ The root has been used for upset stomach and constipation and as a tonic to stimulate the appetite. Some of the tribes of American Indians in Canada and the region of the Great Lakes believed that the root could be used both to prevent smallpox and to reduce the scarring caused by the disease. A British surgeon corroborated this use of the plant in a paper delivered in 1861. There was considerable controversy in the medical profession and the treatment was eventually rejected.

Saururus cernuus
Lizard's-tail*

414

*Also Water Dragon, Swamp-lily
Flowers white, tiny, fragrant, in a long dense terminal cluster, to 15 cm. long (6 in.), nodding at tip; leaves alternate, heart-shaped, to 15 cm. long (6 in.), with long petioles sheathing the stem; stems often branched, jointed at nodes, smooth, to 120 cm. high (4 ft.).

June–August

A perennial with a slender creeping, aromatic rhizome. The flowers have neither sepals nor petals, the white color of the flower cluster resulting from the long white filaments of the stamens. Odor is important in attracting insect pollinators for some plants, and the fragrance of the flower cluster of *S. cernuus* may serve this function. The fruit has a warty surface and 3 or 4 persistent coiled styles and encloses 3 or 4 seeds. It is eaten by the Wood Duck and probably other birds that visit wetlands.

This is a genus of only 4 species, one native to North American and 3 native to eastern Asia. The genus name is derived from Greek words that mean "lizard tail", referring to the shape of the flower cluster. The specific name is a Greek word that means "drooping" and alludes to the nodding tip of the cluster.

6 cm

Lizard's-tail is found in open swamps and pond margins and in shallow water from Quebec to Minnesota, south to Florda and Texas.

℞ The boiled and crushed rhizomes were applied as a poultice by the Choctaw Indians. In herbology, an extract of the rhizome is described as a sedative, a muscle relaxant and an astringent.

Wetlands

Chrysosplenium americanum
Golden Saxifrage*

415

*Also Water-mat, Water-carpet

A perennial with tufts of fibrous roots at nodes of the prostrate stem. The heart-shaped, 2-lobed seed capsule splits across the top to release numerous tiny, black, glandular-hairy seeds. The genus name is a

Greek word meaning "gold spleen." This may refer to an early medicinal use of the plant, although no reference to it can be found in modern herbology.

There are 2 species of this genus in eastern North America. The other one, *C. iow-*

352

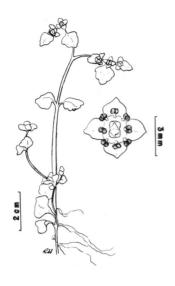

Flowers yellow or greenish, at the tips of branches; petals none, sepals 4; stamens 4–8, in the notches of a central disk, anthers red; leaves opposite, upper ones often alternate, succulent, roundish, with irregular blunt teeth; stems creeping, branched, often forming mats, to 20 cm. long (8 in.).

April–June

ense, not illustrated, has alternate leaves and occurs in Asia and arctic America, extending southward to Iowa.

Golden Saxifrage is found in shaded wet soil along stream margins, around springs and rills, and in wet woods from Quebec to Saskatchewan, south to Virginia and Indiana, and in the mountains to Georgia.

✘ *C. americanum* and *C. iowense* are both edible as cooked greens or fresh in salads. Unlike many other edible wild plants, they do not become bitter with age and thus can be used throughout the growing season. European species of this genus are used in a similar way.

Parnassia glauca
Grass-of-Parnassus

416

Saxifragaceae
Saxifrage Family

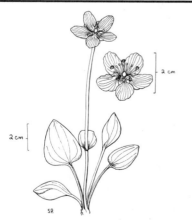

Flowers white with green veins, to 2.5 cm. wide (1 in.), solitary on long stalks; sepals 5, petals 5; fertile stamens 5, between petals, a 3-pronged cluster of sterile stamens opposite each petal, pistil 1; stem leaf 1, roundish, clasping at about middle of stem; basal leaves in a rosette, long stalked, heart-shaped; flowering stem smooth, to 50 cm. high (20 in.).

July–September

not liquid at all. It has been suggested that the function of these false nectaries is to attract the attention of insects and guide them to the true nectar glands. The ruse seems to work since the flowers are visited by flies, bees, and butterflies. The seed capsule splits into 4 sections, releasing numerous light brown, flattened seeds with spongy wings that aid in dispersal by wind or water.

A showy perennial with leaf veins unbranched from base to apex. Each of the clusters of sterile stamens has 2 small nectar glands at its base and is tipped with knobs that resemble drops of nectar but are

In a related species, *P. palustris*, Northern Grass-of-Parnassus, not illustrated, with an even more elaborate system of false nectaries than in *P. glauca*, self-pollination occurs about as often as cross-pollination by insects. This casts some doubt on the hypothesized function of the sterile stamens as false nectaries.

The genus name is derived from Mount Parnassus in Greece and refers to a member of this genus that was called Grass-of-Parnassus by Dioscorides. *P. glauca* is one of 6 species that occur in eastern North Amer-ica. Two related species, *P. asarifolia*, Kidney-leaved Grass-of-Parnassus, with leaves broader than long, and *P. grandifolia*, Large-leaved Grass-of-Parnassus, with sterile stamens longer than fertile ones, neither species illustrated, extend the range of the genus from Virginia and West Virginia along the mountains to Georgia.

Grass-of-Parnassus (*P. glauca*) is somewhat rare and is found in swamps, bogs, and wet soil, especially in limestone areas, from New Brunswick and Quebec to Saskatchewan, south to Virginia and Illinois.

Saxifraga pensylvanica Swamp Saxifrage*   **417**   *Saxifragaceae* Saxifrage Family

*Also Wild-beet
Flowers greenish white, in terminal clusters on a long, mostly leafless stalk; leaves mostly basal, lanceolate, slightly toothed or entire, to 25 cm. long (10 in.); flowering stems thick, soft, hairy, to 1 m. high (40 in.).

May–June

A perennial with a thick, shallow, horizontal rhizome. In some species of this genus self-pollination is avoided by the maturation and shedding of the pollen from the anthers before the stigma becomes receptive. The 2-beaked seed capsule splits into 2 sections, releasing numerous tiny seeds that may become airborne.

This is a large genus mostly of temperate and sub-arctic regions. The genus name is derived from Latin words that mean "stone-breaker" and there are two explanations for its application to this genus. By one account, it refers to several species that grow in rock crevices. Another account is based on the small bulblets produced by some European species. These apparently reminded an early herbalist of kidney stones and thus, by the doctrine of signatures, these

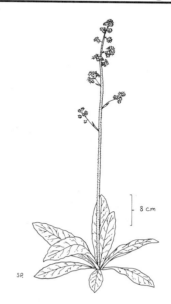

8 cm

plants should be useful in treating the affliction.

Swamp Saxifrage is found in swamps, bogs, seepage areas, and wet meadows from Ontario to Minnesota, south to Virginia, Indiana, and Missouri.

354

✂ The leaves of this species and several other large leaved members of the genus such as *S. micranthidifolia*, Lettuce Saxifrage, not illustrated, can be added to salads, cooked as greens or prepared as a wilted lettuce salad.

Chelone glabra
Turtlehead*

418

Scrophulariaceae
Figwort Family

Also Snakehead, Balmony

A perennial with a fibrous root system. Pollination is usually accomplished by bumblebees which have learned to force open the lips of the corolla. After a visit to one flower with its rich supply of nectar, they search widely for flowers of the same type. The roundish seedpod splits from the top to release many black seeds, each bordered by a gray, corky or spongy wing which aids in dispersal by wind or water.

This is a highly variable species with several varieties based on leaf shapes and de-

Flowers white, sometimes tinged with pink, 2-lipped, upper lip arching over lower giving a shape resembling a turtle's head; flowers in dense clusters at ends of stems and branches; leaves opposite, lanceolate, finely toothed, with prominent midrib; stems smooth, often branched in upper part, to 120 cm. high (4 ft.).

August–September

gree of coloration of flowers. A related species, *C. obliqua*, Red Turtlehead, not illustrated, has red or purple flowers and is common along the coastal plain and in swamps and wet woods from Florida and Alabama, north to Maryland and Tennessee.

Turtlehead is found in open swamps, wet meadows, roadside ditches, and wet woods and along stream margins from Newfoundland to Minnesota, south to Georgia and Alabama.

℞ In herbology, the fresh plant is reported to be useful in the treatment of liver problems and for intestinal worms in children. An ointment made from the leaves has been used for sores and inflamed breasts and the itching and irritation of hemorrhoids. There is some evidence that it was used as a laxative by some tribes of American Indians.

Mimulus ringens
Monkey Flower

419

Scrophulariaceae
Figwort Family

Flowers blue, on long stalks on axils of upper leaves; corolla 2-lipped, upper lip 2-lobed, lower lip 3-lobed; leaves opposite, sessile, lanceolate, finely toothed, upper ones progressively smaller; stems smooth, square, to 1 m. high (40 in.).

June–September

A perennial with a shallow horizontal rhizome and spreading runners which produce roots at intervals. The stigma has 2 lobes that close like the pages of a book when they are touched. After a period of time they will reopen. When they receive pollen from another member of the same species they close and do not reopen. Cross-pollination is effected by bees. The cylindrical seed capsule splits into 2 sections to release many tiny, black-tipped, windblown seeds.

A related species, *M. alatus,* Winged Monkey Flower, not illustrated, is similar but has a winged stem and leaves with petioles and coarse teeth. It occurs over much of the range of *M. ringens.* The flower of these plants supposedly looks like a monkey's face, but the comparison requires imagination.

This is a large genus with species on every continent but most numerous in western North America. Several of the 5 species that

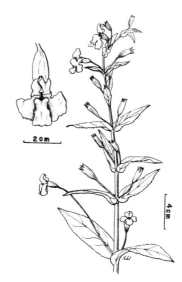

occur in the Northeast are natives of the West which were introduced as ornamentals and have escaped cultivation. *M. ringens* is the most abundant species of the genus native to eastern North America.

Monkey Flower is found in swamps, wet shores and meadows, swales, and roadside ditches from Nova Scotia to Manitoba, south to Georgia, Alabama, and Texas.

Veronica anagallis-aquatica
Water Speedwell*

420

Scrophulariaceae
Figwort Family

*Also Brook-pimpernel

A perennial native to North America, Europe, and Asia. Most of the species of this genus have flower features that attract insects as possible pollinators. The nearly round seed capsule splits into two sections, releasing oval, granular, light brown seeds that are small enough to become windborne in a slight breeze.

A very similar and closely related species, *V. americana,* American Brooklime, not illustrated, has toothed and stalked leaves that do not clasp the stem. It is a native species that occurs in the same types of habitats as *V. anagallis-aquatica* but has a wider geographic distribution in North America.

Water Speedwell is found in brooks, ditches, wet shores, swamps, and shallow water, especially in limestone areas, from Maine to Washington, south to North Carolina and Texas.

Flowers blue or violet, in long clusters, to 10 cm. long (4 in.), from upper leaf axils; corolla 4-lobed, upper lobe larger than other 3, stamens 2; leaves opposite, sessile, often clasping; stems often prostrate with erect branches, succulent, rooting at nodes of prostrate sections, to 90 cm. long (3 ft.).

May–September

✗ Water Speedwell and American Brooklime are both edible but somewhat bitter. They can be added to salads, eaten on sandwiches, or cooked as greens. They are said to be high in Vitamin C and to have been favored as potherbs by some tribes of western Indians.

℞ A similar but introduced species, V. beccabunga, not illustrated, has the same common name as V. americana and it is recommended in herbology as a preventative and cure for scurvy and for disorders of the urinary tract.

Sparganium eurycarpum
Giant Bur-reed

421

Sparganiaceae
Bur-reed Family

Flowers whitish or greenish, in dense globular unisexual clusters, on upper part of stem, upper clusters staminate, lower ones pistillate; the staminate clusters soon wither, the pistillate clusters become many-seeded, bur-like fruits; stem leaves alternate, in 2 rows, narrow, ridged on underside, to 40 cm. long (16 in.); stems thick, zigzagged toward the tip, to 1.2 m. high (4 ft.).

July–September

A perennial with a shallow horizontal tuber-bearing rhizome and fibrous roots. The staminate flowers at the tip of the stem re-

lease pollen which falls or is carried by air currents to the lower pistillate flowers. In germination studies the top-shaped, angular, light brown seeds retained 84% viability after storage for 5 years in water at 1–3°C.

This is an important wildlife food genus. The seeds are eaten by several species of ducks and shorebirds and comprise up to 25% of the diets of mallards and Whistling Swans. These plants do not form extensive mats, as do some aquatics, but they do produce local colonies that are good cover for marsh birds and waterfowl. The leaves and stems provide a source of food that is favored by deer and muskrats.

This is a family of only one genus with 10 species native to the Northeast. *S. eurycarpum* is the most abundant but several others, including *S. americanum*, *S. angustifolium*, and *S. chlorocarpum*, not illustrated, are similar in appearance and occur in the same types of habitats. They are distinguished from one another mainly by characteristics of the mature fruit.

Giant Bur-reed is found in swamps, on the edges of ponds and streams, and in wet soil, especially in limestone areas, from Nova Scotia and Quebec to British Columbia, south to Virginia, Missouri, and California. ✗ Several species of this genus produce small, scattered tubers in the fall. These can be prepared like potatoes, but they are small and hard to collect. The Klamath Indians of Oregon valued them as a food source.

Typha angustifolia
Narrow-leaved Cattail*

422

Typhaceae
Cat-tail Family

---

*Also Nailrod

Flowers tiny, inconspicuous, unisexual, densely crowded in cylindrical terminal clusters, staminate flowers at tip of long stalk, usually separated by a few centimeters from the pistillate flowers beneath them; diameter of mature cluster about 12 mm. (½ in.); leaves long and narrow, slightly rounded on one side, typically no more than 12 mm. wide (½ in.), sheathing the stem; basal leaves and flowering stalk to 2.4 m. high (8 ft.).

May–July

A perennial with a thick horizontal rhizome, of North America, Europe, Asia, and Africa. It is wind-pollinated and although a considerable quantity of pollen is produced, it is not a cause of hayfever. For a given plant the staminate flowers mature before the pistillate ones, thus assuring cross-pollination. After pollen is shed the staminate flowers wither but the brown cylinder of seed-like fruits usually persists into the winter months. It eventually disintegrates and the tiny seeds are spread by wind, often aided by a parachute of hairs.

This is a highly variable species of wide distribution of North America. It is less abundant that *T. latifolia*, but it is particularly common along the Atlantic coast. A

6 cm

hybrid of this species and *T. latifolia* grows to a height of 3 m. (10 ft.) or more and is recognized as a distinct species, *T. glauca*, by some botanists. A related species, *T. domingensis*, not illustrated, is very similar but may reach a height of 3.6 m. (12 ft.) or more, and it may be of hybrid origin also.

Narrow-leaved Cattail is found in fresh and brackish water swamps and marshes from Nova Scotia and Quebec to Minnesota, south to Florida and Texas. It also occurs

358

along the Pacific coast from Washington to California.

✗ The cattails have a firmly established history as food plants. The core of the young shoots can be used in salads or cooked as asparagus. The young flower spikes, before the yellow pollen is visible, can be boiled and eaten like corn on the cob. These have a bland flavor, so they are not likely to replace corn on the cob in the near future but they are nutritious.

Pollen can be collected from these plants, especially *T. latifolia*, quickly and in great quantities by simply shaking the staminate flower cluster in an open bag. It is almost

32% carbohydrate and can be mixed with wheat flour or used in pure form to make bread.

In fall, winter and early spring the rhizome is very rich in starch. This can be extracted by peeling away the outer covering and crushing the core in water, allowing the starch to settle out. This has been found to have a carbohydrate content of almost 57% and can be used as flour.

The plants of this genus have great potential as a commercial source of food for the future.

℞ See *T. latifolia*.

*Also Flag

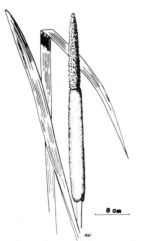

A perennial with a thick creeping rhizome, of North America, Europe, and Asia. Pollination is by wind but, although a great quantity of pollen is produced, this is usually not considered a hayfever plant. Cross-pollination is assured by the shedding of the pollen of a given plant before its stigmas become receptive. After the pollen is released the staminate flowers wither but the

Similar to *T. angustifolia* but there is usually no space between the staminate and pistillate flower clusters; the mature flower cluster about 2.5 cm. (1 in.) in diameter; leaves bluish-green, nearly flat, to 2.5 cm. wide (1 in.); leaves and flowering stems to 3 m. high (10 ft.).

May–July
brown velvety cylinder of seed-like fruits usually persists into the winter months. It eventually disintegrates and thousands of minute seeds, with parachutes of hairs, become windborne.

This is a family of only one genus. The species most common in North America are *T. angustifolia* and *T. latifolia*. These are sometimes difficult to distinguish because there is considerable variation in leaf width. In one form of *T. latifolia* there is a space between staminate and pistillate flower clusters. Probably the most reliable way to distinguish between them is to examine the pollen grains with a microscope. The grains of *T. latifolia* are always in clusters of 4's, most often as flat plates. The pollen of *T. angustifolia* always occurs as single grains.

Broad-leaved Cattail is found in swamps,

pond margins, roadside ditches, and wet waste areas throughout the United States and southern Canada.

✖ See *T. angustifolia.*

℞ In colonial times the crushed rhizome was used as a poultice for skin irritations and burns.

The dried leaves of cattails are highly valued in making braided rush-bottomed chairs. The leaves of *T. latifolia* are used for caulking in the wood barrel industry. Cattail fluff is used for stuffing pillows.

Boehmeria cylindrica
False Nettle*

424

Urticaceae
Nettle Family

*Also Bog-hemp

Flowers greenish, inconspicuous, in dense cylindrical clusters, in upper leaf axils; leaves opposite, long stalked, with 3 prominent veins, tapering to a point, rounded at base, toothed, to 12.5 cm. long (5 in.); stems smooth, unbranched, to 90 cm. high (3 ft.).

July–September

A perennial with a tuft of dense fibrous roots. Pistillate and staminate flowers are usually on different plants assuring cross-pollination. It is wind-pollinated but is usually not considered a serious hayfever plant. Each pistillate flower produces a single tiny, light brown, slightly winged, seed-like fruit. The dead stems with clusters of seeds may persist into autumn or early winter, providing food for the Common Goldfinch, Slate-colored Junco and Swamp Sparrow.

This is a genus mostly of tropical and sub-tropical regions with a single species native to the Northeast. It resembles Clearweed (*Pilea pumila*) and Stinging Nettle (*Urtica dioica*) but does not have a translucent stem

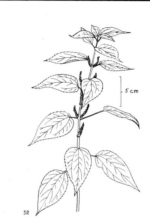

5 cm

SR

or stinging hairs. A related species, *B. nivea*, Ramie, not illustrated, is the source of an important fiber and is sometimes grown as an ornamental in southern United States because of the woolly-white undersides of its leaves.

False Nettle is found in shady swamps and wet woods from Quebec and Ontario to Minnesota, south to Florida and Texas.

Plants of Beaches, Dunes, and Sandy Soils

In the northeastern United States, sand dunes are associated with the Atlantic coast and the eastern shores of the Great Lakes. Many areas of the Northeast include fossil beaches and sand plains associated with glaciation. Sand is a unique substrate for plant growth in a number of ways. In comparison with other soil types that support plant life, it is very coarse textured and has a low water holding capacity. In addition, sandy substrates are usually low in organic matter and in mineral nutrients. Added to this, under the influence of wind sand is constantly being moved about, so that plants growing here must also contend with sand abrasion and frequent burial. If they survive these threats, high surface temperatures of the sand in summer may cause dessication and death. Therefore, sand dunes and sand plains are harsh environments.

Despite these rigors, many plant species are adapted for survival in sand. For example, pioneer species that establish themselves on sand dunes have extensive root systems that tap deep supplies of water and mineral nutrients. This network of roots often serves to anchor the moving sand and to initiate the accumulation of organic matter. The organic matter provides a medium for microorganisms and the beginning of soil development. With these modifications, the dune environment may be sufficiently ameliorated to permit the invasion and establishment of a few hardy trees and shrubs. In time, as the dune becomes increasingly stabilized and the soil more mature, climax species will invade and eventually replace sub-climax species.

If at any point in this process the vegetative cover is removed and the sand substrate exposed, a blow-out will occur and the dune will again begin to move with the prevailing wind. Activities of man, such as clearing for cottage construction or dune buggy operation, are often responsible for the initiation of blow-outs. A case in point are the sand dunes of Cape Cod, Massachusetts. When the pilgrims arrived in Massachusetts in 1620 they recorded the dunes of Cape Cod as wooded to the shore. Settlement by man with the attendent cutting, burning, and grazing created a disturbance that continues to this day among the coastal dunes. The situation was so out of control at one point that the entire town of Provincetown was threatened with burial by sand. Sand dune ecosystems are, thus, very delicately balanced and highly vulnerable to the activities of man.

Beaches

The plants included in the following section are most often found in exposed sandy areas or on sand dunes in the early stages of succession.

The following species are sometimes observed on sand dunes and beaches and in open sandy soil, but they also occur in other types of habitats. These plants are described and illustrated in the section that describes the habitat where they are believed to be most common.

Aizoaceae, the Carpet-weed Family
   Mollugo verticillata, Carpet-weed
Asclepiadaceae, the Milkweed Family
   Asclepias tuberosa, Butterfly-weed
Asteraceae, the Aster Family
   Ambrosia artemisiifolia, Common Ragweed
   Aster linariifolius, Stiff Aster
   Bidens cernua, Nodding Bur Marigold
   Bidens bipinnata, Spanish Needles
   Solidago bicolor, Silverrod
   Sonchus oleraceus, Common Sow Thistle
   Xanthium strumarium, Cocklebur
Caesalpiniaceae, the Caesalpinia Family
   Cassia fasciculata, Partridge-pea
Caryophyllaceae, the Pink Family
   Arenaria lateriflora, Grove Sandwort
Clethraceae, the White Alder Family
   Clethra alnifolia, Sweet Pepperbush
Crassulaceae, the Orpine Family
   Sedum acre, Mossy Stonecrop
Droseraceae, the Sundew Family
   Drosera filiformis, Thread-leaved Sundew
   Drosera intermedia, Spatulate-leaved Sundew
Ericaceae, the Heath Family
   Vaccinium macrocarpon, Large Cranberry
   Vaccinium oxycoccus, Small Cranberry
Gentianaceae, the Gentian Family
   Centaurium umbellatum, Centaury
Hypericaceae, the St. John's-wort Family
   Triadenum virginicum, Marsh St. John's-wort
Iridaceae, the Iris Family
   Sisyrinchium montanum, Blue-eyed Grass
Lamiaceae, the Mint Family
   Scutellaria parvula, Small Skullcap
   Teucrium canadense, American Germander
Orchidaceae, the Orchid Family
   Pogonia ophioglossoides, Rose Pogonia
Plantaginaceae, the Plantain Family
   Plantago aristata, Buckhorn
Polygonaceae, the Smartweed Family
   Polygonum aviculare, Knotweed
Rubiaceae, the Madder Family
   Diodia teres, Buttonweed
Solanaceae, the Nightshade Family
   Solanum carolinense, Horse-nettle
   Solanum nigrum, Black Nightshade

Artemisia caudata
Tall Wormwood

425

Flower heads green, nodding, crowded, in a branched terminal cluster; basal leaves 2 or 3 times pinnately divided, stem leaves alternate, smaller and less dissected; stems very leafy, smooth, usually unbranched, to 1 m. high (40 in.).

July–October

A biennial that produces a dense basal rosette of leaves and a substantial taproot during the first year of growth. Pollination is by wind and the pollen may be a significant cause of hayfever in some areas. Each flower head bears 14–25 flowers but only the outer ones are fertile. The seed-like fruits are small enough to become airborne in a moderate breeze.

Of the approximately 14 species of this genus in eastern North America, 4 are native to the Northeast. *A. stelleriana*, Dusty Miller, not illustrated, a native of northeastern Asia and Japan, is a perennial with pinnately lobed leaves that are white-woolly on both sides. It was introduced as an ornamental for its white leaves but has escaped to sandy sea beaches from Quebec to Virginia.

Tall Wormwood is found on beaches and dunes from Quebec and New Brunswick,

south along the Atlantic coast to Florida, and inland along the Great Lakes.

✘ Some of the species of this genus are occasionally used as seasoning when bitter aromatic herbs are required. *A. absinthium*, Wormwood, not illustrated, a species introduced from Europe, is the source of a flavoring for the liqueur absinthe and is an ingredient in the flavoring of vermouth.

℞ In herbology several European species of this genus are recommended for intestinal worms and for use as tonics.

Beaches

Solidago sempervirens
Seaside Goldenrod

426

Flower heads with yellow ray flowers, showy, to 12 mm. high (½ in.), ray flowers 8–12, flower heads clustered at end of stem or upper axillary branches; leaves alternate, numerous, succulent, entire, basal leaves spatula-shaped, to 40 cm. long (16 in.); stems smooth, erect or arching, to 2.5 m. high (8 ft.).

August–December

A perennial with a short persistent stem base and fibrous roots. Since it is self-incompatible, cross pollination is necessary for seed production and is effected mainly by bees and flies. Seed dispersal by wind is facilitated by a tuft of hairs on the seed-like fruits.

The Goldenrods include about 100 species, confined mainly to North America and concentrated in the Northeast. As wildlife food, their uses are low in comparison to their abundance. Among the documented consumers are the Ruffed Grouse, Common Goldfinch, Junco, Cottontail Rabbit, Wood Rat, and White-tailed Deer.

Seaside Goldenrod is found on sandy beaches and dunes and in saline and brackish areas along the Atlantic Coast from Newfoundland to Florida, Texas, and Mexico. ℞ Although it is insect-pollinated Seaside Goldenrod sheds large quantities of pollen, which may cause a minor hayfever problem locally.

Several species of *Solidago* have been used medicinally. Specifically *S. virgaured* (Eurasia), *S. rigida, S. odora,* and *S. sempervirens* are reported to be useful in the treatment of a variety of ailments including kidney stones, hemorrhaging, diptheria, dysentery, and wounds. In the Ozark Mountains, the roots of some species are chewed as a treatment for toothache. Some species may cause allergic reactions on contact with sensitive individuals. Several species appear to be toxic to sheep when eaten green or dried in hay. For additional information see *S. caesia.*

The flower of this genus is a source of yellow or gold dye for wool.

Solidago tenuifolia / Slender-leaved Goldenrod — **427** — Asteraceae or Compositae / Aster Family

Ray flowers yellow, very small, in branched, flat-topped, terminal clusters; leaves alternate, with one vein, numerous, very narrow, resin dotted, with clusters of smaller leaves in axils, fragrant, to 7.5 cm. long (3 in.); stems slender, to 80 cm. high (32 in.).

August–October

and vegetation are eaten sparingly by several species of birds and small mammals.

The genus name is derived from a Latin word meaning "to make whole," referring apparently to supposed healing properties. It is a very complex genus and hybridization is not uncommon. Although most species are native to North America, a few occur in South America, Europe, and Asia, and one in the Azores.

A perennial with a creeping branched rhizome. The leaves on the lower part of the stem are usually shed early, leaving the upper ones all about equal in size. Each flower head produces 13–20 hairy seeds with tufts of hair that aid in wind dispersal. The seeds

Slender-leaved Goldenrod is found in dry or moist sandy soil and on shores, beaches, and the edges of salt marshes along the Atlantic Coast and inland from Nova Scotia to Florida and Louisiana.

For additional information see *S. caesia* and *S. Sempervirens.*

Cakile edentula
Sea Rocket

428

Brassicaceae or Cruciferae
Mustard Family

Flowers pale purple to white, to 6 mm. wide (¼ in.), in long clusters at the ends of branches and stems; sepals 4, petals 4, stamens 6, 2 shorter than others, pistil 1; leaves alternate, with wavy-toothed margin or pinnately lobed, succulent, to 12.5 cm. long (5 in.); stems erect, smooth, often bushy-branched, to 40 cm. high (16 in.).

June–September

A fleshy annual with a deeply penetrating root system. The seedpod consists of 2 unequal divisions; the lower one smaller, one-seeded or seedless; the upper separating from the lower at maturity, usually one-seeded. The two sections may be adapted for different types of dispersal; the lower for dispersal in the near vicinity of the parent plant; the upper for dispersal by water. The lower section is sometimes seedless.

This is a small genus mainly of the Northern Hemisphere with only one species native to the Northeast. A related species, *C. maritima*, European Sea Rocket, not illustrated, is similar but has deeply pinnately dissected leaves and seedpods with pro-

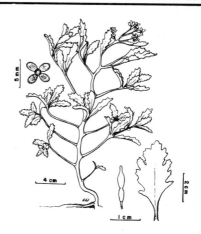

jecting wings. It is a native of Europe and occurs along the northern Atlantic beaches.

Sea Rocket is found on sandy beaches and dunes along the Atlantic coast from Labrador to South Carolina and inland along the Great Lakes.

✄ The young stems and leaves have the flavor of horseradish and may be cooked as greens or added raw to salads.

Beaches

Opuntia compressa
Prickly Pear

429

Cactaceae
Cactus Family

Flowers yellow, to 7.5 cm. wide (3 in.), sometimes with reddish centers, solitary or a few; numerous sepals, petals and stamens; fruit a large, egg-shaped, purplish berry, to 5 cm. long (2 in.); leaves brown, small, shedding early; stems composed of broad, flat, fleshy joints; surface of flat sections bearing tiny clusters of reddish-brown barbed bristles; to 30 cm. high (12 in.).

May–July

A perennial with prostrate clumps of stems and fibrous roots. Numerous seeds are produced without fertilization by each showy flower. These are known to pass unharmed through the human digestive tract, and they probably also survive ingestion by

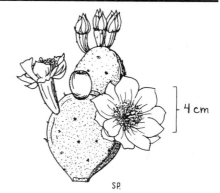

other consumers and are dispersed by them. This is a plant that must be handled with caution. The tiny barbed bristles penetrate the skin and are released from the plant at the slightest touch.

This is a very large genus native to North and South America. In North America it is most commonly associated with the deserts of the Southwest. *O. compressa* is the only widespread cactus east of the Mississippi River. Plants that occur west of the Appalachian Mountains sometimes have long spines. This species is classified by some botanists as *O. humifusa* or *O. rafinesquii.*

Prickly Pear is found in open sandy or rocky areas from southern New England and Ontario to Minnesota, south to Georgia, Mississippi, and Oklahoma.

✘ The Indians of the Southwest make wide use of this genus for food. The pulp of the fruit is flat tasting, although sweet and somewhat juicy. The ground seeds are used as a thickener for soup and stew dishes, and the flat sections of stem are peeled and cooked as vegetables. The reddish-purple fruits of Prickly Pear are usually available in late summer at markets in the East.

℞ The plants of this genus have been used in home remedies for a variety of ailments. The flat stems have been used as poultices on bruises and on the breasts of mothers to increase the flow of milk. The flowers have been used to make a tea for kidney ailments and the roots have been used in a treatment for dysentery.

One member of the Cactus Family, *Lophophora williamsii*, Peyote, not illustrated, is well known as a hallucinogen. It has been used for centuries in religious ceremonies by Indians of the Southwest and more recently by the quarter of a million members of the Native American Church.

Arenaria serpyllifolia
Thyme-leaved Sandwort

430

Caryophyllaceae
Pink Family

Flowers white, tiny, in terminal clusters; sepals 5, petals 5, rounded, shorter than sepals, stamens 10; leaves opposite, oval, sessile, about 6 mm. long (¼ in.); stems rough and minutely hairy, profusely branched, to 20 cm. high (8 in.).

May–August

An annual with a branched taproot, native to Europe and Asia. The oval or flask-shaped seed capsule opens by 6 teeth, releasing numerous kidney-shaped, dull black seeds with concentric rings of small tubercles. These are small enough to become wind-borne in a slight breeze. It sometimes invades cultivated areas but is easily overcome by healthy crop plants.

The genus name is derived from a Latin word for sand and refers to the typical habitat of some of the species. The plants of the genus are small, white flowered herbs that often grow in dense tufts or mats. *A. caro-*

liniana, Pine-barren Sandwort, not illustrated, has flowers to 16 mm. wide (⅔ in.) and very narrow, stiff leaves. It occurs in dry sand along the Atlantic coastal plain from Rhode Island to Florida.

Thyme-leaved Sandwort is inconspicuous and easily overlooked but is common in sandy or rocky areas, fields, and roadsides from Nova Scotia and Quebec to British Columbia, south to Florida and Missouri.

Honkenya peploides — Seabeach Sandwort* — **431** — *Caryophyllaceae* Pink Family

Also Sea-chickweed, Sea Purslane
Flowers white, in terminal clusters or solitary in upper axils; stamens usually 10, petals 5, sepals 5; leaves opposite, bright yellow-green, succulent, oval, pointed, to 2 cm. long (1 in.); stems succulent, branched, spreading over the ground, to 30 cm. high (1 ft.), forming mats to 2 m. across (6½ ft.).

June–July

A perennial with a horizontal rhizome and runners that develop roots and new plants at intervals. The stamens in the pistillate flowers are usually small and non-functional; thus the flowers are mostly unisexual. This is an adaptation to promote cross-pollination, but there is some evidence that seeds are produced without fertilization. The resolution of this problem will have to await further study. The one-celled seedpod contains only a few pear-shaped, brownish seeds.

This genus, which was named for G. A. Honkenya, a German botanist, has only one species. It is highly variable and is circumpolar in distribution. In *Gray's Manual of Botany* this species is classified as *Arenaria peploides*.

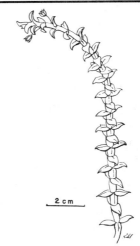
2 cm

Seabeach Sandwort is found on sand dunes and seabeaches along the Atlantic Coast from Newfoundland to Virginia, and along the Pacific Coast from Alaska to California.

✑ Young tender plants may be added to salads, cooked as a potherb, or made into pickles. Eaten raw it has a flavor that is somewhat similar to cabbage. In Iceland it is used to make a fermented beverage.

Beaches

Salsola kali
Russian Thistle*

432

*Also Saltwort, Russian Tumble Weed

An annual with a branched taproot, of North America and Asia. Pollination is by wind, and it has been reported as a very important hayfever plant in Arizona, Colorado, Oklahoma, and Oregon. The seed-like fruit is enclosed by an enlarged, winged calyx that aids in dispersal by wind. In addition, the stem breaks at the base and the entire plant rolls in the wind as a tumbleweed, spreading seeds as it goes.

This is a highly variable species with both native and introduced varieties. The illustration is of the introduced variety tenui-

Flowers greenish or pink, solitary, in axils of leaves; petals none, calyx 5-parted, stamens 5; leaves alternate, first ones formed fleshy, awlshaped, later ones shorter, stiff, with a sharp spine at the tip; stems profusely branched and bushy, to 80 cm. high (32 in.)

July-October

folia. It was introduced accidentally in a shipment of flax seeds in South Dakota in 1886. In years of low rainfall when forage was scarce, prairie farmers cut the plant before its spines hardened and used it as hay. The native variety, Saltwort, occurs mostly along the Atlantic coast.

The variety *tenuifolia*, Russian Thistle, has become an important food plant for several species of game birds, songbirds, small mammals, and hoofed browsers mostly in western and southwestern North America. However, 7 states including the 4 listed above have seed laws that declare it a noxious weed. It is spreading rapidly along eastern and southern coasts.

Russian Thistle or Saltwort is found along the Atlantic Coast from Newfoundland to Louisiana, along the Great Lakes, and throughout western North America.

✘ The young plants up to 12 cm. high (4½ in.) can be cooked as a vegetable and are said to be very good.

Hudsonia ericoides
Golden Heather

433

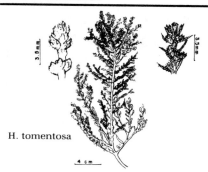

H. tomentosa

Flowers yellow, numerous, each at the tip of a short branch; petals 5, stamens 10–30; leaves needle-like, not overlapping on stem, hairy but not densely so, less than 1 cm. long (½ in.); stems prostrate, with dense erect branches, forming thick greenish mats, to 20 cm. high (8 in.) and 90 cm. across (3 ft.).

May–July

A low-growing evergreen shrub. The many small flowers usually open only in sunshine and give a bright touch of color to their spring environments. The one-celled ovary splits to the base, releasing a few finely pitted, brown seeds.

Hudsonia tomentosa, False Heather (Beach Heath, Poverty Grass), is similar to *H. ericoides*, but with leaves scale-like, overlapping, covering stem, densely hairy. False Heather is found along the Atlantic Coast from the Gaspé Peninsula to North Caolina. It extends inland to Alberta along the shores of the Great Lakes and in sandy prairies, with one outpost on Panther Knob, West Virginia.

The genus was named in honor of William Hudson, an English botanist who, in the eighteenth century, wrote a book on the plants of England in which he introduced the Linnaean system for naming plants. There are 3 species in North America; the third, *H. montana*, not illustrated, occurs in the mountains of North Carolina.

Golden Heather is capable of surviving the harsh environment of sterile sand and salt spray on exposed dunes and beaches along the Atlantic Coast from Newfoundland to Virginia and North Carolina.

Corema conradii
Broom Crowberry

434

Empetraceae
Crowberry Family

Corolla none, stamens purple, flowers in terminal clusters; leaves alternate, very narrow, about 6 mm. long (¼ in.); stems freely branched, forming dense clumps, to 40 cm. high (16 in.).

April–May

A low-growing, evergreen shrub. Cross-pollination is assured since staminate and pistillate flowers are produced on different plants. The pollen grains are compound (in fours), like those of the Heath family. The fruit is dry, brownish-black, and about the size of a pinhead and usually contains 3 seeds.

This is a genus of only 2 species, the other one in Spain and the Canary Islands. They are of limited value as food for wildlife and of little economic importance. They are difficult to maintain but are sometimes cultivated as ornamentals or novelties.

2 cm

Broom Crowberry is found on dunes, sandy beaches, and sandy or rocky soils along the Atlantic Coast, and rarely inland, from Newfoundland to New Jersey. It attains its greatest height and is most abundant in the northern part of its range.

Arctostaphylos uva-ursi
Bearberry*

435

*Also Kinnikinick, Mealberry, Hog
Cranberry*

An evergreen perennial shrub of North
America, northern Europe, and Asia. It is
mainly self-pollinated, and each fruit contains 5 seeds. The seeds are dispersed by
numerous birds that eat the fruits, including Wild Turkey and Ruffed Grouse. The
plant is heavily browsed by Mule Deer and
White-tailed Deer, and the fruit is said to be
favored by bears.

The mats formed during normal growth
are good stabilizers of sand and other sterile
soils. Bearberry is found in open or shaded,
sandy or rocky areas ranging in North Amer-

*Flowers white to pale pink, bell-shaped,
nodding; fruit a red berry; leaves alternate,
entire, leathery, paddle-shaped, to 3 cm.
long (1½ in.); stems freely branched, trailing, forming dense mats.*

May–June

ica from Labrador and Alaska, south to
Virginia.

The western species are tall shrubs or
trees which are called Manzanita. Several of
these, including *A. pungens* and *A. pringlei*,
are common in the evergreen chaparral vegetation of the Southwest.

✖ The berries are edible as survival food.
There are conflicting reports concerning
their tastiness, but they appear to be more
palatable when cooked and served with
cream and sugar.

℞ A tea made from the leaves was used by
early settlers of eastern North America for
diseases of the kidney and inflammation of
the urinary tract. Some American Indians
used a tea made from the berries for weight
control. Other tribes mixed the dried leaves
with tobacco to make a mixture for smoking
called Kinnikinick.

Euphorbia polygonifolia
Seaside Spurge*

436

Also Shore Spurge

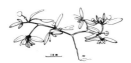

An annual with a deep, branching root
system. The seed capsule is 3-lobed and 3-
celled with one smooth, gray seed in each
cell. The seeds and vegetation are eaten by
several species of upland game birds, songbirds, and small mammals. In most of the
species of this family the seeds have appendages that have been interpreted as adaptations for dispersal by ants.

*Petals and sepals absent, flowers small and
inconspicuous, solitary, on stalks arising
from cup-shaped structures in axils of
leaves; leaves opposite, pale green, oblong,
to 15 mm. long (⅗ in.); stems red, freely
branched, prostrate, to 20 cm. long (8 in.);
plants with milky juice.*

July–October

This tiny, ground-hugging plant may be
overlooked unless one is specifically looking
for it. A hardy pioneer, it is capable of surviving in a harsh environment of temperature extremes and shifting sand. It is found
on sandy beaches and dunes along the Atlantic Coast from Quebec to Georgia and
around the Great Lakes.

☠ Several species of this genus have been used medicinally, mainly as emetics and cathartics. Zuni Indian women used a small quantity of ground spurge leaves in cornmeal mush to promote the flow of milk. However, the milky sap of these plants contains toxic compounds, and they should not be used in home remedies. If it comes in contact with the skin the milky juice may cause dermatitis in sensitive individuals.

Lathyrus maritimus — Beach Pea — 437 — *Fabaceae or Leguminosae* Bean Family

Flowers purple, to 2.5 cm. long (1 in.), on stiff axillary stalks, 5–10 per cluster; fruit a pea-like pod, to 5 cm. long (2 in.); leaves alternate, pinnately compound, with 6–12 oval leaflets; leaf subtended by a pair of arrow-shaped stipules and terminated by a tendril; leaflets and stipules to 5 cm. long (2 in.); stems creeping, usually erect at tip, to 100 cm. long (40 in.).

June–August

A perennial of North America and the cooler parts of Europe and Asia. At maturity the seedpods split longitudinally into 2 sections, releasing several smooth, globular, olive green seeds. These are somewhat resistant to salt water and under certain circumstances may be dispersed by sea water. This species is classified by some botanists as a variety of *L. japonicus.*

There are 2 varieties of this species in the Northeast. The plants that grow around the Great Lakes are hairless, while those along the Atlantic coast have hairs on their calyces and on the undersides of their leaves. It is a hardy successional pioneer capable of surviving in a harsh environment of temperature extremes and shifting sand. It helps to stabilize the sandy substrate as nitrogen fixing nodules on its roots add enrichment. Often occurring in clumps or dense patches, its foliage and clusters of Sweet-pea-like flowers are very showy.

Beach Pea is found on sandy beaches and dunes along the Atlantic coast from Labrador to New Jersey, along the Pacific coast, and inland along lakes Champlain and Oneida and the Great Lakes.

✖ The very young tender seeds are reported by some writers to be indistinguishable from garden peas. Other writers claim they are dry and have a slightly disagreeable taste. All agree that older seeds are dry and tasteless. They were eaten in large quantities as a famine food in England in 1555. The Iroquois Indians cooked, as greens, the young shoots under 25 cm. high (10 in.). *Caution:* Instances of poisoning, known as lathyrism, in humans and livestock have resulted from the ingestion of large quantities of the seeds of some species of this genus. See *L. latifolius.*

Beaches

Lupinus perennis
Wild Lupine

438

Fabaceae or Leguminosae
Bean Family

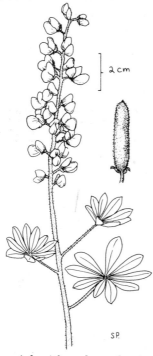

2 cm

S.P.

Flowers blue, pea-like, to 16 mm. long (⅔ in.), in long terminal clusters, to 25 cm. long (10 in.); leaves alternate, long stalked, palmately compound, with 7–11 leaflets; leaflets widest above middle, to 5 cm. long (2 in.); stems erect, hairy, branched, often tinged with purple, to 60 cm. high (2 ft.).

May–July

In North America this genus is concentrated in the western states with about 90 species along the Pacific coast. The genus name is derived from the Latin word for "wolf" and refers to the once held but mistaken belief that these plants would "wolf" or monopolize the mineral nutrients of the soil. *L. perennis* is the only native species in the Northeast. It thrives in the very poorest of sandy soils, which it enriches by the action of nitrogen-fixing nodules on its roots.

Wild Lupine is found in the dry sandy soils of fields, pastures, and open woods from Maine and Ontario to Minnesota, south to Florida and Louisiana.

A perennial with a deep, horizontal, creeping rhizome. The very showy, sweet scented flowers are pollinated by honeybees and bumblebees. The hairy seedpod, to 5 cm. long (2 in.), splits into 2 sections, releasing 4–7 oval, smooth, white or mottled seeds. This is an important wildlife food genus in the western states. A species native to the Southwest, *L. texensis*, Texas Bluebonnet, not illustrated, is the state flower of Texas.

✖ There are ancient Egyptian and Roman records of using lupine seeds for human food. However, some species are known to contain toxic compounds that cause a condition known as lupinosis. Animals vary with regard to their susceptibility to lupinosis but major losses have resulted from ingestion of the plants in western rangelands. Consumption of sub-lethal amounts by pregnant cows and sheep may cause malformed offspring. Poisoning most often occurs in sheep from eating the pods and seeds.

Trichostema dichotomum
Blue Curls*

439

Lamiaceae or Labiatae
Mint Family

*Also Bastard Pennyroyal

An annual species of a genus that includes only one other species in the eastern United States. Each flower produces 4 gray to black, coarsely pitted seeds. The long,

blue, downward curling stamens are responsible for the common name Blue Curls.

A closely related species *T. setaceus*, Narrow-leaved Blue Curls, not illustrated, has very narrow leaves and ranges from Con-

*Flowers blue, corolla tube extending be-
yond the 2-lipped calyx, with 5 unequal
lobes, the lower one bending downward;
stamens 4, extending beyond the upper
corolla lobes and curling downward; leaves
opposite, simple, entire, tapering at base;
stems erect, freely branched, with glandular hairs, to 70 cm. high (28 in.).*

August–October

necticut to Pennsylvania, south to Florida
and Louisiana.

Blue Curls is found in sandy soils of old
fields and dry open woods from Maine to
Michigan, south to Florida and Texas.

Smilacina stellata
Star-flowered Solomon's Seal   **440**   *Liliaceae*
Lily Family

*Flowers white, in a long, several-flowered,
unbranched, terminal cluster, to 5 cm.
long (2 in.); fruit at first green then becoming almost black or with black stripes;
leaves 7–12, alternate, pointed, clasping the
stem, to 15cm. long (6 in.); stems erect,
zigzagging between leaves, to 60 cm. high
(2 ft.).*

May–August

A perennial with a slender, creeping,
branched rhizome. The berries, each containing 1–2 oval, brown, lightly wrinkled and
pitted seeds, persist into autumn. They are
a minor source of food for the Ruffed Grouse
and a few species of songbirds and mammals that probably contribute to seed
dispersal.

This is a genus of 20–25 species native to
North America and eastern Asia. The 3 species in the Northeast are all native and are
easily distinguished from one another. *S.
stellata* is less abundant and usually not as
tall as *S. racemosa*. It has fewer and larger
star-shaped flowers in an unbranched
cluster.

Star-flowered Solomon's Seal is found on
shores, beaches, protected dunes, and sandy
meadows from Newfoundland to British
Columbia, south to Virginia, Indiana, Missouri, and California.

✄ The very young shoots can be prepared like asparagus or, before flowering, cooked as greens. However, since the plant is even less abundant than False Solomon's Seal, it should not be considered as a food plant.

Rhexia virginica
Meadow Beauty*

441

*Also Deergrass

Flowers pink or purple, to 3.8 cm. wide (1½ in.), in terminal clusters; petals 4, stamens 8, conspicuously yellow; leaves opposite, hairy, particularly on upper surface, sessile, rounded at base, with 3 prominent veins, to 6.3 cm. long (2½ in.); stems 4-sided, with slightly winged angles, bristly at nodes, to 60 cm. high (2 ft.).

July–September

are small enough to be windblown in a moderate breeze.

This is a family of mostly tropical plants with only one genus of about 10 species native to Eastern North America. *R. virginica* is the most widespread species in the Northeast. A related species, *R. mariana*, Maryland Meadow Beauty, not illustrated, has pale pink flowers, leaves narrowed at the base, a round stem, and surface runners. It is found on damp sand of the Atlantic coastal plain from Massachusetts to Florida.

Meadow Beauty is found in wet sandy areas and moist meadows from Nova Scotia to Wisconsin, south to Florida, Louisiana, and Texas.

✄ The leaves of *R. virginica* and *R. mariana* can be eaten as a field nibble and in salads or cooked as greens. The tubers of *R. virginica* are edible raw or cooked. These beautiful plants are usually not abundant and should be collected for food only in emergencies.

A perennial with tuberous roots and no surface runners. The style extends beyond the stamens and usually makes the first contact with visiting insects, thus decreasing the possibility of self-pollination. The flowers are visited by honeybees and the yellow butterfly *Colias philodice*. The 4-celled, urn-shaped seed capsule contains many rough, snail-shell-shaped seeds that

Polygonella articulata
Sand Jointweed*

442

Polygonaceae
Smartweed Family

*Also Heather

Flowers pink or white, tiny, in several loose, elongate, terminal clusters, petal-like sepals 5; fruit a smooth 3-angled nutlet; leaves alternate, thread-like, shedding early, to 2.5 cm. long (1 in.); stems slender, erect or prostrate; freely branched, to 40 cm. high (16 in.).

August–October

An annual with a slender, often coiled taproot. The tiny nodding flowers of this species are very dainty and, although individually inconspicuous, are usually present in sufficient numbers to make an attractive cluster. The reddish-brown tinged stems often stand out, especially when growing in white sand. Populations occurring along coastal dunes and in blowout areas have prostrate stems and may not exceed 30 cm. in length (1 ft.). A form with dark purple-red flowers is observed occasionally.

A related species, *P. polygama*, October-flower, not illustrated, is somewhat similar but is a perennial. It occurs in sandy soil and in the pine barrens along the Atlantic coastal plain from Virginia to western Florida. *P. articulata* and *P. polygama* are the only species of this North American genus that occur in the Northeast. *P. articulata* is sometimes erroneously called Heather.

Sand Jointweed is found in dry sandy soil and on dunes and beaches along the Atlantic coastal plain from Maine to North Carolina and inland from Ontario to Minnesota and Iowa.

Potentilla anserina
Silverweed*

443

Rosaceae
Rose Family

*Also Goose-grass, Argentine

Flowers yellow, solitary, on leafless stalks, to 2.5 cm. wide (1 in.), petals 5; leaves alternate, pinnately compound, leaflets 7–25, toothed, increasing in size toward tip of leaf, with silvery silky hairs on underside; stems very short, sending out one or more long creeping runners which root at intervals and produce clusters of leaves.

June–August

A perennial with fleshy fibrous roots, of North America, Europe, and Asia. There are variations in the chromosome numbers among the populations of this species. Some populations with high chromosome numbers are seed-sterile but reproduce vigorously by runners. In other populations with lower chromosome numbers, pollination takes place and fertile seeds are produced. Each flower produces several finely wrinkled seeds with corky bases. These are very

376

buoyant and have been known to float in streams for 15 months.

This is a large genus of the North Temperate and Arctic Zones. The genus name is derived from Latin words that mean "little potent one." This name was originally applied to *P. anserina* because of supposed medicinal properties. The specific name, meaning "of geese," and the name Goosegrass are apparently based on the belief that this plant is favored by geese. Evidence to support this contention is scanty or nonexistent.

Silverweed is circumboreal in distribution, and in North America it is found in wet sandy or gravelly beaches, on the bor-

ders of salt marshes, and inland around areas of elevated salt concentration from Newfoundland to Alaska, south to New England, the Great Lakes region, New Mexico, and California.

✖ The fleshy roots can be eaten raw or cooked as a vegetable. They can be used in stews or casseroles and are reported to taste like parsnips or sweet potatoes.

℞ All parts of the plant contain tannin, which is probably responsible for its astringent properties. It has been used as a gargle for ulcers in the mouth and sore throat, taken internally for diarrhea, and applied externally for sunburn.

Gratiola aurea
Golden Hedge Hyssop*

444

Scrophulariaceae
Figwort Family

*Also Golden Pert

A perennial with a thick rhizome and purple tinted runners that root at intervals. The fruit is a roundish capsule that splits into 4 sections to release numerous tiny, pitted seeds.

This is a genus of about 20 species restricted to the Northern hemisphere. There are 6 species native to eastern North America, usually found in wet or damp soils. In contrast to the bright yellow ones of *G. aurea*, the other species have white or whitelobed flowers. The genus name is derived

Flowers yellow, on long axillary stalks; corolla appears to be 4-lobed, actually it is 2-lipped, lobes of upper lip united, lower lip 3-lobed; leaves opposite, sessile, usually entire, to 2.5 cm. long (1 in.); stems freely branched, often creeping, to 30 cm. high (1 ft.).

June–September

from a Latin word which means "grace" or "thanks" and refers to supposed medicinal properties possessed by some member(s).

A related specied, *G. neglecta*, not illustrated, is somewhat similar but has white lobed flowers with yellow tubes and toothed leaves to 4 cm. long (1½in.). This species has a wider inland distribution than *G. aurea*.

Golden Hedge Hyssop is found on moist sandy or gravelly shores and in wet depressions from Newfoundland and Quebec, south along the Atlantic Coast to Florida, and at several inland stations in New York, Ontario, and Illinois.

⚘ A species of southern Europe, *G. officinalis*, is reputed to possess medicinal properties. *The Dispensatory of the United States of America* (1851) states, "It is a drastic cathartic and emetic, possessing also diuretic properties, and is employed on the continent of Europe in dropsy, jaundice, worms, chronic hepatic affections, scrofula, and various other complaints." There are no known remedies that make use of American species. In early French and German manuals, and in some modern herbals, the genus is listed as poisonous.

Beaches

Plants of Salty Soils

Most plants are very sensitive to environmental salt and cannot grow in areas where there is more than 0.1% salt in the substrate. Some plants however, called Halophytes, have characteristics that permit them to survive in higher salt concentrations. The amount of salt that can be tolerated by these plants varies, with some species preferring higher concentration than others. Thus, to a certain extent, the degree of salinity of the substrate is indicated by the presence of certain species.

Although many halophytes achieve maximum growth only under saline conditions, it is also true that most of them can survive in non-saline habitats. Some, however, can complete their life cycles only in the presence of elevated salt concentrations. Therefore, although the presence of salt tolerant plants in a natural ecosystem is not a guarantee that the substrate is saline, their presence indicates that it may be.

Halophytes occur naturally in two types of areas in the northeastern United States: salt marshes along the Atlantic Coast and inland areas where salt concentrations appear near the surface. The latter are more common in arid or semi-arid portions of the western United States but are not unknown in the Northeast. For example, Central New York State was an important commercial source of salt for colonial America because of its salt springs. Today there are local concentrations of Halophytes in that area.

In early America, salt marshes were common along the Atlantic Coast from Northern Florida to Canada. In terms of productivity (the amount of carbohydrate produced per unit area, per unit of time) they outrank well cultivated and fertilized fields of corn or wheat.

The salt concentration in the substrate of a salt marsh depends on the duration of inundation by tidal salt water. The greatest salinity occurs in those areas which are under water daily at high tide. In more elevated zones which are flooded only during Spring tides (about every two weeks) there is a decrease in salinity. Above this zone of high tides the salt concentration drops rapidly. Over an extended period of time silt and sediments carried by incoming tides are trapped by the intertidal vegetation. The result is a gradual build-up of the level of the marsh and a gradual migration seaward of the zones of salinity and vegetation.

Unfortunately, the salt marshes of the Atlantic Coast are being destroyed at an alarming rate. These unique ecological areas have been filled for commercial development, dredged for boat marinas and waterways, and used as dumping grounds for garbage and waste. As our level of technology has advanced, the rate of destruction has accelerated. For example, at the end of World War II Massachusetts had sixty thousand acres of salt marsh; thirty years later it had about forty thousand acres left. This rate of destruction is not confined to Massachusetts; the coastal salt marsh is rapidly becoming an endangered habitat.

The following species are sometimes observed in brackish water or saline areas, but they also occur in other types of habitats. These plants are described and illustrated in the section that describes the habitat where they are believed to be most common.
Asteraceae, the Aster Family
    Bidens cernua, Nodding Bur Marigold
    Bidens laevis, Larger Bur Marigold
    Mikania scandens, Climbing Hempweed
    Solidago sempervirens, Seaside Goldenrod
    Solidago tenuifolia, Slender-leaved Goldenrod
    Xanthium strumarium, Cocklebur
Chenopodiaceae, the Goosefoot Family
    Salsola kali, Common Saltwort
Haloragaceae, the Water-milfoil Family
    Myriophyllum spicatum, Water-milfoil
Najadaceae, the Pondweed Family
    Potamogeton crispus, curly Pondweed
Polygonaceae, the Smartweed Family
    Polygonum arifolium, Halberd-leaved Tearthumb
    Polygonum aviculare, Knotgrass
Rosaceae, the Rose Family
    Potentilla anserina, Silverweed
Typhaceae, the Cattail Family
    Typha angustifolia, Narrow-leaved Cattail
Verbenaceae, the Vervain Family
    Phyla lanceolata, Fog Fruit

Aster tenuifolius Perennial Salt-marsh Aster 445 *Asteraceae or Compositae* Composite Family

Flower heads with pale purple rays, to 2.5 cm. wide (1 in.), few to several, scattered or solitary; leaves alternate, few, fleshy, very narrow, to 15 cm. long (6 in.); stems smooth, with few or no branches, often zig-zagged between leaves, to 60 cm. high (2 ft.).

August–October

A perennial with fibrous, rooted, creeping surface runners. Each flower head produces numerous gray, slightly hairy, seed-like fruits with parachutes of reddish brown hairs. Forms with whitish ray flowers are occasionally observed. This small attractive aster often goes unnoticed among the tall salt-marsh grasses with which it grows.

A related species, *A. subulatus*, Annual Salt-marsh Aster, not illustrated, has many smaller flower heads, to 13 mm. wide (½ in.), very short rays, and leaves to 20 cm. long (8 in.), and reaches a height of 90 cm. (3 ft.). It often occurs in the same type of habitat as A. tenuifolius but has a broader geographic range occurring from New Brunswick to Florida and Louisiana and in inland saline areas in central New York and southeastern Michigan.

3 cm

SP.

Perennial Salt-marsh Aster is found in brackish water and salt marshes along the Atlantic coast from New Hampshire to Florida and Louisiana.

Pluchea purpurascens Salt-marsh Fleabane 446 *Asteraceae or Compositae* Aster Family

An annual with a faint camphor-like odor. The inner flowers of each flower head produces pollen but have non-functional pistils, while the outer flowers are fertile and pistillate. This is an adaptation to reduce the possibility of self-pollination. Each of the numnerous seed-like fruits of the outer flowers has a parachute of hairs that aids in dispersal by wind.

This is a genus of about 30 species mainly

382

Flower heads pink or purple, in flat-topped terminal clusters, ray flowers absent; leaves alternate, numerous, lanceolate or oval, toothed, to 15 cm. long (6 in.); stems stiff, sticky, to 90 cm. high (3 ft.).

August–September

of warm or tropical regions, with 3 native to the Northeast. The species of the Northeast are similar in having pinkish flowering heads, an odor of camphor, and a tolerance for salt or brackish water. Of the three, *P. purpurascens* has the most widespread distribution. It is a colorful contributor to the autumnal aspect of the salt marsh. Its showy heads are often used in dried flower decorations.

Salt-marsh Fleabane is found in saline and brackish marshes along the Atlantic Coast from Maine to Florida, the West Indies, and Mexico, and in inland areas of high salt concentrations.

Spergularia marina
Salt Marsh Sand Spurry

447

Caryophyllaceae
Pink Family

Flowers white or pink, in branching terminal clusters; petals 5, sepals 5, larger than petals; leaves opposite, succulent, very narrow, to 4 cm. long (1½ in.), with small triangular stipules; stems erect or prostrate, freely branched, to 35 cm. long (14 in.).

June–September

An annual of uncertain origin. Some botanists think it is native to North America; others believe the North and South American plants were introduced from Europe or Asia. The seed capsule splits from the top into 3 sections, releasing several tiny, brown, wingless seeds. These may be dispersed on the feet of waterfowl, and they are small enough to become windborne in a moderate breeze.

This genus includes 5 species in the Northeast. Most of these occur in salty habitats and are distinguished from one another mainly by seed characteristics. *S. rubra*, Sand Spurry, not illustrated, is similar to *S. marina* but has non-succulent leaves and slightly smaller seeds. These two species are the most widespread representatives of the genus in eastern North America.

Salt Marsh Sand Spurry is found in brackish and salt marshes from Quebec to British Columbia, south along the Atlantic coast to Florida and Texas, and inland on saline substrates.

Atriplex patula
Orach*

448

Chenopodiaceae
Goosefoot Family

*Also Spearscale

Flowers green, unisexual, numerous in scattered clusters at the tips of upper branches; leaves opposite, alternate, or both, often triangular or arrow-shaped; stems erect or prostrate, freely branched, to 150 cm. long (5 ft.).

June–November

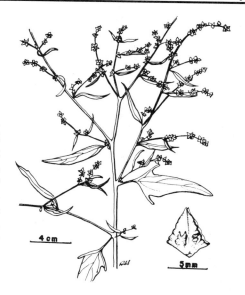

4 cm

5 mm

A highly variable annual of North America, Europe, and Asia. It is wind-pollinated but usually does not occur in sufficient numbers to be an important hayfever plant. The hard, shiny, black seeds are spread by birds and mammals that feed on the plant.

There are about 60 species of this genus in the United States, most of which are found in the western states, where they go by the common names of saltbush and shadscale. They often occur in association with sagebrush and are of fair importance as wildlife food. They may also be more important in the West as hayfever plants.

A related species in the Northeast, *A. arenaria*, Seabeach Orach, not illustrated, has alternate, oval-oblong, silvery leaves and occurs on sea beaches and the edges of salt marshes.

A. patula is found in salt marshes along the Atlantic Coast and in waste spaces and inland saline areas throughout most of the United States and southern Canada.

✗ The young tips can be cooked as greens and according to several sources they are better than Lamb's-quarters, domestic spinach, beets, or chard. Some tribes of American Indians used the seeds of plants in this genus to make an edible meal which they mixed with cornmeal.

℞ The newly ripened seeds, bruised and soaked in alcohol for six weeks, produce a tincture that is reported to be a mild laxative, a cure for headaches, and a treatment for the first attacks of rheumatism.

Salty Soils

Chenopodium rubrum
Coast Blite

449

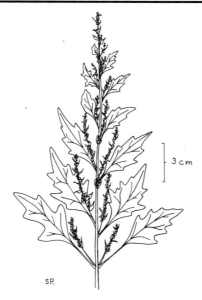

3 cm

S.P.

An annual, native to North America, Europe, and Asia. Pollination is by wind, but it is not known to be an important hayfever plant. A single plant may produce as many as 175,000 seeds. The plants persist until

Flowers reddish, tiny, in dense axillary and terminal clusters; seeds dark brown, glossy, lens-shaped; leaves mostly alternate, reddish, the larger ones oval to triangular, with 1 or more conspicuous teeth on each side, tapering at base; stems erect or prostrate, often freely branched, to 80 cm. long (32 in.).

July–October

late in the season and the seeds are relished by many species of songbirds.

This is a genus of at least 100 species of worldwide distribution with about 6 species native to the Northeast.In *C. rubrum* there are 2 contrasting growth forms: tall erect plants with large leaves, and low-growing, often prostrate plants with smaller leaves. The low-growing forms are recognized by some botanists as a separate species, *C. humile*.

Coast Blite is found in salt marshes and in saline soil along the Atlantic coast from Newfoundland and Nova Scotia to New Jersey and, perhaps more commonly, inland in saline areas to Minnesota and Iowa, and on the Pacific coast.

Salicornia europaea
Glasswort*

450

*Also Samphire, Saltwort, Chickenclaws, Pigeonfoot

A shallow-rooted annual of North America, Europe, and Asia. Pollination is by wind, but ordinarily it does not release pollen in large enough quantities to be considered a hayfever plant. Each flower produces a single brown seed covered with fine white hairs.

In autumn the stems turn red, yellow, or orange and are eaten by several species of waterfowl, especially the Canada Goose and

Pintail Duck, which probably serve as agents of dispersal. This species normally grows just above the high tide level, but it may be flooded during monthly Spring tides. It is the most common species of the genus and is occasionally observed in the waste areas of pickle factories and salt works.

Glasswort is found on tidal flats and in salt marshes along the Atlantic coast from Quebec to Florida, and inland in areas of high salt concentration.

Flowers green or colorless, small, imbedded in upper joints of stem, usually 3 in a cluster, the center one attached above and extending beyond the other two; leaves reduced to minute opposite scales; stems erect, green, freely branched, jointed, joints longer than wide, succulent, salty to the taste when crushed, to 40 cm. high (16 in.).

August–November

✂ The young tender stems can be cooked as a vegetable, but a change of water may be necessary to reduce the saltiness. They may also be pickled in vinegar or added fresh to salads.

℞ All species of the genus are rich in sodium carbonate. The ashes of these plants were formerly used in the manufacture of soap and glass; thus the name glasswort.

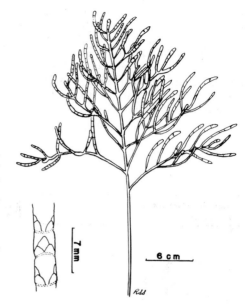

Salicornia virginica
Perennial Saltwort*

451

Chenopodiaceae
Goosefoot Family

*Also Woody glasswort, Leadgrass

Flowers green or colorless, inconspicuous, imbedded in upper joints of stem, usually 3 in a cluster, attached at the same level in a horizontal row, all about the same height; leaves reduced to minute opposite scales; main stems woody, prostrate, rooting at joints, forming mats, with erect, succulent, jointed, flower-bearing branches, to 30 cm. high (1 ft.).

August–October

A perennial with a woody rhizome, of North America, West Indies, western Europe, and northern Africa. Pollination and seed dispersal are similar to that in *S. europaea*. This species is easily distinguished from *S. europaea* by its unbranched or slightly branched, erect flowering stems. These two species often occur together in the margins of salt marshes along the Atlantic coast, but *S. virginica* does not extend inland.

The genus name is derived from Latin

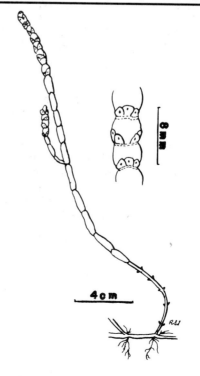

Salty Soils

words which mean "salt horn," alluding to the saline habitat of these plants and their horn-like branches. There are 4 species native to the Northeast and an additional 4 in the southern and western states. There has been some investigation into the use of these species as crop plants in areas of saline ground water in the western states.

Perennial Saltwort is found in the margins of salt marshes along the Atlantic coast from Massachusetts to Texas, and along the Pacific coast from Alaska to California.

For food uses and other information, see *S. europa.*

Suaeda linearis **Sea Blite**	**452**	*Chenopodiaceae* **Goosefoot Family**

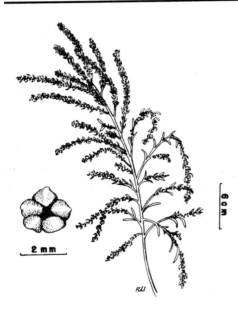

2 mm

RLI

An annual or perennial of eastern North America and the West Indies. Pollination is by wind, but it is not ordinarily considered an important hayfever plant. Each pistillate flower produces a single smooth, glossy black seed. The whole plant often breaks at

Flowers green, in axillary clusters, stamens 5, petals none; calyx lobes 5, completely enclosing the fruit at maturity, upper 2 or 3 lobes with noticeable keels; leaves very narrow, succulent, salty when crushed, lower surface rounded, to 5 cm. long (2 in.), progressively shorter toward top of plant; stems erect, freely branched, to 90 cm. high (3 ft.).

August–October

the base and rolls with the wind like a tumbleweed spreading seeds as it goes.

This is a genus of worldwide distribution with about 20 species in North America. The 5 species native to the Northeast are all salt marsh or saline soil plants, and *S. linearis* is the most widespread. A related species, *S. maritima*, not illustrated, is similar but the upper 3 lobes of the calyx are not prominently keeled. It is much smaller, often with prostrate stems, and occurs in Europe and along the Atlantic coast from Quebec to New Jersey.

S. linearis is found in salt marshes from Maine to Florida and Texas.

✕ The young leaves and stems can be cooked as greens in two or three changes of water to reduce the saltiness.

Hibiscus palustris
Swamp Rose-mallow*

453

*Also Sea-hollyhock

Flowers pink,to 15 cm. wide (6 in.), usually borne singly in upper axils, calyx 5-lobed, petals 5, stamens united into a tube around style, anthers numerous, style branches 5; leaves alternate, broadly rounded at base, pointed at tip, toothed, to 20 cm. long (8 in.); stems hairy in upper part, to 2 m. high (6½ ft.).

August–September

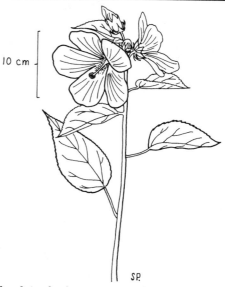

10 cm

A perennial with a dense cluster of fibrous roots. The spiny pollen grains stick together in clumps which are transferred from one flower to another by hummingbirds or bees. The seed capsule splits into 5 sections, releasing several dark, oval, warty seeds. The seed capsules may persist on the dead stems throughout the winter months.

This is a genus of about 200 species mainly of the warm regions of the world. Of the more than 20 species that occur in eastern North America, 4 are native to the Northeast. *H. palustris* is included by some botanists as a variety of the very similar *H. moscheutos*. Two related introduced species, *H. syriacus*, Rose-of-Sharon, and *H. rosa-sinensis*, Rose-of-China, neither illustrated, are widely cultivated as ornamentals.

Swamp Rose-mallow is found in brackish or salt marshes along the Atlantic coast from Massachusetts to North Carolina and in-land in fresh water marshes, to Ontario, Michigan, and Illinois.

✗ According to E. Gibbons and G. Tucker, the young roots of *Hibiscus* species can be fried or boiled as a vegetable. They are rich in mucilage and the cooking water can be whipped like gelatin. A related cultivated species, *H. esculenthus*, Okra or Gumbo, not illustrated, has a very high mucilage content and the young seedpods are used as a thickener for soup and stew dishes.

Salty Soils

Ruppia maritima
Ditchgrass*

454

*Also Widgeon Grass

An aquatic perennial with tufts of basal leaves and flowering stems arising from long prostrate runners. This is one of a very few species in which pollination is accomplished by the pollen floating on the surface of the water from the anthers to the stigmas. Since this is a two-dimensional pollination medium, as compared with wind pollina-tion, less pollen is required for seed production. During flowering, the pollen grains spread over the surface film like droplets of oil. The anthers of the flower mature before pistils; thus self-pollination is avoided.

Each flower usually produces 4 long-stalked, oval, green to black, seed-like fruits. This is one of the most valuable waterfowl food plants, comprising up to 25% of the

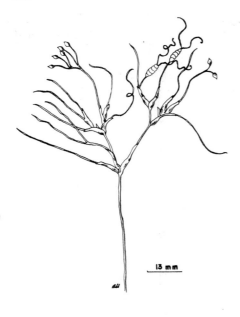

13 mm

Flowers tiny, on long, coiled, axillary stalks that rise to the surface of water at flowering, petals and sepals none; leaves alternate, thread-like, sheathing the stem, to 12 cm. long (4½ in.); stems submerged, often branched, to 80 cm. long (32 in.).

June–September

diet of the American Brant, Coot, the Baldpate, Gadwell and Redhead Ducks, the Greater and Lesser Scaups, and the Canada Goose. The seeds, leaves, stems, and runners are all consumed.

Ditchgrass is found along the Atlantic and Pacific coasts and in saline and brackish waters inland. It is abundant in alkaline lakes, ponds, and streams in western North America and is widespread in Europe. It rarely occurs in fresh water.

| *Limonium nashii* Sea Lavender* | **455** | Plumbaginaceae Leadwort Family |

*Also Marsh Rosemary

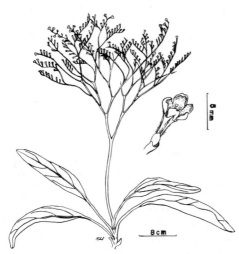

8 cm

9 mm

A perennial with a thick woody root. Each of the many flowers produces a single dark brown, shiny, faintly ridged seed, enclosed by the persistent calyx. The specific name

Flowers pale purple or lavender, about 3 mm. wide (⅛ in.), in a profusely branched terminal cluster on a leafless stalk; calyx 5-lobed, hairy on lower half, persistent, petals 5; leaves basal, lanceolate, widest above middle, tapering to long stalks, to 25 cm. long (10 in.); flowering stems erect, to 60 cm. high (2 ft.).

July–October

honors George V. Nash, a botanist who was an authority on the grasses and who became curator of the New York Botanical Garden in 1899.

This is a genus of primarily desert and salt-marsh plants. It has a world-wide distribution and includes species that are grown in rock gardens and greenhouses and used in dried flower arrangements. The genus is represented by two native species in the Northeast. *L. carolinianum*, not illustrated, is very similar to *L.nashii* but has a calyx devoid of hairs. It occurs in tidal marshes from southern New York to Texas.

Sea Lavender (*L. nashii*) is found in salt marshes along the Atlantic coast from Labrador to northeastern Mexico.

℞ The root has a very bitter and astringent taste and in the nineteenth century was used for diarrhea and dysentery in eastern North America. An ointment made from the powdered root has been used as a soothing treatment for hemorrhoids, chronic gonorrhea, and inflammation of body orifices.

Ranunculus cymbalaria — Seaside Crowfoot — 456 — *Ranunculaceae* Crowfoot Family

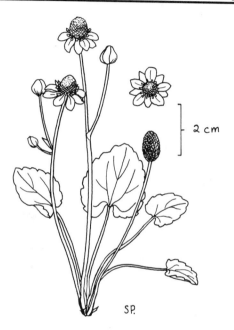

2 cm

S.P.

Flowers yellow, solitary or in few-flowered clusters on leafless stalks, about 12 mm. wide (½ in.); petals 5, slightly longer than sepals; leaves basal, long stalked, rounded at tips, heart-shaped at bases, toothed, to 6 cm. long (2½ in.); flowering stems smooth, to 25 cm. high (10 in.).

June–September

A perennial with thread-like runners that root at intervals, producing new plants with slender fibrous roots. The flower receptacle enlarges and becomes cylindrical in fruit, bearing up to 200 flattened veined, nutlets. The mature fruiting receptacle is sometimes red.

This is a genus of at least 250 species most abundant in northern and Arctic regions and at high altitudes. It includes species in a wide range of habitats from dry sand and open fields to salt marshes, fresh water swamps, and shady woods. Seed production patterns vary from species that require cross-pollination to those that produce seeds without fertilization.

Seaside Crowfoot is found in salt marshes along the Atlantic coast from Labrador to New Jersey and inland in saline areas from Alaska to Iowa and Texas and in the western states. It is rarely observed in fresh water habitats.

☠ This species has the same toxic properties as other members of the genus. See *R. acris*.

Glossary

Actinomorphic. Radially symmetric, as a flower (illustration in key).

Alternate (leaf). Arranged so that only one leaf is attached at a given level on a stem (illustration in key).

Anemophily. Pollination by wind.

Annual. A plant that completes its life cycle, from seed to mature plant to seed, in one growing season.

Anther. The usually enlarged part of the stamen that produces pollen (illustration in key).

Apocarpus. Having several separate pistils (illustration in key).

Axil. The angle formed between a leaf and the stem.

Basal (leaf). Attached at or below the ground level. Not attached to the aerial stem (illustration in key).

Biennial. A plant that requires two growing seasons to complete its life cycle, usually producing a basal rosette of leaves the first year and a flowering stem the second year.

Bipinnate. Twice pinnately divided.

Blade. The flat, usually expanded part of a leaf.

Bloom. A white powdery coating on a leaf.

Bract. A greatly reduced leaf, usually located at the base of a flower or cluster of flowers.

Bulb. An underground bud, usually with thick fleshy leaves.

Calyx. The sepals of a flower (illustration in key).

Cauline. Pertaining to the stem.

Compound (leaf). Having a blade made up of two or more leaflets (illustration in key).

Cordate (leaf). Heart-shaped, at least at the base.

Corolla. The usually showy part of a flower, the petals. They may be separate or united into a tube or cup (illustration in key).

Decumbent. Lying on the ground at the base (usually a stem) with erect tips.

Dentate. With marginal teeth.

Diaphoretic. A substance that promotes sweating.

Disk (flower). Occurring in the center part of flower head of many members of the Aster family.

Diuretic. A substance that increases the flow of urine.

Emetic. A substance that causes vomiting.

Entire (leaf). Having a smooth margin.

Entomophily. Pollination by insects.

Epigynous. With the sepals, petals and stamens attached at the top of the ovary (illustration in key).

Expectorant. A substance that promotes discharge of mucus from the respiratory tract.

Fertilization. The union of male and female sex cells to form a zygote.

Filiform. Long and thin, threadlike.

Filament. The thin part of the stamen that supports the anther (illustration in key).

Floret. A small individual flower of a dense cluster, as in the flower head of the Aster family.

Fruit. A ripened ovary or a cluster of ripened ovaries.

Glaucous. Having a bloom.

Head. A dense cluster of flowers without pedicels (illustration in key).

Hypogynous. With sepals, petals and stamens attached below the ovary (illustration in key).

Inflorescence. A cluster of flowers (illustration in key).

Imperfect (flower). With staminate or pistillate parts present but not both.

Lanceolate. Shaped like a lance, longer than broad, wide at base, tapering to a point.

Lateral. On the side.

Leaflet. A division of a compound leaf.

Linear. Long and narrow with parallel sides.

Monoecious. Pistillate flowers and staminate flowers on the same plant.

Nodding. Bending downward, as a flower.

Node. The place on a stem where a leaf is attached.

Opposite (leaves). Two at a node.

Ovary. The lower usually enlarged part of the pistil bearing the young seeds.

Palmate (leaf). With lobes, leaflets or veins radiating outward from a point like the fingers from the palm of the hand (illustration in key).

Panicle. A branched flower cluster (illustration in key).

Pedicel. The stalk of an individual flower in an inflorescence.

Penduncle. A stalk that supports a solitary flower or an inflorescence.

Peltate (leaf). With the stalk attached to the underside of the blade, somewhat umbrella-like.

Perennial. A plant that usually produces flowers and seeds every year and lives for several years.

Perfect (flower). Having both pistil and stamens.

Perfoliate (leaf). With the leaf surrounding the stem (illustration in key).

Perianth. The calyx and corolla. Often used when one is missing or when they are very similar.

Perigynous. With the sepals, petals and stamens attached to the margin of a cup-shaped structure that surrounds the ovary or ovaries (illustration in key).

Petal. A unit of the corolla, the inner, often showy, floral envelope (illustration in key).

Petiole. The stalk of a leaf.

Pinnate (leaf). With lobes, leaflets, or veins arising from two sides of a midrib or stalk (illustration in key).

Pistil. The central seed bearing organ of a flower consisting of an ovary, style, and stigma (illustration in key).

Pollen. Spores produced by the anther that bear the male gametes.

Pollination. The transfer of pollen from the anther to the stigma.

Polypetalous. Having two or more separate petals (illustration in key).

Ray. The marginal floret of the flowerhead of some members of the Aster family, with a petal-like corolla (illustration in key).

Reflexed. Bent abruptly downward or backward (illustration in key).

Rhizome. An underground, usually horizontal, stem.

Rosette. A dense cluster of basal leaves attached in a circular pattern.

Sagittate. In the shape of an arrow-head.

Sepal. One of the units making up the calyx, usually green (illustration in key).

Serrate (leaf). Margin with sharp, forward directed teeth.

Sessile. Without a stalk.

Simple (leaf). Blade not divided into leaflets.

Spadix. A thick, fleshy column bearing flowers at its base as in the Arum family (illustration in key).

Spathe. A sheath or bract enclosing a cluster of flowers as in the Arum family (illustration in key).

Spike. An elongate unbranched cluster of sessile flowers (illustration in key).

Stamen. A pollen producing structure consisting of a filament and an anther (illustration in key).

Stigma. The part of the pistil, usually the tip, that receives pollen (illustration in key).

Stipules. Reduced leaf-like structures attached at the base of a petiole (illustration in key).

Style. The part of the pistil, often long and slender, that attaches the stigma to the ovary (illustration in key).

Succulent. Soft, thick, and fleshy.

Sympetalous. Petals united into a one piece corolla (illustration in key).

Syncarpous. Having a single pistil (illustration in key).

Ternate. Divided into threes.

Tonic. A substance that invigorates, refreshes or stimulates the body.

Umbel. A cluster of flowers with pedicels radiating from the same point (illustration in key).

Whorled (leaves). With three or more leaves attached at a node (illustration in key).

Winter Annual. A plant that completes its life cycle in one growing season. The seeds germinate in autumn and flowering occurs the following spring.

Zygomorphic (flower). Having bilateral symmetry (illustration in key).

Additional Sources of Information

The following books and papers are the main sources of information used by the author. This list is not intended to be a complete bibliography.

Allen, O. N., and E. K. Allen. 1981. *The Leguminosae, A Sourcebook Of Characteristics, Uses, and Nodulation.* The University of Wisconsin Press, Madison.

Appalachian Mountain Club. 1964. *Mountain Flowers of New England.* Appalachian Mountain Club, Boston.

Bailey, L. H. 1949. *Manual of Cultivated Plants Most Commonly Grown In The Continental United States and Canada.* Rev. Ed. The Macmillan Company, New York.

Borror, D. J. 1960. *Dictionary of Word Roots and Combining Forms.* Mayfield Publishing Company, Palo Alto, California.

Brainerd, E. 1921. *Violets of North America.* Free Press Printing Co., Burlington, Vermont.

Clarkson, R. B., W. H. Duppstadt, and R. L. Guthrie. 1980. *Forest Wildlife Plants of the Monongahela National Forest.* The Boxwood Press, Pacific Grove, California.

Coon, N. 1979. *Using Plants for Healing.* Rodale Press, Emmaus, Pa.

Curtis, J. T. 1959. *The Vegetation of Wisconsin, An Ordination Of Plant Communities.* The University of Wisconsin Press, Madison.

Cuthbert, M. J. 1948. *How to Know The Fall Flowers.* Wm. C. Brown Company, Dubuque, Iowa.

Cuthbert, M. J. 1949. *How To Know The Spring Flowers.* Wm. C. Brown Company, Dubuque, Iowa.

Dampier, Sir W. C. 1949. *A History of Science*, 4th Ed. Cambridge University Press, Cambridge.

Dressler, R. L. 1981. *The Orchids, Natural History and Classification.* Harvard University Press, Cambridge, Massachusetts.

Faegri, K., and L. Van Der Pijl. 1966. *The Principles of Pollination Ecology.* Pergamon Press, New York.

Fassett, N. C. 1976. *Spring Flora of Wisconsin*, 4th Ed. Revised by O. S. Thomson. The University of Wisconsin Press, Madison.

Fassett, N. C. 1957. *A Manual of Aquatic Plants*, revised by E. C. Ogden. The University of Wisconsin Press, Madison.

Fernald, M. L. 1950. *Gray's Manual of Botany*, 8th ed. D. Van Nostrand Company, New York.

Fernald, M. L., and A. C. Kinsey. 1958. *Edible Wild Plants of Eastern North America*, revised by R. C. Collins. Harper and Row, Publishers, New York.

Fryxell, P. A. 1957. "Mode of Reproduction in Higher Plants." Bot. Rev. 23:135–233.

Gibbons, E. 1962. *Stalking The Wild Asparagus*. David McKay Company Inc. New York.

Gibbons, E., and G. Tucker. 1979. *Euell Gibbons' Handbook of Edible Wild Plants*. A Unilaw Library Book, Virginia Beach/Norfolk.

Gleason, H. A. 1952. *The New Britton and Brown Illustrated Flora of The Northeastern United States and Adjacent Canada*. (in 3 volumes) Hafner Press, New York.

Gleason, H. A., and Cronquist, A. 1963. *Manual of Vascular Plants of Northeastern United States and Adjacent Canada*. D. Van Nostrand Company, New York.

Grant, V. 1971. *Plant Speciation*. Columbia University Press, New York.

Grieve, Mrs. M. 1971. *A Modern Herbal*. (Two volumes.) Dover Publications Inc., New York.

Harborne, J. B. 1982. *Introduction to Ecological Biochemistry*. Academic Press, New York.

Hartmann, H. T., and D. E. Kester, 1959. *Plant Propagation, Principles and Practices*. Prentice-Hall Inc., Englewood Cliffs, New Jersey.

Haslam, S. M. 1978. *River Plants*. Cambridge University Press, New York.

Heinrich, B. 1975. "Bee Flowers: A Hypothesis on Flower Variety and Blooming Times." Evolution 29:325–334.

Heiser, Jr., C. B. 1969. *Nightshades, The Paradoxical Plants*. W. H. Freeman and Company, San Francisco.

Heywood, V. H. Editor. 1978. *Flowering Plants Of The World*. Mayflower Books Inc., New York.

Hinds, H. R., and W. A. Hathaway. 1968. *Wildflowers of Cape Cod*. The Chatham Press, Inc. Chatham, Mass.

Hyypio, P., and E. Cope. "Giant Hogweed, *Heracleum mantegazzianum*." A Cornell Cooperative Extension Publication, Cornell University, Ithaca.

Kingsbury, J. M. 1964. *Poisonous Plants of the United States and Canada*. Prentice-Hall, Inc., Englewood Cliffs, New Jersey.

Kozlowski, T. T. (Editor) 1972. *Seed Biology. Volume 1. Importance, Development and Germination*. Academic Press, New York.

Krochman, Arnold and Connie Krochman. 1973. *A Guide To The Medicinal Plants Of The United States*. Quadrangle/The New York Times Book Co., New York.

Lawrence, G. H. M. 1951. *Taxonomy of Vascular Plants*. The Macmillan Company, New York.

Lewis, W. H., and M. P. F. Elvin-Lewis. 1977. *Medical Botany, Plants Affecting Man's Health*. John Wiley and Sons, New York.

Lust, J. B. 1974. *The Herb Book.* Bantam Books, New York.

Magee, D. W. 1981. *Freshwater Wetlands, A Guide To Common Indicator Plants Of The Northeast.* University of Massachusetts Press, Amherst.

Martin, A. C., H. S. Zim, and A. L. Nelson. 1951. *American Wildlife And Plants, A Guide To Wildlife Food Habits.* Dover Publications, Inc., New York.

Mathews, F. S. 1927. *Field Book Of American Wild Flowers.* Rev. ed. G. P. Putnams' Sons, New York.

Meeuse, B. J. D. 1961. *The Story of Pollination.* The Ronald Press Company, New York.

Meyer, J. E. 1960. *The Herbalist.* Meyerbooks, Glenwood, Illinois.

Mitchell, R. S., and E. O. Beal. 1979. *Magnoliaceae Through Ceratophyllaceae of New York State.* New York State Museum Bulletin, Albany.

Mitchell, R. S., and J. K. Dean. 1982. *Ranunculaceae (Crowfoot Family) Of New York State.* New York State Museum Bulletin, Albany.

Mitchell, R. S., and J. K. Dean. 1978. *Polygonaceae (Buckwheat Family) Of New York State.* New York State Museum Bulletin, Albany.

Montgomery, F. H. 1977. *Seeds and Fruits of Plants of Eastern Canada and Northeastern United States.* University of Toronto Press, Toronto.

Morgan, A. H. 1930. *Field Book of Ponds and Streams.* G. P. Putnam's Sons, New York.

Muenscher, W. C. 1944. *Aquatic Plants of The United States.* Cornell University Press, Ithaca, New York.

Muenscher, W. C. 1940. *Poisonous Plants of the United States.* The Macmillan Company, New York.

Newcomb, L. 1977. *Newcomb's Wildflower Guide.* Little, Brown and Company, Boston.

Niering, W. A. 1979. *The Audubon Society Field Guide to North American Wildflowers, Eastern Region.* Alfred A. Knopf, Inc., New York.

Nygren, A. 1954. "Apomixis in Angiosperms II." Bot. Rev. 20:577–649.

Palmer, E. L. 1975. *Fieldbook of Natural History.* 2nd ed. Revised by H. S. Fowler. McGraw-Hill Book Company, New York.

Parrish, J. A. D., and F. A. Bazzaz. 1979. "Difference in Pollination Niche Relationships in Early and Late Successional Plant Communities". Ecology 60:597–610.

Percival, Mary. 1965. *Floral Biology.* Pergamon Press, New York.

Peterson, L. 1978. *A Field Guide to Edible Wild Plants.* Houghton Mifflin Company, Boston.

Peterson, R. T., and M. McKenny. 1968. *A Field Guide to Wildflowers of Northeastern and North-central North America.* Houghton Mifflin Company, Boston.

Van Der Pijl, L. 1972. *Principles of Dispersal in Higher Plants.* 2nd ed. Springer-Verlag, New York.

Rickett, H. W. 1965. *Wildflowers of The United States, The Northeastern States* (in 2 parts). McGraw-Hill Book Company, New York.

Shosteck, R. 1964. *Flowers and Plants, An International Lexicon With Biographical Notes.* Quadrangle/The New York Times Book Company, New York.

Strausbaugh, P. D., and E. L. Core. 1964. *Flora of West Virginia* (in 4 volumes). West Virginia University Bulletin, Morgantown, West Virginia.

United States Department of Agriculture. 1971. *Common Weeds of The United States.* Dover Publications, Inc., New York.

United States Department of Agriculture. 1961. *Seeds, The Yearbook of Agriculture.* Government Printing Office, Washington, D. C.

Weiner, M. A. 1980. *Earth Medicine—Earth Food.* Rev. Ed. The Macmillan Company, New York.

Wheelwright, E. G. 1974. *Medicinal Plants and Their History.* Dover Publications Inc., New York.

Wodehouse, R. P. 1935. *Pollen Grains.* McGraw-Hill Book Company, Inc., New York.

Wood, G. B. and F. Bache. 1851. *The Dispensatory Of The United States Of America,* Ninth Edition. Lippincott, Grambo, and Co., Philadelphia.

Index of Plant Names

In the following index, heavy boldface type is used for entry numbers. Page numbers follow in standard roman type.

408